Introduction to Exercise Science

Duane V. Knudson, PhD

Texas State University

EDITOR

HUMAN KINETICS

Library of Congress Cataloging-in-Publication Data

Names: Knudson, Duane V., 1961- editor.
Title: Introduction to exercise science / Duane V. Knudson, PhD, Texas
  State University, Editor.
Description: Champaign, IL : Human Kinetics, [2024] | Includes
  bibliographical references and index.
Identifiers: LCCN 2022039360 (print) | LCCN 2022039361 (ebook) | ISBN
  9781718209954 (paperback) | ISBN 9781718209961 (epub) | ISBN
  9781718209978 (pdf)
Subjects: LCSH: Sports sciences--Research. | Exercise--Physiological
  aspects--Research. | Kinesiology. | Interdisciplinary research. |
  Physical education and training--Research. | Performance.
Classification: LCC GV558 .I58 2024 (print) | LCC GV558 (ebook) | DDC
  796.01/5--dc23/eng/20220908
LC record available at https://lccn.loc.gov/2022039360
LC ebook record available at https://lccn.loc.gov/2022039361

ISBN: 978-1-7182-0995-4 (print)
ISBN: 978-1-7182-1389-0 (loose-leaf)

The web addresses cited in this text were current as of November 2022, unless otherwise noted.

Acquisitions Editor: Diana Vincer; **Developmental and Managing Editor:** Amanda S. Ewing; **Copyeditor:** Marissa Wold Uhrina; **Proofreader:** Leigh Keylock; **Indexer:** Rebecca L. McCorkle; **Permissions Manager:** Laurel Mitchell; **Senior Graphic Designer:** Nancy Rasmus; **Graphic Designer:** Dawn Sills; **Cover Designer:** Keri Evans; **Cover Design Specialist:** Susan Rothermel Allen; **Photograph (cover):** gorodenkoff/iStock/Getty Images; **Photographs (interior):** © Human Kinetics, unless otherwise noted; **Photo Asset Manager:** Laura Fitch; **Photo Production Manager:** Jason Allen; **Senior Art Manager:** Kelly Hendren; **Illustrations:** © Human Kinetics, unless otherwise noted; **Printer:** Walsworth

Printed in the United States of America   10 9 8 7 6 5 4 3 2 1

The paper in this book was manufactured using responsible forestry methods.

Human Kinetics
1607 N. Market Street
Champaign, IL 61820
USA

United States and International
Website: **US.HumanKinetics.com**
Email: info@hkusa.com
Phone: 1-800-747-4457

Canada
Website: **Canada.HumanKinetics.com**
Email: info@hkcanada.com

E8447 (paperback) / E8675 (loose-leaf)

CONTENTS

PART III   Related Professional Subdisciplines   209

PREFACE

Introduction to Exercise Science presents the exciting world of the subdisciplines of evidence-based exercise prescription. Students and instructors will find an engaging blend of influential past discoveries and future possibilities.

Goals

The primary goal of *Introduction to Exercise Science* is to provide students with an overview of the major subdisciplines of this important field of knowledge. Second, the text introduces students to the basics of research and knowledge in these subdisciplines of exercise science. Third, the text illustrates how interdisciplinary collaboration and applied research in exercise science–related professions contribute to the performance and health of all people. Evidence-based exercise prescription blends scientific and professional knowledge with client values to promote the best possible results. We hope the book inspires students to continue their study of exercise science and join the authors in promoting evidence-based exercise prescription for the benefit of all.

Organization

The book is organized into three major sections. Part I has three important chapters laying out preliminary knowledge in the study of exercise and physical activity. Chapter 1 presents the diverse scope of exercise science and kinesiology as fields of scientific study, key subdisciplinary sciences, and sources of exercise science knowledge, and it shows how this knowledge is used in evidence-based exercise prescription. Chapter 2 introduces students to key concepts of human musculoskeletal anatomy, a major prerequisite area of study that, when integrated with exercise science, can explain how human movement is created. Chapter 3 illustrates how measurement and statistics are essential to exercise science and its evidence-based application.

Part II reviews five of the most important subdisciplines of exercise science: biomechanics, exercise physiology, motor behavior, sport and exercise psychology, and physical activity epidemiology. These chapters describe the goals and benefits of that area of exercise science knowledge, major historical research advancements, and common research methods. Each of these chapters also gives examples of real-world applications from research in that subdiscipline. Readers will come to understand how these subdisciplines are influential and relate to exercise science professions.

Part III focuses on major related professional subdisciplines critical to the application of exercise science in specialized, real-world settings. Chapter 9 introduces students to principles of scientific research and how knowledge is applied in the field using principles of evidence-based practice. This knowledge is especially relevant to exercise science majors who participate in research as well as future professionals who must be continual consumers of scientific research on exercise. Chapters 10 and 11 highlight interdisciplinary, professional research areas related to exercise science. Students interested in sport performance will learn about knowledge creation in strength and conditioning, sport nutrition, and sport analytics in chapter 10. In chapter 11 students interested in medicine and allied health will learn about knowledge generation in applied research in several health-related professions.

Features

Introduction to Exercise Science has many features to help students organize and understand essential knowledge about exercise science.

- Chapter objectives highlight the key themes of the chapter.
- An opening scenario provides an engaging introduction to the chapter content.
- Key point boxes reiterate important information.
- Glossary terms are set in bold throughout the text for easy identification.
- Full-color figures and photos bring the text to life.
- References to cutting-edge and seminal research in the field provide relevance.

In addition, unique sidebar content provides insight into different areas:

- Exercise Science Colleagues sidebars introduce readers to inspirational professionals.
- Professional Issues in Exercise Science features highlight current issues and research studies.
- Research and Evidence-Based Practice in Exercise Science sidebars make evidence-based practice come to life.
- Hot Topic features highlight the latest exciting developments in the field.
- Career Opportunity sidebars detail new and interesting career paths.

Students should take advantage of these features to stimulate their thinking and reinforce what they have learned by reading these chapters.

Student Resources on HK*Propel*

Students have access to a variety of online resources in HK*Propel*:

- Guided notes ensure students are retaining the most pertinent information.
- Flash cards with a self-test quiz students on the key terms.
- Key point review questions test student comprehension of key concepts.
- Journal article research activities help students develop reading comprehension and become comfortable reviewing scholarly literature.
- Scenario activities allow students to evaluate a chapter-specific situation and select the best response to that situation.
- Video activities reinforce chapter content.
- Organization research activities allow students to become more familiar with the myriad exericse science-related organizations.

Instructor Resources on HK*Propel*

A variety of instructor resources are available online within the instructor pack in *HKPropel*:

- *Presentation package:* The presentation package includes more than 380 slides that cover the key points from the text, including select figures and tables. Instructors can easily add new slides to the presentation package to suit their needs.
- *Image bank:* The image bank includes most of the figures and tables from the book, separated by chapter. These items can be added to the presentation package, student handouts, and so on.
- *Instructor guide:* The instructor guide includes a sample class calendar and syllabus as well as chapter-specific files with robust lecture outlines, suggested assignments, ideas for topics and guest speakers, and answers to the chapter review questions.
- *Test package:* The test package includes 330 questions in true-false and multiple-choice formats. These questions are available in multiple formats for a variety of instructor uses and can be used to create tests and quizzes to measure student understanding.
- *Chapter quizzes:* These LMS-compatible, ready-made quizzes can be used to measure student learning of the most important concepts for each chapter. Ten questions per chapter are included in multiple-choice format.

Adopting instructors receive free instructor ancillaries, including an ebook version of the text that allows instructors to add highlights, annotations, and bookmarks. Please contact your sales manager for details about how to access instructor resources in HK*Propel*.

PART I
Foundational Knowledge

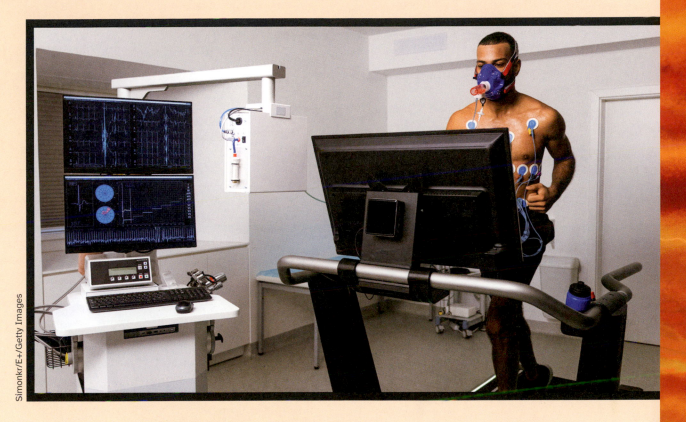

Movement is essential to human life. From basic fundamental movement patterns required for mobility, manipulating objects, and self-care to passionate practice of exercise and sport, movement is essential to a long, healthy life. Exercise science professionals use a variety of equipment and protocols to measure key variables to evaluate the health and performance status of people from all walks of life and levels of ability. While the folk wisdom to exercise regularly has been around for thousands of years, what does science say about the importance of exercise and physical activity to people's mental and physical well-being? Questions like these are explored by the academic discipline of exercise science. The evidence that has been and is being collected confirms and even extends the hypothesized benefits of physical activity and exercise.

The chapters in part I introduce you to the disciplines of exercise science and kinesiology (chapter 1). What are these areas of study, and how are they different? What is the difference between exercise and physical activity? Chapter 2 summarizes one of the most important prerequisite areas of study to exercise science: human musculoskeletal anatomy. Chapter 3 introduces measurement and statistics, a foundational area of exercise science knowledge, and evidence-based application to specific clients and special populations. Examples of unique populations include intercollegiate athletes, tactical athletes, construction workers, children, and the elderly.

Following part I, you will be ready to begin your study of major subdisciplines of the discipline of exercise science (part II) and how this knowledge can be applied and integrated with clinical sciences in professional evidence-based practice (part III). Your exercise science knowledge will put you in position to prescribe perhaps one of the most powerful factors in promoting healthy performance and longevity: a lifestyle of regular exercise and physical activity.

CHAPTER 1

The Scope of Exercise Science and Kinesiology

Duane V. Knudson

CHAPTER OBJECTIVES

In this chapter, we will

- help you appreciate the pervasiveness of human movement, physical activity, and exercise;
- introduce you to ways of defining and thinking about physical activity and exercise;
- define the disciplines of exercise science and kinesiology;
- familiarize you with the types of knowledge about exercise science that are acquired through exercise experience, scholarly research, and professional experience in exercise program implementation; and
- help you gain a preliminary understanding of what a profession is and of the many career possibilities springing from exercise science.

The majority of injuries that pull tactical athletes (fire, military, and police) out of service are musculoskeletal. Sue is an athletic trainer working with Nate to recover neuromuscular skills and strength lost due to a shoulder injury. Exercise science knowledge, professional knowledge, and client goals and values all inform the rehabilitative exercise program she provides Nate to help him return to duty.

Human life involves movement, and more and more people are seeking a variety of exercise science professionals to improve their movement performance or regain lost physical activity ability after injury or disease. The strength and conditioning coach uses knowledge of exercise technique, training, and motivation to help athletes reach their performance potential. The athletic trainer, physical therapist, or exercise physiologist uses knowledge of exercise science with additional clinical skills to recover healthy and effective movement. Elite and high-level athletes rely on exercise science to monitor nutrition, performance, and training in order to prescribe exercise programs that match their competitive schedule.

The low levels of exercise and physical activity in many peoples' lives contribute to numerous diseases, so exercise science professionals are increasingly called on to provide preventive programming to promote physical activity and exercise through community, fitness center, public health, and school programs. The wide variety of expressions of physical activity and exercise means that numerous opportunities exist for exercise science professionals.

The author acknowledges the contributions of Timothy Brusseau to this chapter.

You might not have fully appreciated it until now, but performing physical activity consumes most of your daily life. Even if you do not go to the gym or athletic field or engage in hard labor on a given day, you will probably get out of bed, walk to the bathroom, brush your teeth, get dressed, eat breakfast, and make your way to class. After your morning classes, you will probably eat lunch, visit the library, go back to your room, and do some organizational and cleaning chores. All of these are forms of physical activity.

You might think it silly to ask what is meant by the term *physical activity*. Everybody knows what it is, so why waste time defining it? However, definitions are important, especially in scientific and professional fields in which terms are often defined precisely and differently than they are in everyday language. These technical definitions ensure that people working in a particular profession or science share a common understanding of key concepts and terms on which to build a body of knowledge. This book's glossary includes many of these exercise science terms.

Scientific fields require a definition that is neither too inclusive (e.g., all human movement) nor too exclusive (e.g., only human movement related to sport). The definition of **physical activity** used in this text takes its cue from Karl Newell's (1990) formulation of physical activity as "intentional, voluntary movement directed toward achieving an identifiable goal."

Notice three things about Newell's definition: First, it does not stipulate the energy requirements of the movements used to produce the activity. The highest levels of energy are typically required for large-muscle activities, such as swimming, lifting weights, and running marathons. However, Newell's definition does not limit physical activity to such activities. Indeed, smaller-scale activities such as typing, handwriting, sewing, and surgery are every bit as much forms of physical activity as are large-muscle activities such as chopping wood.

Second, the setting in which physical activity takes place is irrelevant. Surely, shooting a basketball is a form of physical activity, and so is tossing a piece of paper into a wastebasket. Pole-vaulting is a physical activity, and so is jumping over a fence. Lifting weights is a physical activity, and so is swinging a sledgehammer. In other words, just as physical activity takes place in many settings, it also takes many forms. Wrestling and skiing differ greatly from typing and performing sign language, but all are forms of physical activity.

The third point to note about Newell's definition is that simply moving your body does not constitute physical activity. Rather, the movements must be directed toward a purposeful end. This distinction can be confusing, especially when we consider that

the term *kinesiology* is derived from the Greek words *kinesis* (movement) and *kinein* (to move). Human movement consists of any change in the position of one's body parts relative to each other. Of course, physical activity requires that we move our bodies or body parts, but movement by itself does not constitute physical activity. One way to think about the relationship between movement and physical activity is this: movement is a necessary but not sufficient condition for physical activity.

Only movement that is intentional and voluntary—purposefully directed toward an identifiable goal—meets the technical definition of physical activity. This formulation excludes all involuntary reflexes and all physiological movements controlled by involuntary muscles, such as peristalsis, swallowing, and reflexes such as eye blinking. It also excludes voluntary movements that people perform without having a goal in mind. For example, a thoughtless scratch of the head, absentminded twirling of one's hair, and the repetitive movements of a compulsive-obsessive psychiatric patient constitute human movement but fall outside of the technical definition of physical activity because they are not designed to achieve a goal.

If you take a moment to reflect on how physically active you are, you will see that your life involves an endless variety of physical activity. You walk, reach, run, lift, leap, throw, grasp, wave, push, pull, move your fingers and toes, adjust your head for a better line of vision, adjust your posture, and perform thousands of other movements as part of living a normal human existence. As a result, physical activity is essential in your work, whether you perform hard physical labor or precise tasks with low metabolic energy expenditure. We also use physical activity to express ourselves in gesture, music, art, and dance. Sometimes we don't appreciate the importance of physical activity until it is taken away from us by injury, disease, or aging.

Indeed, physical activity is part of human nature. It is an important means by which we explore and discover our world. Linking movement with complex cognitive plans helps us define ourselves as human beings. A significant part of our lifetime is spent in learning to master a broad range of physical activities, from the earliest skills of reaching, grasping, and walking to enormously complex skills such as hitting a baseball, performing a somersault, or playing the piano. Most of us master a broad range of physical activities at a moderate level of competence. Others concentrate on a limited number of skills, and this focus can lead to extraordinary performances. For example, NBA star Steph Curry's ability to sink three-point shots consistently from 30 feet (9 meters) away is the result of intense practice and motivation, as is Yo-Yo Ma's skill in positioning the bow on the cello's strings or a pilot's ability to land a fighter jet on the runway of an aircraft carrier that is being tossed by the sea.

KEY POINT

Physical activity is voluntary human movement performed intentionally in order to achieve a goal. A wide variety of physical activity is essential to a healthy and long life.

In this chapter, we talk about physical activity and human movement in relation to the academic fields of exercise science and kinesiology. Taking time to read the chapter carefully will help you appreciate the complexity and diversity of physical activity as well as its importance to human life. The chapter will also help you understand how the scientific field of exercise science is organized and related to kinesiology. If you've been physically active throughout your life, you already have some knowledge of physical activity. This background will be of enormous benefit to you as you roll up your sleeves and begin to probe the depths of knowledge of exercise science. However, prior experiences can also hinder your understanding, especially when you are required to think about those experiences in new ways. At times, therefore, you will have to set your assumptions aside so that you can examine physical activity from a fresh and exciting point of view. This endeavor sometimes can be more challenging than you might imagine.

The Disciplines of Exercise Science and Kinesiology

Because people are now more aware than ever of the importance of physical activity, enrollment in college and university curriculums devoted to the study of physical activity and exercise are on the rise. The number of undergraduate students majoring in exercise science and kinesiology has increased dramatically from the early 2000s to 2016 (Stevens et al., 2018; Wojciechowska, 2010), which made them some of the fastest-growing majors in the recent history of higher education (Nuzzo, 2020). In some universities the exercise science or kinesiology department is one of the largest academic units on campus. This surge in interest has resulted from two major reasons. First, career opportunities have expanded greatly for college-trained professionals who possess in-depth knowledge of the scientific and humanistic bases of physical activity. Before the 1990s most departments of exercise science or kinesiology (then referred to as "physical education") were designed primarily for preparing physical education teachers and coaches. Now, however, exercise science and kinesiology serve as the academic base for a diverse assortment of careers such as physical education teaching, athletic training, cardiac rehabilitation, coaching, fitness leadership and management, kinesiotherapy, occupational therapy, physical therapy, public health, sport management, and more (Spittle et al., 2021).

The growth in exercise science and kinesiology also derives from increasing awareness of the importance of physical activity and exercise to health, learning, and longevity. Many people also realize that exercise science and kinesiology deserve to be studied just as seriously and systematically as do other disciplines in higher education such as biology, psychology, and sociology (Henry, 1964). No doubt you have heard the word *discipline*, but you might not fully understand what it means in this context. A **discipline** is a body of knowledge organized around a certain theme or focus (see figure 1.1); it embodies knowledge that learned people throughout the world consider worthy

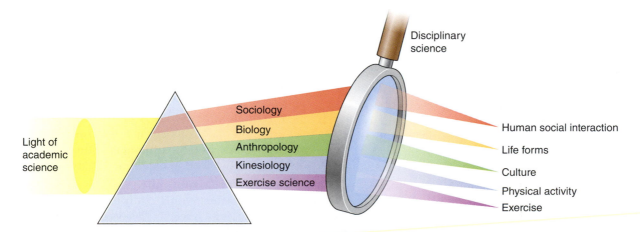

Figure 1.1   The disciplines of science each focus on knowledge of one topic. Kinesiology focuses on the wider topic of human physical activity, and exercise science focuses on the evidence-based exercise prescribed by physical activity professionals.

of study. The focus of a discipline identifies what is studied by those who work in the discipline. For example, biology focuses on life forms, psychology on the mind and mental and emotional processes, and anthropology on cultures.

You probably noticed the terms *exercise science* and *kinesiology* have been used together so far, so it is time to note the subtle differences. Recall that for hundreds of years scholars studying human movement at universities typically defined this discipline as physical education. Physical education programs in universities in the United States in the late 1800s were founded by men and women physicians (long before coeducational universities) interested in the health-promotion effects of exercise in students enrolled in colleges and universities. It would take almost 100 years for research documenting the numerous health, performance, and cognitive benefits to catch up to these visionaries who pioneered this discipline. **Kinesiology** is the larger discipline focusing on all aspects of physical activity, while **exercise science** is the discipline focusing on evidence-based exercise prescribed by physical activity professionals.

Kinesiology

In the early 20th century, the term *physical education* became more associated with the required primary and high school–based programs throughout the world than its original, more holistic meaning similar to exercise science and kinesiology. That and changes in higher education in the late 1960s initiated debate for a new name for the scientific discipline of physical education in higher education (Henry, 1964; Newell, 1990; Thomas, 1987). Gradually, the consensus for describing the whole discipline or body of knowledge focused on physical activity became kinesiology in North America (American Kinesiology Association, 2021; Hoffman, 2009; Mahar & Crenshaw, 2015; National Academy of Kinesiology, 2021; Newell, 1990).

The American Kinesiology Association (AKA), an association of academic departments of kinesiology at North American universities, defines kinesiology as "an academic discipline which involves the study of physical activity and its impact on health, society, and quality of life" (2021). Kinesiology draws on several sources of knowledge, including

- actual physical activity experience,
- scholarly study and research on physical activity, and
- results of professional experience in implementing physical activity programs.

Although the disciplines are most often associated with scholarly study and research, AKA recognizes that the body of knowledge of kinesiology is informed by and defined by the other two sources as well.

Ultimately, the uniqueness of kinesiology as a discipline derives from its embrace and integration of subdisciplinary study and application of physical activity, including biological, medical, philosophical, psychological, social-humanistic, and numerous other subdisciplinary as well as professional perspectives. This integration of many disciplinary perspectives into a unique field has been described as interdisciplinary (Harris, 1993; Knudson, 2013, 2021; Zeigler, 1990), meaning the knowledge from subdisciplines must be integrated, not just looked at separately from several unique perspectives or disciplines. Students may take specific subdisciplinary courses (e.g., biomechanics and exercise physiology) in their academic programs, but the whole disciplines of kinesiology and exercise science are holistic in nature. This chapter provides a preview some of the subdisciplines of exercise science explored throughout the book. You will find many of the authors note examples of how their subdisciplinary knowledge is integrated with other subdisciplines of the field.

However, one should be aware of other less recognized and borrowed uses of the term *kinesiology*. For example, the founding of physical therapy was primarily based on the corrective exercise research from physical education in the late 19th and early 20th centuries, so "clinical kinesiology" is still used sometimes to refer to functional anatomy courses taught in occupational and physical therapy programs. Similarly, "dental kinesiology" applies virtually the same techniques in anatomically and mechanically analyzing the motions of the jaw and tongue. In addition, several forms of alternative medicine have begun to use the term *kinesiology* in their titles. Here are some examples:

- Behavioral kinesiologists treat patients using biofeedback.
- Spiritual kinesiologists claim to promote healing by uniting body and soul.
- Applied kinesiology is used by some practitioners of chiropractic to refer to an alternative medicine in which manual testing of muscle strength claims to diagnose disease.
- Kinesiology tape (e.g., Kinesiotape) is an elastic tape product, though little exercise science and kinesiology research was used in the development of these products.

All these uses of the term *kinesiology* are rare and not the primary academic meaning of the term recognized worldwide. So far, these alternative meanings of *kinesiology* have not been problematic, given the lack of scientific evidence for their claims as compared with the long-term, robust international body of research in academic kinesiology. Given the more than 150-years, world-wide research on physical education and kinesiology, it is little wonder that products and scientifically questionable alternative health services try to use the term *kinesiology* to encourage sales or services.

While kinesiology is now generally regarded as the discipline that focuses on human physical activity, many university departments specialize in academic programs and research on more focused areas from the great diversity of physical activity. Many international universities focus their mission and name on human movement studies or sport sciences given the great cultural interest in sport (Čustonja et al., 2009). Still other programs focus on the applications of the field with a name like Department of Health and Human Performance. Although individual departments might choose to shape their curriculums and research agendas on the basis of selected areas of kinesiology, such institutional preferences should not be interpreted as reflecting a comprehensive definition of the large, diverse discipline of kinesiology.

Exercise Science and the Definition of Exercise

Perhaps the second most common scholarly focus within kinesiology is exercise science or exercise and sport science. Exercise science is the discipline or body of knowledge focusing on evidence-based exercise prescribed by physical activity professionals. While *physical activity* is the more general term for all voluntary human movement performed intentionally to achieve a goal, exercise is voluntary human movement consisting of evidence-based prescription based on exercise science and kinesiology knowledge. While it might be colloquial wording to say a person went running for 3 miles (5 km) for exercise, it would be more accurate to say they chose the physical activity of running. An exercise prescription for cardiovascular fitness goals that included jogging or running would have much more specific details on warm-up, modification of speed and grade to elicit certain metabolic conditions, cool-down, and perhaps other exercise movements and activities. A passionate runner might use sophisticated sensors and notes tracking their run, but this would not usually constitute the full evaluation, evidence-based guidance,

and monitoring of professional exercise prescription. What is meant by evidence-based professional practice in exercise science is developed throughout this text. Throughout the rest of this text, use of the term *exercise science* will be emphasized, although in many cases it would also be relevant to use the term *kinesiology*, which also uses evidence-based practice in other physical activity contexts.

KEY POINT

Exercise is voluntary human movement consisting of evidence-based prescription based on the specialized knowledge of the field of exercise science or kinesiology.

While dance and sport have physical, social, and aesthetic benefits, exercise might be the most important form of physical activity given its greater potential for a variety of benefits from the evidence-based programming. People engage in guided exercise to improve their appearance, physical performance, and health and longevity, or to regain performance ability that has been reduced as a result of injury, disease, or aging. Because exercise involves many types of prescribed or programmed movements focused on improving movement function, it is helpful to break the concept down into a few subcategories.

Training Exercise

Training exercise consists of programmed movements performed for the express purpose of improving athletic, military, work-related, or recreation-related performance. This exercise is often intense in order to achieve high levels of physical performance necessary to be successful in vigorous competition and work movements. Exercise training is particularly important to exercise science graduates who embark on careers as strength and conditioning specialists for university, Olympic, or professional sport teams (chapter 10). Exercise science graduates can take the test for the Certified Strength and Conditioning Specialist (CSCS) and other advanced credentials affiliated with the National Strength and Conditioning Association (NSCA). Training exercise is now being used in medical settings to help people acquire physical skills and functions considered normal or expected. This type of preventive medical training is called *habilitative therapeutic exercise* (chapter 11 and in the subsequent section on Therapeutic Exercise) and is a new addition to the traditional remedial, rehabilitation, or therapeutic exercise to treat disease or injury common in medicine.

Health-Related Exercise

Health-related exercise is undertaken specifically to develop or maintain a sound working body and reduce the risk of disease for the purpose of healthy longevity. Health-related exercise serves as the primary focus for exercise science graduates who work as fitness leaders or personal trainers and in public health. Recall that this preventive or wellness promotion hypothesis of exercise was what physical education was founded on. Over 100 years of research has shown that this health promotion hypothesis was correct.

In the late 20th century this research on physical activity and exercise accelerated. This research documented the dramatic benefits of physical activity in reducing cardiovascular disease and premature deaths. By the mid-1990s major public health recommendations were announced in the United States, concluding that major health benefits could be obtained by only a moderate amount of physical activity on most, if not all, days of the week (Pate et al., 1995; U.S. Department of Health and Human Services, 1996). More recent research has documented expanded, dramatic positive impacts of exercise and physical activity in reducing morbidity (diseases like cancer, diabetes, and heart disease) and healthy mortality (Bull et al., 2020; Lee et al., 2012; Moore et al., 2016). These health benefits are associated with total volume of physical activity, irrespective of the mix of the intensity, duration, or type of activity (Physical Activity Guidelines Advisory Committee, 2018), so current recommendations also include limiting sedentary time (Bull et al., 2020). The epidemiology chapter (chapter 8) will summarize this large body of research on exercise and physical activity as preventive, wellness-promotion strategies that reduce the burdens of disease and disability. Public health scientists working at the population level often emphasize physical activity of groups of people over individual-specific exercise prescription.

Unfortunately, medical professionals, organizations, and the pharmaceutical industry often have been resistant to accepting this strong evidence (Ekkekakis, 2021). Medical calculations of the global burden of disease related to physical activity make numerous errors in thoroughly accounting for the multidimensional benefits of physical activity (Stamatakis et al., 2021). The American College of Sports Medicine (ACSM) does strive to educate physicians on exercise as a vital sign and promote exercise through an Exercise Is Medicine campaign. These initiatives are positive for educating medical professionals who are not well trained in health promotion and the preventive benefits of exercise and physical activity. Exercise is not just medicine because the benefits of exercise and physical activity are much more than disease remediation, they are powerful preventive interventions (Cheng & Mao, 2016; Hart & Zernicke, 2020), reducing disease and premature mortality. Exercise science must keep documenting and promoting the positive and preventive benefits of physical activity and exercise to ensure that greater emphasis is placed on exercise by public health and medical professionals.

Cosmetic Exercise

Cosmetic exercise is intended to reshape a person's body for aesthetic reasons. This kind of systematic training is used by bodybuilders, models, and people wanting to lose or gain weight for a summer trip. It is sometimes difficult to determine whether trying to achieve, say, a smaller waistline is more a cosmetic effort or a health-related endeavor. Even so, it is important for exercise science professionals to understand the motivations and difficulties involved in exercising to change one's body shape or size because these goals are often difficult to achieve. Motivating clients on this difficult path could be as important as knowing the best exercise and dietary programs to achieve the cosmetic goals.

Therapeutic Exercise

Therapy or **therapeutic exercise** involves specialized and individualized movements performed to restore or develop physical capacities that have been lost due to injury, disease, behavioral patterns, or aging. For example, individualized therapeutic or rehabilitative exercise programs are prescribed and implemented by exercise science graduates who complete additional education as athletic trainers or occupational or physical therapists (see chapter 11). Stroke patients, for instance, sometimes require many sessions of physical or occupational therapy to help them regain the ability to perform important movements of daily living.

As was noted earlier, therapeutic exercise in allied health professions traditionally has been remedial in nature, correcting or accommodating for injury or illness. In contrast, recall we noted physical education in higher education, the precursor to kinesiology and exercise science, was founded on research documenting the proactive, preventive benefits of exercise. Several biomechanical and sports medicine studies have followed this lead and documented that specific, specialized prehabilitation training programs (usually implemented by athletic trainers or strength and conditioning professionals) can reduce one's risk of certain traumatic or overuse injuries (Elliott & Khangure, 2002; Herman et al., 2009; Hewett et al.,

Kevin Guskiewicz

Kenan Distinguished Professor, Department of Exercise and Sport Science, University of North Carolina, Chapel Hill

Have you ever imagined yourself working for the National Football League (NFL) to help understand the causes of concussions and other brain injuries? How about waking up one day to find that you have been given several hundred thousand dollars to pursue that work? That's exactly what happened to Kevin Guskiewicz, Kenan Distinguished Professor of Athletic Training in the Department of Exercise and Sport Science at the University of North Carolina, Chapel Hill. He is an athletic trainer and researcher who has been a longtime scientific advisor to the NFL thanks to his research on sport-related concussions. His work has been so influential that in 2011 he received a MacArthur Foundation Fellowship, or "Genius Grant." These prestigious awards are given to scholars and artists whose work is deemed to show great promise for improving the world; the award's prestige and funding make it essentially equivalent to a Nobel Prize. Dr. Guskiewicz and other exercise scientists are changing the world for the better. Like many other outstanding exercise science scholars, his expertise and leadership also have been rewarded with numerous appointments as deans, provosts, and university presidents. Dr. Guskiewicz serves as chancellor of the University of North Carolina at Chapel Hill.

1999; Noehren et al., 2011). Another highly successful preventive exercise program from exercise scientists is highlighted later in the chapter in the Research and Evidence-Based Practice in Exercise Science feature. These are exciting times for exercise science knowledge and interdisciplinary collaboration with medical professionals who have now begun to adopt a more proactive, wellness-promoting perspective on prescribed exercise. Since injuries are multifactorial in nature, exercise scientists can do considerable research seeking to document when prehabilitation or habilitative therapeutic exercise can reduce the risk of injury.

Collaboration of Exercise Science and Kinesiology

Even though exercise science and kinesiology grew from the same physical education roots in higher education, as governments and universities deal with limited financial resources some academic programs have specialized. Specific faculty, administrators, and college or university mission often drive if a department's mission focuses on the whole discipline of kinesiology or a tighter focus on exercise, sport, or other specific area of physical activity. Despite these trends the curriculum of courses and experiences of kinesiology and exercise science majors are similar, with specific variations designed by faculty to best meet specific fields and careers, all within the university context.

The rest of this text will focus on the discipline of exercise science. Many exercise science majors

and programs place great emphasis on the scientific research that creates the knowledge that drives evidence-based exercise prescriptions. Remember that this discipline is similar to others across the world that are referred to as *health and human performance*, *kinesiology*, or *sport science*. These other programs and majors have instruction in many of the same subdisciplinary sciences, but some place additional emphasis on experiential and professional experience sources of knowledge (see figure 1.2). Exercise science students and faculty can learn much from their kinesiology colleagues, and vice versa. Integration of all relevant knowledge is what professionals seek to provide in evidence-based practice.

Sources of Knowledge of Exercise Science

In your college courses, you might have noticed that disciplines are not all learned or studied in the same way. Art, for example, can be studied through reading, writing, and experimentation with studio projects. People learn history, literature, and philosophy largely through reading, writing, memorization, debate, and discussion. The same activities are important when learning chemistry and biology, but these disciplines also involve active participation in laboratory experiences.

People acquire exercise science knowledge in three different but related ways (see figure 1.2), one of which is through exercise experience. Just as students in art and music learn to appreciate their disciplines

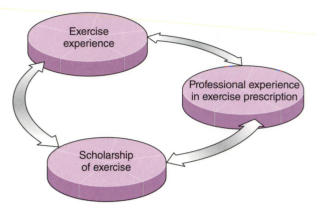

Figure 1.2 Three interrelated sources of knowledge in exercise science.

in part by watching, listening, and performing, you can develop your understanding of exercise science in part through the direct personal experience of observing or performing exercise. You may also learn important exercise testing and monitoring skills through actual physical experience and practice.

A second way of developing an understanding of exercise science is through scholarship of exercise science. This way of learning involves researching, reading about, studying, and discussing with colleagues both theoretical and practical aspects of exercise; it also involves modeling, experiments, and laboratory experiences. These forms of study are necessary in order to master the various subjects included in the exercise science curriculum. Where does the knowledge contained in such subjects come from? Mostly, it comes from work done by scholars in the field of exercise science who have developed and added to the knowledge base through systematic research and scholarship. Scholars who conduct research in biomechanics, exercise and sport psychology, motor learning, and many other subdisciplines develop important foundational knowledge for the scholarly study of physical activity and exercise.

A subdiscipline in exercise science is often related to a broader, more established parent discipline such as psychology, physiology, sociology, biology, history, or physics. For instance, biomechanics draws on physics concepts, exercise physiology draws on basic concepts and theories from physiology, and the study of motor behavior draws on psychology. These relationships mean that exercise science students must develop a working knowledge of the language, theories, and conceptual frameworks of a number of major disciplines and learn to apply them to exercise. Some preprofessional (designed to align with admission to DAT, DPT, and other programs) exercise science degree programs include additional

parent discipline prerequisite coursework because graduate or professional schools require them of all students who apply to their programs. Some exercise science subdisciplines focus on the effects of exercise in particular populations such as elite athletes, older adults, children, or people affected by disease or disability. Exercise science has numerous subdisciplinary sciences, given the diversity of populations and purposes for which exercise is used.

A third way of learning about exercise is through professional experience in exercise science. Here, the focus is placed not so much on learning to perform exercise movements or on learning about them, but on learning by designing and implementing exercise programs for clients in particular professional scopes of practice. For example, professionals such as athletic trainers, cardiac rehabilitation specialists, physical therapists, and strength coaches systematically manipulate and monitor the exercise experiences of students, clients, patients, and others whom they serve in order to help them achieve personal goals.

Part II of this book explores the knowledge gained through study of five major subdisciplines common in university exercise science curriculums. Part III describes three topics in professional experience in exercise science. First, students are introduced to exercise science research methods that generate knowledge and how this is applied in evidence-based practice in professions. The last two chapters introduce research in two professional practice areas where this applied knowledge is integrated with exercise science knowledge in evidence-based practice.

KEY POINT

Knowledge in the discipline of exercise science can be acquired in three ways: experience, scholarship, and professional experience in exercise science.

Knowledge gained through these three sources becomes part of the discipline of exercise science only when it is embedded in a college or university curriculum or is universally accepted and used by exercise scientist scholars in their research. This caveat is given in order to clarify precisely what is considered part of the "official" discipline and what is not. Many people experience exercise (e.g., jogging, playing golf, weight lifting), study it informally (e.g., read *Sports Illustrated* or popular websites on fitness or sport), or engage in some form of exercise leadership (e.g., volunteering as a coach) outside the confines of the university curriculum. However, these activities do not constitute the practice of exercise science, and the people engaging in them are not exercise scientists in the strict definition used in this text. Exercise scientists are usually university faculty who have earned the highest doctoral degrees (PhD) offered in disciplines and then contribute to that body of knowledge through their own peer-reviewed research. Professional degrees (recently called doctorates like DAT, DPT, MD, DO, etc.) are not research degrees since they focus on knowledge and skills of professional practice.

To be sure, exercise and exercise leadership are important and valuable in their own right, but they do not constitute practicing exercise science any more than the use of elementary psychological principles by a businessperson to motivate their sales force constitutes professional practice of psychology. The discipline of psychology remains tied to the college and university curriculum and to the research conducted by psychologists. Similarly, people can use the principles of exercise science outside of the discipline, but exercise science per se remains a function of curricula and the consensus of research usually conducted in colleges and universities.

KEY POINT

Only knowledge about exercise that is included in a college or university curriculum and based on the consensus of research is considered to be part of the body of knowledge of exercise science.

Another reason for limiting our definition of exercise science lies in the fact that the knowledge you acquire in your major curriculum is more highly organized and more scientifically verifiable than the knowledge of exercise held by laypersons. Universities use rigorous methods to organize and monitor the authenticity of the knowledge included in their curricula and in the research conducted by their faculty members. Think about it: Would you have more confidence in recommendations made by a university exercise scientist who specializes in knee rehabilitation research than in recommendations made by someone who had and recovered from a knee injury? Similarly, would you have more confidence in the scientific accuracy of recommendations for exercise programs offered by an exercise physiologist than in recommendations offered by a television fitness model or exercise guru who lacks formal training in exercise science? Given your decision to invest several years of hard work in preparing for a career in the specialized field of exercise science, your answer to these questions is likely to be yes. Additionally,

numerous studies have reported significant improvements in skills, fitness, and lower rates of injury from programs based on exercise science research compared to traditional instruction or training. (See the Exercise Science and Evidence-Based Practice Boxes throughout this text.)

Exercise Science and Professional Careers

Exercise science opens the door to a wide range of professional careers. Because exercise science programs expose you to broad knowledge about prescribing exercise, they provide excellent preparation for careers in fitness leadership and consulting, teaching and coaching, cardiac and neuromuscular rehabilitation, sport management, strength and conditioning, and admission to numerous professional programs like athletic training, physical therapy, occupational therapy, podiatry, chiropractic care, medicine, public health, nutrition, and, of course, graduate study in exercise science and kinesiology.

Many college graduates pursue master's and doctoral degrees in exercise science or kinesiology (see figure 1.3). Some seek this advanced study to improve their qualifications and meet advanced certification requirements. Others pursue graduate work to meet the educational requirements of a medical profession; in fact, a master's degree is increasingly viewed as the minimal requirement for most allied health professions. Sometimes students continue their studies beyond the master's level in order to obtain a PhD in a subdiscipline of exercise science or kinesiology so that they can become college or university faculty members or researchers in other organizations. The greater the level of your education, the higher the typical pay in exercise science–related careers (Wagner, 2021).

Holistic Study of Subdisciplines of Exercise Science

We hope that by the time you complete this introductory study of exercise science, you will be convinced of the holistic nature of exercise and physical activity. Holism refers to the interdependence of mind, emotion, body, and spirit. Although some people think that exercise science deals exclusively with the body and with bodily movement, in reality it spans a much broader range. Exercise and physical activity involves our minds, our emotions, and our souls as much as it does our bodies. It is convenient to speak

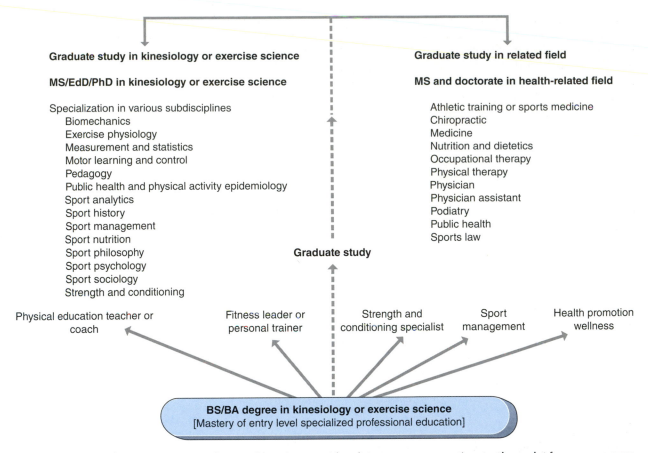

Figure 1.3 An undergraduate degree in exercise science or kinesiology can serve as the starting point for many careers, some of which require additional training in graduate programs.

of physical activity because the physical aspects are so easily observed, yet physical activity is also cognitive activity, emotional activity, and even spiritual activity for many people. In short, it is best to avoid the false dichotomy of separate mind and body in exercise science.

KEY POINT

Although bodily aspects of exercise science receive the most attention, the study and application of subdisciplines of exercise science must account for the fact that human beings are holistic, multidimensional creatures characterized by the interrelated elements of cognition, emotion, body, and soul.

Studying exercise science will take you far beyond the study of the anatomical aspects of exercise. Exercise science also studies the psychological antecedents and outcomes of physical activity and exercise; the dynamics of skill development, performance, and learning; and other biophysical subdisciplinary sci-

ences informing exercise. No introductory exercise science book can cover all the subdisciplines of the field. This book focuses on five of the most common subdisciplines in exercise science programs (see figure 1.4):

- Biomechanics
- Exercise physiology
- Motor behavior
- Sport and exercise psychology
- Physical activity epidemiology

Notice how these subdisciplines have different, historically-driven, preferred titles that emphasize different areas of the larger discipline of kinesiology. For example, physical activity epidemiology focuses on large (population-based) public health efforts to promote health and longevity through the powerful intervention of physical activity.

It is great that the importance of exercise and physical activity draws so much attention from numerous parent disciplines and subdisciplines within our field. The diverse discipline of exercise science with

The holistic nature of human experience—effort, competition, fulfillment—is particularly evident in our passions for physical activity and sport.

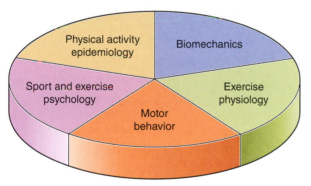

Figure 1.4 The five major subdisciplines of exercise science explored in this book.

many subdisciplines sometimes results in a variety of meanings for some terms that creates potential miscommunication (Knudson, 2019). Remember that all these subdisciplinary scientific perspectives provide important knowledge about exercise and physical activity, but that knowledge is most effective when not used in isolation but integrated in an interdisciplinary fashion with all relevant subdisciplines (Harris, 1993; Knudson, 2013, 2021; Zeigler, 1990).

One way to achieve this integration is evidence-based professional practice.

Exercise Science and Evidence-Based Practice

Many exercise science and health professions require, in addition to bachelor's and graduate degrees, state or national licensure or certifications. Many legally established professions base their services on principles of evidence-based practice. Evidence-based practice involves use of one of several models of integrating three kinds of evidence to treat or serve clients:

- Scientific research
- Professional or clinical experience
- The values and specific characteristics of the client

Formal evidence-based practice models grew out of evidence-based medicine in the late 1980s at Oxford University (Centre for Evidence-Based Medicine, 2022).

Professional Issues in Exercise Science

Ensuring Equity, Evidence-Based Practice, and Expertise in Our Services

Exercise science professionals have ethical requirements to provide effective, high-quality service to clients. Three common areas of this responsibility are equity, evidence-based practice, and expertise. As a result, each chapter of this book illustrates how exercise science knowledge can be applied to make professional decisions related to these issues. All three sources of exercise science knowledge are illustrated, and key references for further study are cited in these features. The book also integrates a variety of subdisciplines of exercise science in order to illustrate the diversity and interdisciplinarity of professional applications.

Professional Issues in Exercise Science sidebars also provide examples of different formats and styles for scholarly citations and references that are common in this field of numerous subdisciplines and professional applications. Many behavioral sciences use the American Psychological Association (APA) format, whereas publications in sports medicine and athletic training often use the American Medical Association (AMA) format. The references in this text follow a modified APA format to save space, only providing doi (digital object identifier is a unique, permanent URL for a publication) numbers when that is the only way to access the publication. At times, exercise science professionals also need to write reports or read research from other scientific disciplines with different reporting styles. For more information about the APA and AMA styles, please see the following sources:

- American Medical Association. (2020). *AMA manual of style* (11th ed.). Oxford University Press.
- American Psychological Association. (2020). *Publication manual of the American Psychological Association* (7th ed.). American Psychological Association.

Research and Evidence-Based Practice in Exercise Science

Earlier we noted that exercise science research has driven innovations that improve human performance and well-being. One example of national-scale success is the RugbySmart program in New Zealand. Dr. Patria Hume and her colleagues first proposed a 10-point strategy to reduce the rate of injuries in one of the most popular sports in that country, rugby. Following proof of concept studies, and with interdisciplinary support from national sport governing bodies and the national health service, the program has substantially reduced the rate of serious spinal injuries in the sport (Quarrie et al., 2007, 2020). The success of this program has spun off into other "Smart" national injury prevention programs in several sports across New Zealand.

Some of your training in exercise science and beyond will be on how to evaluate the quality of different scientific studies and integrate that with experiential knowledge. This aligns with the three-dimensional picture (exercise experience, scholarship of exercise, and professional experience in exercise) of exercise science knowledge. Only through the thoughtful integration of all these sources of exercise science knowledge can you truly provide the best evidence-based exercise prescription to your clients. Notice how evidence-based practice affirms the holistic perspective of sciences and the clients we serve.

Throughout the subdisciplinary chapters of this book (part II) are boxed features (Research and Evidence-Based Practice in Exercise Science, and Professional Issues in Exercise Science) that illustrate how scientific research in exercise science contributes knowledge for evidence-based practice in a variety of professions. Chapter 9 provides more information on exercise science research and how it can be applied in evidence-based practice. The other chapters in part III preview research on professional experience in exercise prescriptions in two major areas: performance (chapter 10) and allied health and medicine (chapter 11).

Wrap-Up

With this introductory knowledge under your belt, you are now ready to dig into the topics of exercise science more deeply. In this chapter, you have learned that exercise science is the study of professionally prescribed or programmed exercise. This discipline is a major part of the larger discipline of kinesiology, which studies physical activity. The terms *exercise* and *physical activity* have technical definitions that are different from their colloquial use. You also have learned that the expressions of exercise studied by exercise scientists vary widely, as do the professional careers related to the field.

In addition, you have learned something about how the disciplines of exercise science and kinesiology are organized by subdisciplines. Exercise science knowledge comes from three main sources: actual experience that comes from participating in exercise, the scholarly study of exercise, and the knowledge that comes from professional experience in implementing exercise programs with clients.

More Information on Exercise Science

Organizations

American College of Sports Medicine (ACSM)

American Kinesiology Association (AKA)

American Kinesiotherapy Association

British Association of Sport and Exercise Sciences (BASES)

International Council for Health, Physical Education, Recreation, Sport, and Dance (ICHPER-SD)

International Organization for Health, Sports, and Kinesiology (IOHSK)

National Academy of Kinesiology (NAK)

National Association for Kinesiology in Higher Education (NAKHE)

National Strength and Conditioning Association (NSCA)

Journals

Adapted Physical Activity Quarterly

European Journal of Sport Science

Exercise and Sport Sciences Reviews

German Journal of Exercise and Sport Research

ICHPER-SD Journal of Research

International Journal of Exercise Science

International Journal of Kinesiology in Higher Education

International Journal of Sports Physiology and Performance

Isokinetics and Exercise Science

Journal of Aging and Physical Activity

Journal of ICHPER · SD

Journal of Physical Activity and Health

Journal of Science and Medicine in Sport

Journal of Sports Medicine and Physical Fitness

Journal of Sports Sciences

Kinesiology

Kinesiology Review

Measurement in Physical Education and Exercise Science

Medicine & Science in Sports & Exercise

Pediatric Exercise Science

Quest

Research Quarterly for Exercise and Sport

Scandinavian Journal of Medicine & Science in Sports

Sport Nutrition and Exercise Metabolism

Women in Sport and Physical Activity Journal

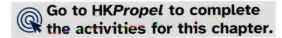 **Go to HK*Propel* to complete the activities for this chapter.**

Review Questions

1. What is the difference between movement and physical activity? Give an example of an instance in which human movement does not meet the technical definition of physical activity.
2. What is the difference between physical activity and exercise?
3. What is meant when exercise science and kinesiology are described as holistic disciplines?
4. What forms or categories of exercise are studied in exercise science?
5. What are some major subdisciplines of exercise science?
6. What are the three sources of knowledge of exercise science?
7. How should exercise science knowledge be applied in prescribing exercise?

Suggested Readings

Berryman, J.W., & Park, R.J. (Ed.). (1992). *Sport and exercise science: Essays in the history of sports medicine*. University of Illinois Press.

Knudson, D. (2013). *Qualitative diagnosis of human movement: Improving performance in sport and exercise* (3rd ed.). Human Kinetics.

Knudson, D.V., & Brusseau, T. (Eds.). (2022). *Introduction to kinesiology* (6th ed.). Human Kinetics.

Massengale, J.D., & Swanson, R.A. (1997). *The history of exercise and sport science*. Human Kinetics.

McCrory, P. (2006). What is sports and exercise medicine? *British Journal of Sports Medicine, 40,* 955-957.

References

American Kinesiology Association. (2021). *About AKA*. Retrieved August 23, 2022, from www.americankinesiology.org/SubPages/Pages/About

Bull, F.C., Salih, S.A., Biddle, S., Borodulin, K., Buman, M.P., Cardon, G., Carty, C., Chaput, J.-P., Chastin, S., Chou, R., Dempsey, P.C., DiPietro, L., Ekelund, U., Firth, J., Friedenreich, C.M., Garcia, L., Gichu, M., Jago, R., Katzmarzyk, P.T., . . . Willumsen, J.F. (2020). World Health Organization 2020 guidelines on physical activity and sedentary behaviour. *British Journal of Sports Medicine, 54*(24), 1451-1462.

Centre for Evidence-Based Medicine. (2022). *CEBM*. Retrieved October 28, 2022, from www.cebm.net

Cheng, S., & Mao, L. (2016). Physical activity continuum throughout the lifespan: Is exercise medicine or what? *Journal of Sport and Health Science, 5,* 127-128.

Čustonja, Z., Milanovic, D., & Sporiš, G. (2009). Kinesiology in the names of higher education institutions in Europe and the United States of America. *Kinesiology, 41*(2), 136-146.

Ekkekakis, P. (2021). Why is exercise underutilized in clinical practice despite it is effective? Lessons in pragmatism from the inclusion of exercise in guidelines for the treatment of depression in the British National Health Service. *Kinesiology Review, 10*(1), 29-50.

Elliott, B.C., & Khangure, M. (2002). Disk degeneration and fast bowling in cricket: An intervention study. *Medicine & Science in Sports & Exercise, 34*(11), 1714-1718.

Harris, J.C. (1993). Using kinesiology: A comparison of the applied veins in the subdisciplines. *Quest, 45*(3), 390-412.

Hart, D.A., & Zernicke, R.F. (2020). Optimal human functioning requires exercise across the lifespan: Mobility in a 1g environment is intrinsic to the integrity of multiple biological systems. *Frontiers in Physiology, 11,* 156. https://doi.org/10.3389/fphys.2020.00156

Henry, F.M. (1964). Physical education: An academic discipline. *Journal of Physical Education, Recreation, and Dance, 35*(7), 32-33, 69.

Herman, D.C., Oñate, J.A., Weinhold, P.S., Guskiewicz, K.M., Garrett, W.E., Yu, B., & Padua, D.A. (2009). The effects of feedback with and without strength training on lower extremity biomechanics. *American Journal of Sports Medicine, 37*(7), 1301-1308.

Hewett, T.E., et al. (1999). The effect of neuromuscular training on the incidence of knee injury in female athletes. A prospective study. *American Journal of Sports Medicine, 27*(6), 699-706.

Hoffman, S.J. (2009). What's in a name? *Kinesiology Today, 2*(1), 1-2.

Knudson, D. (2013). *Qualitative diagnosis of human movement: Improving performance in sport and exercise* (3rd ed.). Human Kinetics.

Knudson, D. (2019). Kinesiology's tower of babel: Advancing the field with consistent nomenclature. *Quest, 71*(1), 42-50.

Knudson, D. (2021). *Fundamentals of biomechanics* (3rd ed.). Springer Science.

Lee, I.-M., Shiroma, E.J., Lobelo, F., Puska, P., Blair, S.N., Katzmarzyk, P.T., & Lancet Physical Activity Series Working Group. (2012). Effect of physical inactivity on major non-communicable diseases worldwide: An analysis of burden of disease and life expectancy. *Lancet, 380* (9838), 219-229.

Mahar, M., & Crenshaw, J.T. (2015). *AKA Salary Survey 2015*. American Kinesiology Association.

Moore, S.C., Lee, I.-M., Weiderpass, E., Campbell, P.T., Sampson, J.N., Kitahara, C.M., Keadle, S.K., Arem, H., Berrington de Gonzalez, A., Hartge, P., Adami, H.-O., Blair, C.K., Borch, K.B., Boyd, E., Check, D.P., Fournier, A., Freedman, N.D., Gunter, M., Johannson, M., . . . Patel, A.V. (2016). Association of leisure-time physical activity with risk of 26 types of cancer in 1.44 million adults. *Journal of the American Medical Association Internal Medicine, 176*(6), 816-825.

National Academy of Kinesiology. (2021). *Kinesiology: The discipline and related professions*. Retrieved August 23, 2022, from https://nationalacademyofkinesiology.org/SubPages/Pages/What%20is%20Kinesiology

Newell, K.M. (1990). Kinesiology: The label for the study of physical activity in higher education. *Quest, 42*(3), 269-278.

Noehren, B., Scholz, J., & Davis, I. (2011). The effect of real-time gait retraining on hip kinematics, pain, and function in subjects with patellofemoral pain syndrome. *British Journal of Sports Medicine, 45*(9), 691-606.

Nuzzo, J.L. (2020). Growth of exercise science in the United States since 2002: A secondary data analysis. *Quest, 72*(3), 358-372.

Pate, R.R., Pratt, M., Haskell, W.L., Macera, C.A., Bouchard, C., Buchner, D., Ettinger, W., Heath, G.W., & King, A.C. (1995). Physical activity and public health: A recommendation from the Centers for Disease Control and Prevention and the American College of Sports Medicine. *Journal of the American Medical Association, 273*, 402-407.

Physical Activity Guidelines Advisory Committee. (2018). *Physical activity guidelines advisory committee scientific report*. U.S. Department of Health and Human Services. https://health.gov/sites/default/files/2019-09/PAG_Advisory_Committee_Report.pdf

Quarrie, K., Gianotti, S.M., Hopkins, W.G., Hume, P.A. (2007). Effect of nationwide injury prevention programme on serious spinal injuries in New Zealand rugby union. *British Medical Journal, 334*(7604), 1150-1153.

Quarrie, K., Gianotti, S., Murphy, I., Harold, P., Salmon, D., & Harawira, J. (2020). RugbySmart: Challenges and lessons from the implementation of a nationwide sports injury prevention partnership programme. *Sports Medicine, 50*(2), 227-230.

Spittle, M., Daley, E.G., & Gastin, P.B. (2021). Reasons for choosing an exercise and sport degree: Attractors to exercise and sport science. *Journal of Hospitality, Leisure, Sport & Tourism Education, 29*, 100330. https://doi.org/10.1016/j.jhlste.2021.100330

Stamatakis, E., Ding, D., Ekelund, U., & Bauman, A.E. (2021). Sliding down the risk factor rankings: Reasons for and consequences of the dramatic downgrading of physical activity in the Global Burden of Disease 2019. *British Journal of Sports Medicine*, https://doi.org/10.1136/bjsports-2021-104064

Stevens, C.J., Pluss, M., Nancarrow, S., & Lawrence, A. (2018). The career destination, progression, and satisfaction of exercise and sports science graduates in Australia. *Journal of Clinical Exercise Physiology, 7*(4), 76-81.

Thomas, J.R. (1987). Are we already in pieces or just falling apart? *Quest, 39*(2), 114-121.

U.S. Department of Health and Human Services. (1996). *Physical activity and health: A report of the surgeon general*. U.S. Department of Health and Human Services, Centers for Disease Control and Prevention, National Center for Chronic Disease Prevention and Health Promotion. https://health.gov/sites/default/files/2019-09/paguide.pdf

Wagner, D.R. (2021). Salaries of exercise science professionals in the United States. *ACSM's Health & Fitness Journal, 25*(1), 36-42.

Wojciechowska, I. (2010, August). A quickly growing major. *Inside Higher Education*, www.insidehighered.com/news/2010/08/11/kinesiology

Zeigler, E.F. (1990). Don't forget the profession when choosing a name! In C.B. Corbin & H.M. Eckert (Eds.), *The evolving undergraduate major: American Academy of Physical Education Papers No. 23* (pp. 67-77). Human Kinetics.

CHAPTER 2

Musculoskeletal Anatomy

Duane V. Knudson

CHAPTER OBJECTIVES

In this chapter, we will

- introduce you to qualitative anatomical descriptions of structure and motion,
- explore anatomical planes of motion and their axes,
- introduce how anatomy hypothesizes muscle contributions to motion,
- introduce three kinds of muscle actions, and
- illustrate how anatomy must be integrated with subdisciplinary exercise science knowledge to determine how muscles contribute to motion.

Brent and Victor are good friends and are both passionate about basketball. Although they play in different leagues, they often shoot around together and share their love for the game. Frequent discussions relate to the best technique and body contributions to shooting success. While they share the same anatomy, Brent and Victor have different beliefs about what exercises and techniques help them to be better shooters in basketball.

Coaches and athletic trainers working with basketball players need knowledge of the structure of the body to help players like Brent and Victor train for play, deal with injuries, and communicate with other medical and exercise science professionals working with them. The human body is an amazing, complex, and living system. The study of the structure of that system is called *anatomy*. Human tissues are fairly consistent in general structure; however, some people are born with variations in anatomical structure. Anatomy is an essential prerequisite area of knowledge that must be combined with subdisciplinary knowledge from exercise science to explain actual musculoskeletal function. Understanding of shooting technique differences between Brent and Victor requires anatomical knowledge to be integrated with biomechanics and motor behavior knowledge. Additional exercise science knowledge can be integrated from sports analytics, conditioning, and other exercise science subdisciplines to help players like Brent and Victor.

The study of exercise and physical activity cannot be pursued without prerequisite structural knowledge of the human body. Like medicine, exercise science requires that students take at least one course in gross human anatomy. The study of the structure of the human body is **anatomy**. Anatomy provides a somewhat standardized qualitative language to describe the structure of the body and relative movements. This common language is essential not only for the study of human motion in exercise science but also for communicating with other medical and allied health professionals.

This chapter introduces you to basic anatomical terms, key elements of gross musculoskeletal anatomy that are emphasized in exercise science majors, and why this requires knowledge from other subdisciplines of the exercise sciences to actually explain the function of anatomical tissues in creating movement. While anatomy is not a specific subdiscipline of exercise science, it is an essential prerequisite course and is related to success in exercise science (Esmat & Pitts, 2020; Vitali et al., 2020). Despite the great importance of anatomy to the study of exercise science, there continues to be some inconsistency in its teaching across exercise science programs (Lafave & Tomkins-Lane, 2014).

Anatomy and Exercise Science

Anatomy describes the structure of living things. Historically, it has been primarily based on dissection and surgery performed on animals and people. This gross anatomy based on unassisted visual observations has been expanded to smaller structures. The advent of microscopes and other imaging technologies has created the study of microanatomy, or histology. For example, the typical 206 bones of the body are studied in gross anatomy (figure 2.1a), while study of the contractile substructures of muscle cells (muscle fibers) requires histology (figure 2.1b). Other areas also make up anatomical study (e.g., comparative anatomy, embryology, pathology). This chapter provides a preview of key concepts of gross musculoskeletal anatomy of importance to the human movement focus of exercise science.

Most of exercise science focuses on the production of effective and safe human movement; thus, the required anatomical courses of study typically focus on gross cardiovascular and musculoskeletal anatomy, with less coverage of the nervous and other anatomical systems. This is not to say that digestive, lymphatic, nervous, reproductive, and other anatomical systems are not important; however, the numerous structures of bones, muscles, joints, and qualitative position and motion descriptions tend to dominate most exercise science professional practice. Most graduate training in allied health includes additional neuroanatomy, so comprehensive knowledge of innervation and spinal levels of various sensory and motor nerves helps professionals diagnose likely sources of pain or injury.

Some majors might even take anatomy or an anatomical kinesiology course in the kinesiology or exercise science department to ensure adequate coverage of this important musculoskeletal knowledge. One founding subdiscipline of exercise science from the 19th century that grew out of anthropology and efforts to expand from anatomy is anthropometrics. **Anthropometrics**, or kinanthropometry, is the science of physical measurements of the properties (e.g., length, mass, volume) of the human body and their relationship to movement (Pheasant, 1990). Exercise scientists look to anthropometrics to see if certain body types and dimensions have advantages in certain activities and sports, while engineers use anthropometrics to ensure good fit for human interaction with clothes, tools, and equipment. For example, people with tall and thin body types (**ectomorph**) have advantages in activities like basketball and high jumping. These people are at a disadvantage compared to **mesomorph** athletes (medium to small and muscular) in activities like gymnastics and wrestling.

Anatomy creates a common qualitative descriptive system for the human body. Efforts are ongoing to standardize anatomical terminology (Greathouse et al., 2004). The International Federation of Associations of Anatomists hosts three terminology standards, but errors and inconsistencies remain from multiple languages and translations that complicate the study and communication of anatomy in modern audiences (Kachlik et al., 2015; Strzelec et al., 2017). The study of anatomy can be challenging because of these subtle inconsistencies, the large number of details (structures, substructures, landmarks, tissues, motions, changes with development, and anatomical variations), and the dominant use of Latin and Greek terms. Descriptions and names of structures in the "dead" Latin language do not have the immediate logical and obvious meaning to people now like they did when they were developed hundreds to thousands of years ago. Added to all this is the confusion of some anatomy terms that develop common or colloquial meanings that often conflict with their professional meaning. For example, in anatomy, *flexion* or *flex* refers to a specific joint rotation in a specific plane, not contracting a muscle or muscle group in

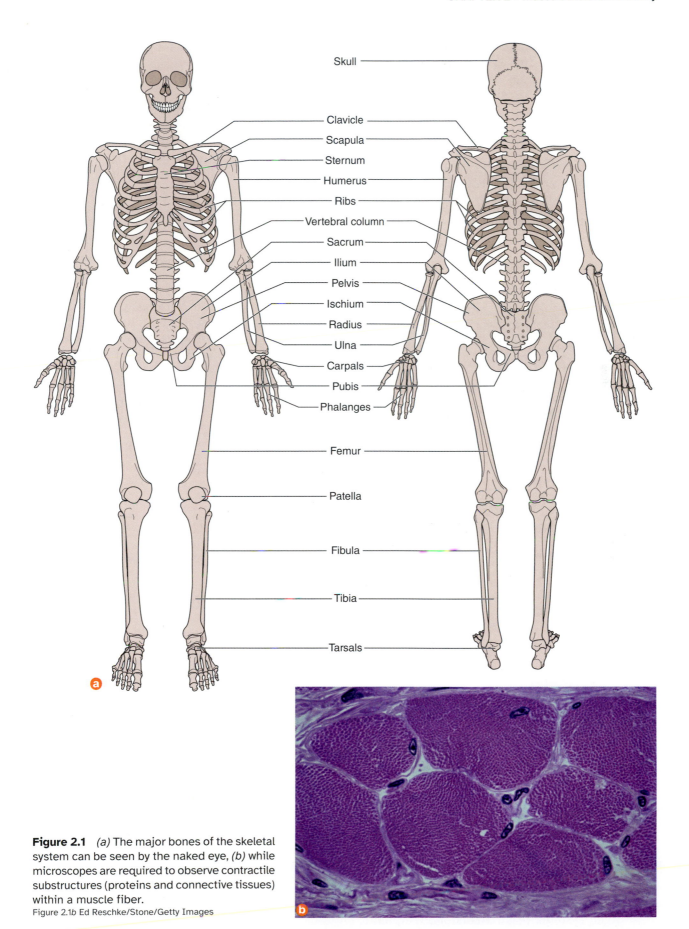

Skull

Clavicle
Scapula
Sternum
Humerus
Ribs
Vertebral column
Sacrum
Ilium
Pelvis
Ischium
Radius
Ulna
Carpals
Pubis
Phalanges

Femur

Patella

Fibula

Tibia

Tarsals

Figure 2.1 *(a)* The major bones of the skeletal system can be seen by the naked eye, *(b)* while microscopes are required to observe contractile substructures (proteins and connective tissues) within a muscle fiber.
Figure 2.1b Ed Reschke/Stone/Getty Images

the gym. It is, therefore, essential that exercise science majors and professionals continually review and reinforce their knowledge of gross anatomy to best serve their clients or patients and to communicate clearly with colleagues.

KEY POINT

Anatomy is the study of the structure of the human body and provides an important, common language for exercise science and other health professions.

Professional Issues in Exercise Science

Ensuring Evidence-Based Practice and Expertise

Expertise in most exercise science professions requires knowledge and lifelong learning that is often required to maintain licensure and certifications. Anatomical knowledge also continues to advance (see the list of anatomical journals at the end of the chapter), and a national survey of physical therapists and anatomical educators documented the need for more continuing education programs in anatomy (Wilson et al., 2018). The professionals responding were particularly interested in the latest advances in knowledge and how it integrates into their clinical and educational responsibilities. Professionals have an ethical responsibility to be continuous students to provide the best, evidence-based care to their clients.

Citation style: APA
Wilson, A.B., Barger, J.B., Perez, P., & Brooks, W.S. (2018). Is the supply of continuing education in the anatomical sciences keeping up with the demand? Results of a national survey. *Anatomical Sciences Education, 11*(3), 225-235. https://doi.org/10.1002/ase.1726

Structure of the Body

The human body is organized at four main levels: cells, tissues, organs, and organ systems.

- Cells are the fundamental units of life.
- Many cells that work together to perform a particular task make up tissues.
- Several tissues that work together to perform physiological or physical function constitute an organ (e.g., brain, heart, kidney).
- Organs are grouped by major functions into organ systems (e.g., circulatory, lymphatic, muscular, nervous, skeletal).

Anatomy classifies tissues into four categories: connective, epithelial, muscular, and nervous.

- *Connective* tissues consist of bone; blood; lymph; and cartilage, **ligaments**, and **tendons**.
- *Epithelial* are lining tissues and contribute to skin, body cavities, and glands.
- *Muscles* are excitable tissues that create force and motion. Muscles are further classified as cardiac, smooth, or skeletal (or striated) muscle.
- *Nervous* tissues are also excitable but are used for biological communication.

Skeletal Anatomy Preview

Gross anatomy instruction often begins with the skeletal system. The skeletal system is composed of bones, cartilage, and connective tissues that act as rigid levers for motion, protect internal organs, and produce and store nutrients and blood cells. The skeletal system is broken up into the axial skeleton (skull, spinal column, ribs, and sternum) and the appendicular skeleton (upper and lower extremities). Figure 2.1*a* illustrates some of the major bones of the skeletal system.

The structure of a typical lone bone (figure 2.2) illustrates several key gross structural features. The shafts of long bones are called the *diaphysis* and the

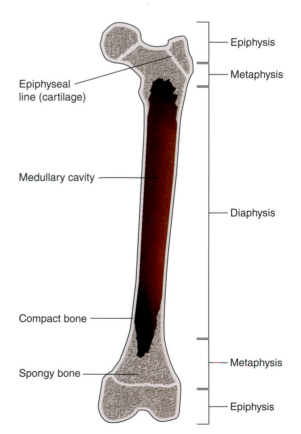

Figure 2.2 The structural regions of a long bone.

ends, the *epiphyses*. Cartilage growth plates allow long bones to grow in length as children mature. Proper diet and physical activity ensure bones will remodel with greater mineral deposits to increase their mechanical strength. Adults replace their skeleton over several years, but poor diet, inactivity, or physical disorders can result in loss of bone mass (osteopenia) that can escalate to dangerous levels (osteoporosis).

Bones are strong, composite structures with a dense outer region (compact or cortical bone), with greater internal space conserved by a three-dimensional lattice structure of spongy, cancellous bone. These truss-like beams are called *trabeculae* and develop along common mechanical stress areas. Much of the importance of regular physical activity involves healthy levels of force applied to bones to stimulate growth and remodeling of cortical and cancellous bone. Bones are living tissues with complex architecture with blood and nerve supplies that run throughout the structure of trabeculae.

Joints are the articulations between bones and are classified into three anatomical types based on mobility.

- *Fibrous* joints are immovable; examples include the fusion of bones forming the skulls of adults or the interosseous membrane between the tibia and fibula and the ulna and radius.
- *Cartilaginous* joints are strong, slightly moveable joints often with large cartilage structure, such as between vertebrae and the pubic bones.
- *Synovial* joints (e.g., shoulder, hip, elbow, knee) are freely movable with large, fluid-filled capsules that can move in a variety of directions depending on bone and ligament geometry.

KEY POINT

The complexity and many levels of anatomical structures like bones and joints have functional consequences.

Hot Topic

Anatomy and Injury

Injury is an unfortunate fact of life. A common experience of exercise science majors is having friends or family ask about aches, pains, and injuries. Has this happened to you? If so, do not feel bad if you did not have an immediate answer to their questions. Detailed study and continued review of anatomy is a major ethical obligation for exercise science, medical, and allied health professionals. You will likely add to your anatomy knowledge in advanced courses' clinical tests to diagnose musculoskeletal disorders. For example, knowledge of muscle and sensory nerves helps professionals diagnose potential sources of pain or dysfunction. Exercise science and medical researchers use a variety of measurements and study designs to figure out causes of injury and pain, effective treatments, and potential risk reduction. There is a place for you in exercise science if you are passionate about helping others with injuries.

Anatomical Position

Anatomy describes the human body relative to the **anatomical position**. The anatomical position and three anatomical planes (**frontal**, **sagittal**, and **transverse**) are illustrated in figure 2.3. Because of

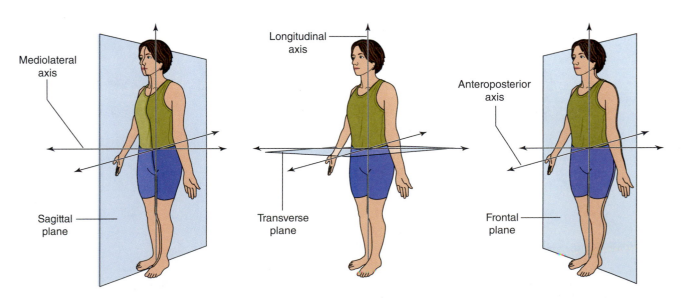

Figure 2.3   The anatomical position, major anatomical planes of motion, and associated axes of rotation.

the standing anatomical position, these planes also correspond to three spatial directions (two horizontal and one vertical).

- Rotations in the frontal plane occur about an anteroposterior axis.
- Rotations in the sagittal plane occur about a mediolateral axis.
- Rotations in the transverse plane happen about a longitudinal axis.

KEY POINT

The anatomical position, planes, and axes of rotation provide key structural knowledge for the qualitative description of body regions and movements.

Anatomical Directions

The major axes of rotations require directional descriptors.

- Directions along the vertical (anatomically longitudinal) axis are usually referred to as **superior** (cephalic) in the direction of the head and **inferior** (caudal) in the direction of the feet. Note this is one example of redundant anatomical terminology, which is discussed below.
- Structures or motion toward the front of the body are described as **anterior**, while structures or motion toward the rear of the body are described as **posterior**.
- Motion or position near the midline of the body is described as **medial**, while structures

or motion away from the midline in either direction (left or right in the frontal plane) are described as **lateral**.

Joint Motions

The major rotations of bones at joints in the three anatomical planes have standard qualitative descriptions in anatomy. Six major joint motion and anatomical plane terms are used (figure 2.4).

Flexion and Extension

Joint flexion and extension movements occur in the sagittal plane about a mediolateral joint axis.

- **Flexion** occurs when anterior surfaces are brought closer together.
- **Extension** occurs when anterior surfaces are moved away from each other. The term **hyperextension** is sometimes used when extension goes beyond the anatomical position or some typical end range of motion.

Abduction and Adduction

Joint abduction and adduction movements occur in the frontal plane about an anteroposterior axis.

- **Abduction** occurs when a joint rotates away from the midline of the body.
- **Adduction** occurs when a joint rotates toward the midline of the body.

Internal and External Rotation

Joint internal and external rotation occurs in the transverse plane about a longitudinal axis.

- **Internal rotation** occurs when the anterior portion of a segment moves toward the midline

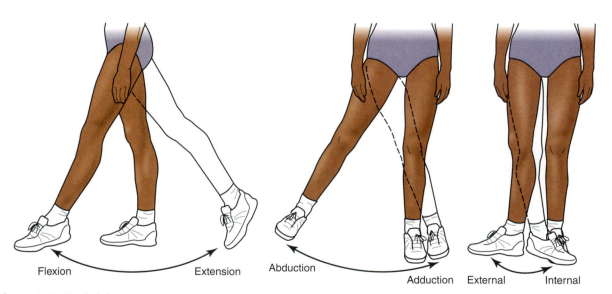

Figure 2.4   Basic joint movements.

of the body. Internal rotation of the forearm is often referred to as *pronation*.

- **External rotation** occurs when the anterior portion of a segment moves away from the midline of the body. External rotation of the forearm is often referred to as *supination*.

Human motion is usually composed of a complex combination of many body segment rotations at joints; thus, special terms often are used to simplify complex movements. Triplanar motion of the rear foot (calcaneus under the talus) in the stance phase of running is sometimes referred to as *pronation* and *supination*. It would best if these terms were not used for both transverse plane forearm movements and triplanar rear foot motions, but it is important to know these redundancies. Context will help professionals reading exercise science publications, and they will also know to use these terms in searching for studies in bibliometric databases. Another example of potential confusion about specialized motion terminology occurs in the transverse plane. The rotation of the upper arm (humerus) toward the midline in a horizontal plane about a vertical axis at the shoulder joint (figure 2.5) is called either *horizontal flexion* or *horizontal adduction*. The axis of rotation is not the longitudinal axis of the humerus even though this is a transverse plane motion, and both flexion and adduction are assigned to other anatomical planes. Notice the difficulty making precise qualitative descriptions of joint motion in a complex human skeletal system as well as the reliance on knowledge of specific bones (calcaneus, humerus, scapula, talus), their positions, and their shared articulations.

Here's another example of often confusing anatomical terminology using fairly common skeletal alignment deviations. Orthopedic medicine has often described a bowlegged person as having genus valgus and a knock-kneed person as having genu varus. Anatomical terminology, however, strives to use *valgus* to refer to the long axis of a segment aligned away from the midline and *varus* as a segment's long axis deviating medially (beyond the midline). Attaching the adjectives *varus* and *valgus* to joints lacks clarity on what specific segment or segments are out of alignment. The uneven pressure in the knee joint (increased compressive force in the lateral half) of a knock-kneed position could cause pain and disability, but how this is treated depends on whether this resulted from a varus thigh alignment from hip adduction or a valgus lower leg alignment (strained medial or tibial collateral ligament) (figure 2.6). This is a complicated problem because a strength or flexibility imbalance that misaligns a segment often contributes to misalignment of adjacent segments. Some in the medical community recommend dropping this terminology used with joints or at least defining it every time it is used to avoid confusion (Houston & Swischuk, 1980).

KEY POINT

Anatomical joint motions are fairly standardized nomenclature for qualitative descriptions of joint rotations.

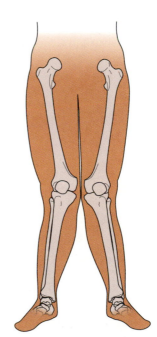

Figure 2.6   A knock-kneed lower extremity could be created by more hip adduction (humerus varus alignment) or tibial valgus alignment. *Varus* (segment aligned toward the midline) and *valgus* (segment aligned away from the midline) adjectives normally should be attached to specific segments, not joints, for accurate description.

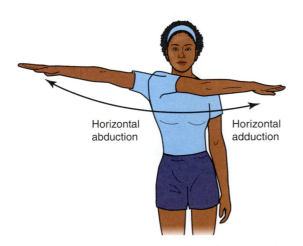

Horizontal abduction

Horizontal adduction

Figure 2.5   Horizontal flexion or adduction and horizontal extension or abduction movements also occur in the transverse plane.

Muscular Anatomy Preview

The structure of muscles is of particular importance to exercise science because these tissues are the effective internal motors, springs, struts, and brakes that create human movement (Dickinson et al., 2000; Kuhman & Hurt, 2019). Structural details from anatomy, however, must be combined with the sciences of biomechanics, physiology, and motor behavior to explain how chemical energy is used to create and regulate our movements.

A muscle cell or fiber (figure 2.7) is a cylindrical structure with multiple nuclei. A muscle fiber has several components:

- Each fiber is covered with connective tissue called **endomysium**.
- An individual muscle is composed of distinct bundles of muscle fibers called **fascicles**.
- Each fascicle is surrounded by a layer of connective tissue (**perimysium**).
- Connective tissue around each individual muscle (**epimysium**) connects with connective tissue of groups of muscles (**septa**).
- These connective tissues within muscles blend together to form **tendons**, which connect muscles to bones.

As we will see later, the passive tension from these connective tissue components within and between muscles has major functional consequences.

Muscle fiber arrangement is classified as either parallel or pennate (figure 2.8).

- **Parallel** muscle fibers are aligned parallel to the long axis or line of pull of the muscle and tendon (e.g., biceps brachii and rectus abdominus).
- **Pennate** muscle fibers are aligned at an angle to a tendon or aponeurosis on the long axis of the muscle. Pennate muscles have a feathered appearance and can have fascicles angled in from one direction (unipennate or tibialis posterior), two directions (bipennate or rectus femoris), or several different directions (multipennate or deltoid).

Muscles with parallel fascicle arrangements (longitudinal, radiate, and fusiform variations) favor range of motion and speed of shortening over tension development. The greater number of muscle fibers and fascicles lined up in series, the greater the muscle excursion and velocity of shortening compared to a pennate arrangement. Pennate arrangements of fascicles (unipennate, bipennate, and multipennate variations) create more muscle tension than parallel arrangement primarily because more fibers can be put in a physiological cross-sectional area. The physiological cross-sectional area is the total area of the muscle at calculated right angles to fascicles. Many parallel architecture muscles (erector spinae, rectus abdominis, sartorius) often accommodate the great range of motion required across multiple joints, while many

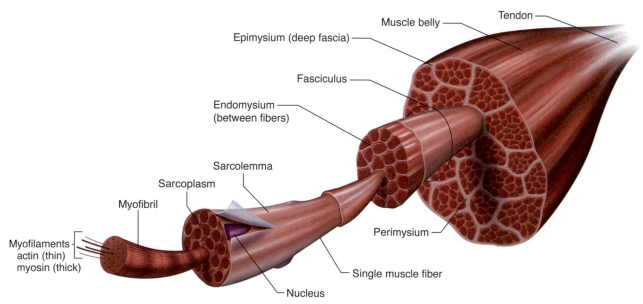

Figure 2.7 Skeletal muscle is a complex structure with interconnected levels of connective tissues throughout that fuse with tendons and bones.

Parallel

Longitudinal   Radiate   Fusiform

Pennate

Unipennate   Bipennate   Multipennate

Figure 2.8   Muscle is classified into parallel or pennate based on the alignment of fibers relative to its tendon and aponeurosis.

pennate muscles (unipennate—semimembranosus, bipennate—rectus femoris, multipennate—deltoideus) contribute greater tension. The amount of tendon a muscle has and the ratio of tendon length to fascicle length also affect the range-of-motion and tension abilities of a muscle. The gross arrangement of muscle fascicles relative to the tendon and aponeurosis (internal tendon) have direct effects on muscle strength, range of motion, and related speed of shortening (Lieber & Friden, 2000).

KEY POINT

Parallel and pennate muscle architecture of muscle fibers strongly influence their range of motion and force potential.

When you get to the biomechanics classes in your exercise science program, you will learn more about the complex interaction between active and passive sources of muscle tension that affect move-

Research and Evidence-Based Practice in Exercise Science

Muscles Within Muscles

Classifications of gross muscle architecture from anatomy change over time. Additional microdissection and interdisciplinary research (e.g., biomechanics and electromyography) sometimes update our understanding of functional subdivisions of complex muscles. Gross anatomy students often learn about three regions of the deltoid muscle (anterior or clavicular, intermediate or acromion, posterior or scapular), the pectoralis major (clavicular, sternal, costal), and the latissimus dorsi (vertebral, iliac, costal, scapular). Electromyography using numerous, high-density microelectrodes, however, indicates that despite apparent attachment-based regions of these three muscles, the central nervous system can activate seven, six, and six sections, respectively (Brown et al., 2007). Biomechanical research has documented how regions of the bipennate rectus femoris can be activated to direct forces toward a specific joint (Watanabe et al., 2021) or side of the leg (Hagio et al., 2012). It is likely that the central nervous system is able to activate motor units within numerous sections of many muscles to fine-tune the forces it creates at their attachments.

ment. **Active tension** in muscle comes from the interaction of microscopic myofilaments (active and myosin) within substructures of muscle fibers and cells called sarcomeres. Active tension is a result of metabolic conversion of high-energy phosphates during activation, which you will learn about in exercise physiology.

The tension muscle creates to make our movements, however, is mediated by the **passive tension** from the stretch and recoil of all the connective tissue components of muscle. This smooths out the transmission of force to the bone, is quite complex, and interacts with mechanical factors of creation of active tension of muscle. In slow, static stretching you experience the discomfort of high muscle passive tension as you stretch a muscle group. Muscle passive tension contributes to all actions of muscles, but in a completely relaxed (unactivated) muscle stretch, the passive tension begins a sharp, progressively steep increase just beyond resting lengths or the midrange of the joint's range of motion. Any movements near the end of a joint's range of motion will be more strongly influenced by passive tension forces in the stretched muscles, even if the muscle is not activated. The creation of muscle tension to modify movement is a complex combination of active and passive forces.

A cross-sectional dissection or view of upper- and lower-extremity body segments shows the muscles are grouped by sheet-like connective tissue (septa) in various structural compartments (anterior, deep, medial, posterior, superficial). These anatomical compartments have different nerve and blood supplies. For example, the upper arm has two major compartments. The anterior compartment bundles the long and short head of biceps brachii, brachialis, and the coracobrachialis muscles, while the posterior compartment surrounds the three heads of the triceps brachii and anconeus. Dissection of the hand shows eleven separate muscular compartments.

KEY POINT

Muscle tension results from both active sources of the interactions of myofilaments using metabolic energy and passive sources using stretch and recoil of connective tissue elements.

Muscle Actions and Muscle Groups

Anatomists observing muscle compartments and the opposing (opposite sides of joints) structure of muscles around joints classify muscles in proposed muscle groups based on their hypothesized actions. A **muscle group** is a set of muscles hypothesized to tend to create the same joint rotation based on anatomy. For example, the four major muscles on the anterior aspect of the thigh are often referred to as the quadriceps muscle group. The quadriceps are classified as extensors of the knee. Anatomists would say that quadriceps are the **agonist** muscle group for knee extension effect of this muscle, and the opposing (**antagonist**) muscle group tending to create knee flexion is the hamstring muscle group.

This section illustrates how these anatomical classifications of tendency to create motion are a good initial system to study muscles, but the muscle actions are much more complex in actual human movement. You will see that when the central nervous system activates a muscle or muscle group, the active and passive tension created can contribute to three muscle actions and can have a wide variety of force and motion effects depending on the biomechanical conditions at the time. **Muscle action** refers to the mechanical effect of activated muscle to contribute to movement by acting either to stabilize (isometric, or act as a strut), shorten (concentric, or act as a motor creating motion), or lengthen (eccentric, or act as a brake on other forces).

Hypothesized Muscle Actions

Most anatomy texts hypothesize actions of specific muscles and muscle groups when they are activated. Anatomical tables often classify a muscle like the vastus lateralis as a knee joint extensor and as part of the quadriceps (four knee extensors: rectus femoris, vastus intermedius, lateralis, and medialis) muscle group. Study of these anatomical tables of the innervations, distal and proximal attachments, and hypothesized actions at joints is an important first step in understanding a small part of what muscles do, but they should be used with caution. We will see in this chapter that many of the muscle actions hypothesized from anatomy (Hellebrandt, 1963) and professionals' subjective observation of movement are not correct when electromyographic and biomechanics measurements are made in actual human movements (Herbert et al., 1993; Bartlett, 2012).

Classifications of muscles as flexors, abductors, or internal rotators at certain joints are based on the logic of a simple mechanical method of muscle action analysis. This logic method examines one muscle's line of action to the axis of rotation one joint at a time (figure 2.9). The joint rotation created by the muscle is then inferred based on assumed attachment stability, orientation, and pulls of the muscle in the anatomical position. For example, the iliacus

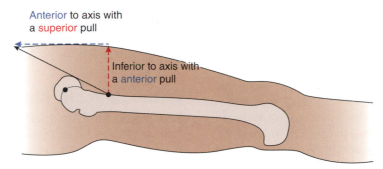

Figure 2.9   The mechanical method of muscle action analysis is used to hypothesize that the iliacus or psoas major flexes the hip joint in the sagittal plane. The vector directions and orientations relative to the hip joint axis logically leading to this motion are illustrated.

or psoas major muscles are classified as hip flexors in the sagittal plane based on several assumptions:

- The least movable attachment (origin) is the hip or lumbar vertebrae.
- The more movable attachment (insertion) is on the lesser trochanter of the femur.
- The muscles' anterior orientation with superior pull and superior orientation and posterior pull to the hip joint axis tend to create hip flexion.

The key point is that while activation of the iliacus or psoas major does tend to flex the hip joint, this analysis does not take into consideration other muscles or external forces. Low iliacus and psoas major forces could be slowing or braking hip extension in a roman chair exercise, or they could be isometrically stabilizing the hip in that exercise. Hypothesized muscle actions from anatomy take into account only a small part of a complex multisegment and muscular system interacting with external forces that affect movement.

The hypothesized muscle actions from anatomy are useful for study, but they paint a limited and sometimes misleading story about how muscles create movement. A major assumption that ignores biomechanical complexity is classifying attachments as origins and insertions. While the tension of activated muscles usually pulls both attachments approximately equally, they often don't. The size of the forces at each attachment might not be the same because of the force transmitted to nearby muscles and extra-muscle connective tissue (Huijing, 1999; Huijing & Baan, 2001; Maas et al., 2004; Maas &

Hot Topic

Anatomical Variation

As more people donate their bodies for anatomical and other scientific study, our knowledge of exercise science and other fields advances. Just like people do not look exactly alike on the outside, so too do many people have anatomical structures that vary from traditional anatomy due to genetic variation and development. Additional muscle bellies or attachments are common in many muscles. For example, the four muscles of the quadriceps group have been found to have up to eight proximal attachments or distinct muscles (Grob et al., 2016; Olewnik et al., 2021; Ruzik et al., 2020), prompting revised anatomic classifications for anterior thigh muscles. Variations in muscle attachments, fascicles, and connective tissue attachments have functional consequences for movement and injury risk. The numbers and sites of attachments of many muscles vary, such as the rhomboid and scalene muscles (Kamibayashi & Richmond, 1998). Variations in musculoskeletal structure are also hypothesized to contribute to risk of injury (Whiting & Zernicke, 2024). The role of variation in muscle and skeletal architecture on muscle actions requires further study (Richmond, 1998). Most introductory anatomy classes do not cover anatomical variations thoroughly. If you are fortunate to have an anatomy course with cadaver dissection or with previously dissected prosections, you will learn that human anatomy is messier than the anatomical models. You also will likely gain some appreciation for anatomical variation.

Finni, 2018). The end-to-end connections within parallel-arranged muscle fascicles and connective tissue connections within muscles make force transmission complex (Patel & Lieber, 1997; Sheard, 2000; Mass & Finni, 2018). For example, the center fibers of the biceps brachii do not shorten uniformly due to differences in the distal and proximal aponeurosis (Pappas et al., 2002). The force within the sarcomeres of a fiber might also not be uniform (Talbot & Morgan, 1996; Morgan et al., 2000), perhaps from activation-related mediation of the passive tension in the molecular spring titin (Herzog, 2019). In short, active and passive tension within muscles varies between and along muscles. Actual muscle actions are quite complex and require full-body modeling and several kinds of biomechanical research evidence for confirmation.

Which attachment of a muscle is more movable, therefore, will depend on many biomechanical factors involving the integration of active and passive forces between all active muscles. Anatomy-in-action books illustrate primarily superficial muscular anatomy. These help readers visualize the muscles in various body positions (figure 2.10). Readers should evaluate carefully whether an illustration clearly shows all muscles that are acting to produce the movement as well as the internal and external forces affecting the motion.

The next section describes the many actions of muscles, and the last section of the chapter previews how anatomical knowledge must be integrated with biomechanics to understand actual muscle actions that create and modify movement. How the neuromuscular system manages the complex muscle control of movement is studied in biomechanics and motor behavior courses in the exercise science curriculum.

KEY POINT

Hypothesized muscle groups and their actions in anatomical tables are often incorrect or incomplete for many movements because of the single muscle, single joint, and many assumptions used.

Actual Muscle Actions

The neuromuscular stimulation of muscle to create tension to move a joint traditionally has been called *contraction*, which implies a shortening of the whole muscle. Several scholars have argued against using this term because of what muscles actually do during movement, which is to shorten, lengthen, and stabilize (Cavanagh, 1988; Faulkner, 2003). Thus, this book uses the term muscle **action** to refer to the neuromuscular activation of muscles and muscle groups in all reflexive and skillful efforts to create or modify body position or movement.

The neuromuscular system activates muscles skillfully and efficiently using three major actions:

- Eccentric
- Isometric
- Concentric

The three main muscle actions require an understanding of two key biomechanical variables:

- **Force** is a linear push or pull effect that acts between two objects and tends to create linear acceleration to modify motion or deform an object. When a force acts off center on an object, it also creates a rotary effect.

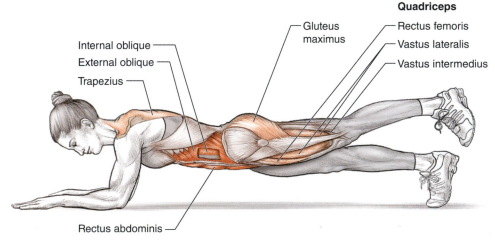

Figure 2.10   Anatomy-in-action illustrations help readers visualize superficial muscles in a movement. What their actual actions are in contributing to the movement requires integration of anatomy and multiple kinds of biomechanical studies.

- **Moment of force** is the tendency of an off-center force to create angular acceleration to modify rotation.

Force and moment of force are **vector** quantities that must specify size, units, and direction. Forces can be illustrated with an arrow, and moments of force can be illustrated with curved arrows about axes of rotations. The lengths of both kinds of arrows represent the size, and direction is represented by the arrowhead.

Let's look more closely at eccentric, isometric, and concentric muscle actions using figure 2.11 as an example.

- An **eccentric** muscle action occurs when a muscle or muscle group's moment of force (M_M) is less than the moment of resistance (M_R). An eccentric muscle action occurs when the person uses the muscles as a brake, slowing motion in the opposite direction. Elbow extension in a standing arm curl exercise is an example of an eccentric muscle action.

- A **concentric** muscle action occurs when the muscle's moment of force (M_M) is larger than the moment of resistance (M_R). The whole muscle or tendon complex shortens to rotate the joint and overcome the resistance. Concentric muscle actions allow people to use muscles as motors to drive joint rotation and combinations of joint rotations to move the body and other objects. Elbow flexion in a standing arm curl exercise is an example of a concentric muscle action.

- An **isometric** muscle action occurs when the muscle's moment of force (M_M) is equal to the moment of resistance (M_R). The muscle acts like a strut or stabilizer. For example, trunk muscles (low back and abdominals) usually isometrically stabilize the spine and upper body while a person performs a standing arm curl or push-up exercise.

You will learn much more about neuromuscular control in future biomechanics and motor behavior courses. Many students are surprised that skillful, effective, and efficient movement usually involves purposeful muscle inaction. Skilled movement means not activating some muscles at a particular point in time because it would be inefficient or even counterproductive to the movement goal. In a sense, remaining inactive might be considered a fourth muscle action. Practice alone allows people's brains to learn in the "school of gravity" how to coordinate muscle actions to overcome the weight forces of gravity with muscles and to tune those actions to cooperate with gravity and other forces when it is effective to do so.

The hypothesized concentric muscle actions common in anatomy are only a small percentage of the many actions and inaction muscles can use to control joint rotation. Notice also the biomechanical variables needed to define muscle action at one joint. The reality of human movement is that we have numerous cooperating (agonist) and opposing (antagonist) muscle forces to consider. Add to this the external forces to consider and the interaction of numerous body segments to know what muscles do, and one can see why anatomy alone is not enough to understand how movement is created. Anatomy is important, but it needs to be integrated with other exercise sciences to understand muscle function. The next section previews how anatomical knowledge is combined with biomechanics (chapter 4) in exercise science so we can help people move more effectively and safely.

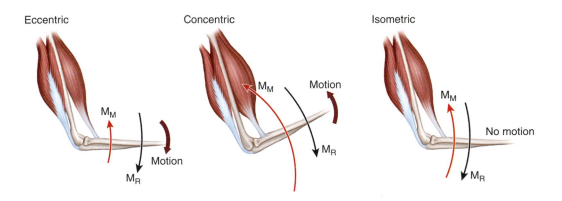

Figure 2.11   Main muscle actions. M_M = moment of force for the muscle group. M_R = moment for the resistance.

Integration of Anatomy and Exercise Science

The structural complexity of the body illuminated by anatomy must be combined with the subdisciplines of exercise science to understand the client, movement, and environment. Knowing your anatomy is not enough to know what to do to help a client with an injury; you also need knowledge of biomechanics, exercise physiology, motor control, and other subdisciplines. This section will preview briefly how anatomical knowledge is integrated with some kinds of biomechanical knowledge to understand how muscles create and modify movement.

The muscle actions noted earlier were the first illustration that hypothesized that concentric actions of muscles from anatomy were often inaccurate (Hellebrandt, 1963). **Biomechanics** extends anatomical knowledge using the branch of physics (mechanics) that precisely measures motion and documents its causes. Biomechanics uses several kinds of research that are integrated to determine the actions of muscles to create and modify movement. Advances in computers, mathematics modeling, and noninvasive imaging and sensor technologies have expanded our understanding of muscle contributions to movement. Three lines of biomechanics evidence are discussed here to illustrate how anatomical knowledge can be integrated with exercise science to understand human movement: electromyography, multisegment modeling, and imaging and sensors.

Electromyography

Electromyography (EMG) is one of the oldest and most extensive bodies of knowledge about neuromuscular control. Since the mid-1900s, skin and needle, or wire, electrodes have been used to detect the electrical potentials of activated muscles. Careful collection, amplification, processing, and interpretation of this activation data with other biomechanical measurements tells us how the central nervous system activates muscles. Extensive EMG research of most of the muscles of the body has documented general patterns of preferential activation of muscles within muscle groups to achieve tasks (Basmajian & De Luca, 1985). This research has also noted significant flexibility in muscle activation within and between people. Much of this variation comes from the muscular redundancy for all joint rotations, so almost identical movements can be created by different muscle activation and subsequent net joint moments of force (Winter, 1984; Patla, 1987; Hatze, 2000). This is why a consensus of EMG and other biomechanics research is needed to confirm manual muscle tests (Kendall et al., 2005), commonly used in clinical orthopedic evaluations (Rowlands et al., 1995; Kelly et al., 1996).

EMG research confirms the neuromuscular activation of muscles really is not as simple as the brain turning on all the muscles in an anatomical muscle group. Many studies indicate that activation of individual muscles is often not representative of all muscles in the same functional group (Bouisset, 1973; Arndt et al., 1998) and these differences are affected by training (Rabita et al., 2000). Recall the structural complexity of muscles, which results in quite sophisticated activation of the motor units within muscles depending on the task and muscle actions it requires (Enoka, 1996; Gandevia, 1999; Gielen,

Rachel Koldenhoven Rolfe

Courtesy of Texas State University.

Dr. Rachel Koldenhoven Rolfe is an athletic training faculty member at Texas State University. She received her PhD from the University of Virginia and has worked as an athletic trainer in a variety of settings from youth sport to the collegiate level. She uses her clinical experiences to guide her research and provide examples during her classes. One of her favorite subjects to teach is human anatomy. She taught cadaver anatomy as a graduate student, and in her current undergraduate anatomy classes students learn from interactive lectures, laboratory activities, online 3D anatomy software, and model demonstrations. She believes that a strong foundation and continued study of human anatomy is essential for all exercise science careers. Anatomy is critical to her clinical biomechanics research on assessing gait and functional activity impairments for individuals with lower-extremity injuries. She treats biomechanical alterations using biofeedback and traditional rehabilitation techniques.

1999; Babault et al., 2001). The muscles used in a task can vary due to variation in joint angle, muscle action (Nakazawa et al., 1993; Kasprisin & Grabiner, 2000), and the amount of body stabilization required (Kornecki et al., 2001). How muscles share loading in a muscle group depends on many factors like fiber type, contractile properties, cross-sectional area, and moment arm (Ait-Haddou et al., 2000). Fatigue can also be addressed by the body by adjusting activation of muscles within a muscle group (Kouzaki et al., 2002). A key principle of muscle contribution coming from this extensive EMG research is a muscle synergy.

A muscle **synergy** is a combination or cooperative activation of several muscles that best achieves a motor task. Sometimes synergies involve muscles in opposing muscle groups. EMG and other biomechanical methods are used in **motor behavior** research given the importance of muscle synergies and force sharing of muscles in creating movement (Herzog,

1996b, 2000; Arndt et al., 1998). Motor behavior research also uses synergy to refer to underlying rules of the neuromuscular system for controlling and creating movements (Bernstein, 1967; Aruin, 2001).

Synergies are just one of several theories of motor control (Loeb, 2021) within motor behavior (chapter 6). One of the first muscle coordination or synergy controversies identified by EMG research was the apparently paradoxical **coactivation** (apparently opposing) of quadriceps and hamstrings in the power stroke (first 180° from top dead center) of cycling, the so-called Lombard's paradox. It turns out this synergy is not counterproductive but is the right combination of leg muscular activation to push with the whole lower extremity on the bicycle pedal effectively, in a rotation-productive direction (van Ingen Schenau et al., 1994; Doorenbosch et al., 1997). In general, though, muscles are usually activated in short bursts (figure 2.12) that are pre-

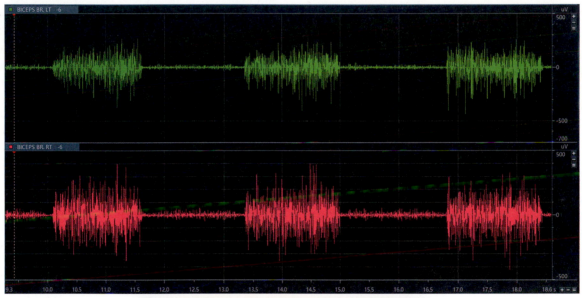

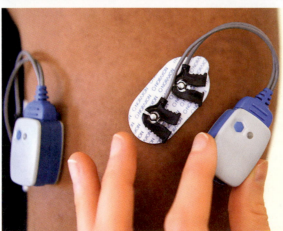

Figure 2.12 Surface EMG activation of left and right biceps brachii muscles using Ultium EMG sensors shows precisely timed bursts to most movements like these repetitive elbow flexions.
Courtesy of Noraxon.

cisely coordinated with other internal and external forces acting on the biomechanical system. This strategy provides a metabolically efficient means of coordinating muscle forces and muscle rest or inaction, with other external forces and multisegment effects to create movement.

EMG research has begun to explore activation of smaller regions of motor units within muscles than have been traditionally studied with a pair of surface electrodes. These recent studies use large numbers of smaller electrodes and special signal analysis. Much of this research focuses on many small segments of large, complex muscles. This work shows that many intramuscular segments are likely controlled by the central nervous system that go beyond simple visual fascicular segments (Wickham & Brown, 1998; Mirka et al., 1997; Paton & Brown, 1994, 1995; Wickham et al., 2004; Brown et al., 2007; Szucs & Borstad, 2013). Microelectrode study of the deltoid muscle indicates at least seven functional activation segments rather than the three anatomical (anterior, middle, and posterior fibers) segments commonly described (Wickham & Brown, 1998). This has functional importance in interpreting previous EMG research on this muscle and how the deltoid really contributes to shoulder stabilization and movement. Activation of small regions in muscles might also be related to risk of injury (Falla & Gallina, 2020). Further advances in microanatomy and high-density EMG recording and analysis research are needed on most muscles of the body given the local pick-up of most two electrode (bipolar) EMG studies (Vieira & Botter, 2021).

Multisegment Modeling

Another area of biomechanics research that is needed to determine the actual actions of muscles creating movement involves various analyses of the kinetics of linked segment models of all relevant segments and forces acting on biomechanical systems. A **linked segment model** is a biomechanical model of a series of rigid body segments linked by joints. **Kinetics** is the branch of mechanics that explains how forces and moments cause motion. Modeling is the use of a mathematical representation of all the motion factors relevant in a biomechanical system. Certain kinds of biomechanical models can be used in simulation to predict how muscle forces can affect movement in complex linked segment systems. Kinetic models of biomechanical systems indicate muscle contributions to movement are much more complex than single-joint hypotheses of anatomy. Several kinds of biomechanical models show how muscle forces and moments create all kinds of muscle actions (not just concentric) and have other motion effects transferred through joints of linked segment systems.

The main method of study of how the body creates movement is using linked segment models with data from anthropometrics and high-speed imaging to calculate net forces and moments in the joints of the body that created movements. This inverse dynamics technique uses Newton's laws of motion and has been used extensively in the study of human motion since the 1930s. Inverse dynamics results of typical movements are consistent with the complex and short bursts of muscle activation measured by EMG (Winter, 1984; Patla, 1987; Hatze, 2000). Advancements of high-speed computers and

Passive Dynamics Demonstration

Demonstrate for yourself how forces can be transferred in a linked-segment system through joints. Create a two-segment system by softly suspending a ruler vertically from its end by your thumb and first index finger. Imagine your forearm and hand are a thigh segment and the ruler is the lower leg in a kick. Push your hand (thumb and finger) horizontally and forward rapidly, and then quickly slow it to a stop. Observe how the slowing of your hand creates a proximal backward force from your hand on the ruler segment that angularly accelerates it. You were able to create a free extension at the hand and ruler joint without any "knee extensor muscles" at that joint. This transfer of forces across joints in linked-segment biomechanical systems is called **passive dynamics**. In fact, the ruler lagged back (model knee flexed) as you accelerated the proximal end of the ruler and then extended when you reversed the forward acceleration of the proximal end. Many times muscles can have more important actions at distal joints through interaction of the segments in the multisegment system than the immediate muscle action at the joints they cross.

videography have made these kinetic studies easier. The first three-dimensional study of the kinetics of the human lower extremity in walking required over 14,000 calculations and 500 hours of graphing by hand (Bressler & Frankel, 1950).

Biomechanics kinetic modeling shows how muscles have strong influences on the movements of joints they do not cross. One multisegment modeling technique called *induced acceleration analysis* has shown that muscles can even accelerate joints in the opposite direction as hypothesized by anatomy because of the interaction of the linked-segment system (Zajac & Gordon, 1989; Zajac, 1991; Hof & Otten, 2005; Kulmala et al., 2016). Another example of the variety of muscle actions at different joints naturally involves multiarticular muscles (Watanabe et al., 2021; van Ingen Schenau et al., 1989; Zajac, 1991). Muscles that cross several joints can directly create moments to modify their rotation, so they often are used differently than monoarticular muscles in the same muscle group (Prilutsky & Zatsiorsky, 1994; van Ingen Schenau et al., 1995; Hof, 2001; Cleather et al., 2015; Flaxman et al., 2017). Many kinds of models and simulations of the use of muscles in animals and humans all consistently indicate complex cooperation of muscles as observed in EMG synergies to achieve movement goals.

KEY POINT

Muscles have actions at joints they do not even cross because forces can be transferred between segments through joints by passive dynamics.

Imaging and Sensors

The third kind of biomechanics research that explains how muscles actually create movement involves the use of advanced imaging and sensors that measure muscle tension, lengths, and other geometry. The majority of these studies often involve invasive surgical techniques and animal models (Biewener, 1998; Dickinson et al., 2000; Griffiths, 1989; Herzog, 1996a, b; Shadwick et al., 1998). These techniques also confirm complex muscle actions during normal movement activities. A study of running turkeys using implanted sensors showed that the gastrocnemius (plantar flexor) muscle acted in essentially an isometric action with stretching and shortening of the tendon (Roberts et al., 1997) in horizontal running. The muscle only performed concentric actions of the whole muscle when the turkey ran uphill.

Do human muscles also have similar complex muscle actions? Yes, and both invasive and recent less invasive techniques confirm this. Biomechanists have gone to great lengths to find this out, some even volunteering themselves as participants in invasive surgery (Komi, 1990). Fortunately, sensors have improved (Komi et al., 1996) and less invasive techniques (fiber optic filaments in tendons) can be combined with bright-mode ultrasound video (Finni, Komi, & Lepola, 2000; Finni et al., 2001; Fukunaga et al., 1997, 2002; Kubo et al., 1999) to produce images of what muscle fibers and tendons are doing under the skin.

Much of this research seeing the motion of soft tissues has only been possible with advances in ultrasound imaging techniques. This kind of research on human muscles also reports complex tendon lengthening while muscle fibers shorten in an isometric action (Ito et al., 1998; Maganaris & Paul, 2000) and lengthening and shortening of muscle fibers, aponeurosis, and tendons in human movements (Finni, 2006; Fukashiro et al., 2006; Kawakami & Fukunaga, 2006; Werkhausen et al., 2019). In slow cycling, Muraoka and colleagues (2001) used ultrasound images to show that as the knee extends after the top position of the pedal cycle, the vastus lateralis fibers shorten more slowly because its tendon first elongates and then shortens before the foot and pedal reach the bottom of the pedal cycle. All these studies indicate that what is an apparently concentric muscle action from anatomical or net joint kinetics from inverse dynamics can be a more complex combination of muscle action because of the interactions of the active and passive tension in muscles.

It is abundantly clear from these three lines of biomechanics research that muscle actions in typical animal and human movements are considerably more complex than simple concentric actions hypothesized by anatomy. We will see other important previews of the theoretical and practical utility of exercise science research in the subsequent subdiscipline chapters of this text.

KEY POINT

Several kinds of biomechanics and motor behavior research methods are required to document the actual muscle actions that contribute to movement.

Professional Issues in Exercise Science

Expanding Anatomical Expertise

Besides using biomechanics to understand how anatomical structures affect movement function, recall that anthropometry, or kinanthropometry, is the scientific field examining the measures of the physical properties of the human body. Besides basic length, density, volume, and other physical variables, research in kinanthropometry includes issues like somatotype, muscle mass, and fat mass, now often using three-dimensional imaging techniques. The International Society for the Advancement of Kinanthropometry is the organization in this field that focuses on measurements of the human body relative to its capacity for movement and a variety of activities and sports. Two journals in this field are the *Journal of Functional Morphology and Kinesiology* and the *American Journal of Physical Anthropology*. Several reference books on kinanthropometry are available to help exercise science professionals (Hume et al., 2018; Norton & Eston, 2019) understand how anthropometric variations influence physical activities and sports.

Wrap-Up

Anatomy is the scientific study of the structure of the human body and is a key prerequisite course for exercise science. Anatomical terminology and qualitative descriptions of the structures of the body also provide an important professional language for exercise science and related medical fields. The hypothesized actions of muscles and muscle groups, however, qualitatively describe only a small percentage of the many ways muscles act to create movement. Muscles are active in complex synergies coordinated with other forces to achieve movement goals efficiently. This is done with interaction of active and passive tension of muscles in primarily concentric, eccentric, and isometric actions. Muscles also are purposely not activated, but when they are activated they affect the motion of other segments in a linked multisegment system beyond the joints they cross. Several kinds of biomechanics research confirm that muscle actions are more complicated than hypothesized by anatomy in normal movements of animals and humans. Anatomical knowledge therefore must be integrated with exercise science knowledge to understand human movement and how it can be improved.

More Information on Musculoskeletal Anatomy

Organizations

American Association for Anatomy

American Association of Clinical Anatomists (AACA)

Anatomical Society

Association Posturologie Internationale (API)

International Federation of Associations of Anatomists (IFAA)

International Society for the Advancement of Kinanthropometry (ISAK)

Journals

American Journal of Physical Anthropology

American Journal of Surgical Pathology

Anatomical Record

Anatomical Science International

Anatomical Sciences Education

Annals of Anatomy

Clinical Anatomy

Developmental Dynamics

European Journal of Anatomy

Journal of Anatomy

Journal of Anthropology of Sport and Physical Education

Journal of Functional Morphology and Kinesiology

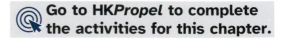
Go to HK*Propel* to complete the activities for this chapter.

Review Questions

1. What is anatomy, and why is it important to exercise science?

2. What are the anatomical terms used to describe the position and motion of the human body?

3. What is the relationship between anatomical planes and the axes of rotation for motion in those planes?

4. How do arrangements of muscle fascicles affect their tension and range of motion capabilities?

5. What are the two major sources of muscle tension, and are they uniform across joint range of motion?

6. Activated muscle can result in what three kinds of muscle actions?

7. Why are hypothesized actions of muscles based on anatomy alone sometimes incorrect?

8. How should anatomy knowledge be used with exercise science subdisciplines to understand muscle actions in creating and modifying human movement?

Suggested Readings

Basmajian, J.V., & De Luca, C.J. (1985). *Muscles alive: Their functions revealed by electromyography* (5th. ed.). Williams & Wilkins.

Gross, M. (2021). *Anatomical kinesiology.* Jones & Bartlett.

Hellebrandt, F.A. (1963). Living anatomy. *Quest, 1,* 43-58.

IFAA. (2002). *Terminologia anatomica: International anatomical terminology* (multilingual ed., 2nd ed.). Theme.

Vigotsky, A.D. (2018). Interpreting signal amplitudes in surface electromyography studies in sport and rehabilitation. *Frontiers in Physiology, 8,* 985.

Zajac, F.E. (2002). Understanding muscle coordination of the human leg with dynamical simulations. *Journal of Biomechanics, 35,* 1011-1018.

References

Ait-Haddou, R., Binding, P., & Herzog, W. (2000). Theoretical considerations on co-contraction of sets of agonist and antagonist muscles. *Journal of Biomechanics, 33,* 1105-1111.

Arndt, A.N., Komi, P.V., Brüggemann, G. P., & Lukkariniemi, J. (1998). Individual muscle contributions to the *in vivo* achilles tendon force. *Clinical Biomechanics, 13,* 532-541.

Aruin, A.S. (2001). Simple lower extremity two-joint synergy. *Perceptual and Motor Skills, 92,* 563-568.

Babault, N., Pousson, M., Ballay, Y., & Van Hoecke, J. (2001). Activation of human quadriceps femoris during isometric, concentric, and eccentric contractions. *Journal of Applied Physiology, 91,* 2628-2634.

Bartlett, R. (2012). *Sports biomechanics: Reducing injury and improving performance* (2nd ed.). Routledge.

Basmajian, J.V., & De Luca, C.J. (1985). *Muscles alive: Their functions revealed by electromyography* (5th ed.). Williams & Wilkins.

Bernstein, N.A. (1967). *The coordination and regulation of movements.* Pergamon Press.

Biewener, A.A. (1998). Muscle function *in vivo*: A comparison of muscles used for elastic energy savings versus muscles used to generate mechanical power. *American Zoologist, 38,* 703-717.

Bouisset, S. (1973). EMG and muscle force in normal motor activities. In J.E. Desmedt (Ed.), *New developments in electromyography and clinical neurophysiology* (pp. 502-510). Karger.

Bressler, B., & Frankel, J.P. (1950). The forces and moment in the leg during level walking. *Transactions of the American Society of Mechanical Engineers, 72,* 27-36.

Brown, J.M.M., Wickham, J.B., McAndrew, D.J. & Huang, X.-F. (2007). Muscles within muscles: Coordination of 19 muscle segments within three shoulder muscles during isometric motor tasks. *Journal of Electromyography and Kinesiology, 17*(1), 57-73.

Cavanagh, P.R. (1988). On "muscle action" vs. "muscle contraction." *Journal of Biomechanics, 21,* 69.

Cleather, D.J., Southgate, D.F.L., & Bull, A.M.J. (2015). The role of biarticular hamstrings and gastrocnemius muscles in closed chain lower limb extension. *Journal of Theoretical Biology, 365,* 217-225.

Dickinson, M.H., Farley, C.T., Full, R.J., Koehl, M.A., Kram, R., & Lehman, S. (2000). How animals move: An integrated view. *Nature, 288,* 100-106.

Doorenbosch, C.A.M., Veeger, D., van Zandwij, J., & van Ingen Schenau, G. (1997). On the effectiveness of force application in guided leg movements. *Journal of Motor Behavior, 29,* 27-34.

Enoka, R.M. (1996). Eccentric contractions require unique activation strategies by the nervous system. *Journal of Applied Physiology, 81,* 2339-2346.

Esmat, T.A., & Pitts, J.D. (2020). Predicting success in an undergraduate exercise science program using science-based admissions courses. *Advances in Physiology Education, 44,* 138-144.

Falla, D. & Gallina, A. (2020). New insights into pain-related changes in muscle activation revealed by high-density surface electromyography. *Journal of Electromyography and Kinesiology, 52,* 102422.

Faulkner, J.A. (2003). Terminology for contractions of muscles during shortening, while isometric, and during lengthening. *Journal of Applied Physiology, 95,* 455-459.

Finni, T. (2006). Structural and functional feature of the human muscle-tendon unit. *Scandinavian Journal of Medicine and Science in Sports, 16,* 147-158.

Finni, T., Ikegawa, S., & Komi, P.V. (2001). Concentric force enhancement during human movement. *Acta Physiological Scandinavica, 173,* 369-377.

Finni, T., Komi, P.V., & Lepola, V. (2000). In vivo human triceps surae and quadriceps femoris muscle function in a squat jump and counter movement jump. *European Journal of Applied Physiology, 83,* 416-426.

Flaxman, T.E., Alkjaer, T., Simonsen, E.B., Krogsgaard, M.R., & Benoit, D.L. (2017). Predicting the functional roles of knee joint muscles from internal joint moments. *Medicine & Science in Sports & Exercise, 49,* 527-537.

Fukashiro, S., Hay, D.C., & Nagano, A. (2006). Biomechanical behavior of muscle-tendon complex during dynamic human movements. *Journal of Applied Biomechanics, 22,* 131-147.

Fukunaga, T., Ichinose, Y., Ito, M., & Kawakami, Y. (1997). Determination of fasicle length and pennation in a contracting human muscle in vivo. *Journal of Applied Physiology, 82,* 354-358.

Fukunaga, T., Kawakami, Y., Kubo, K., & Keneshisa, H. (2002). Muscle and tendon interaction during human movements. *Exercise and Sport Sciences Reviews, 30,* 106-110.

Gandevia, S.C. (1999). Mind, muscles and motoneurons. *Journal of Science and Medicine in Sport, 2,* 167-180.

Gielen, S. (1999). What does EMG tell us about muscle function? *Motor Control, 3,* 9-11.

Greathouse, D.G., Halle, J.S., & Dalley, A.F., II. (2004). Terminologia anatomica: Revised anatomical terminology. *Journal of Orthopaedic & Sports Physical Therapy, 34*(7), 363-367.

Griffiths, R.I. (1989). The mechanics of the medial gastrocnemius muscle in the freely hopping wallaby. *Journal of Experimental Biology, 147,* 439-456.

Grob, K., Ackland, T., Kuster, M.S., Manestar, M., & Filgueira, L. (2016). A newly discovered muscle: The tensor of the vasus intermedius. *Clinical Anatomy, 29,* 256-263.

Hagio, S., & Nagata, K., & Kouzaki, M. (2012). Region specificity of rectus femoris muscle for force vectors in vivo. *Journal of Biomechanics, 45*(1), 179-182.

Hatze, H. (2000). The inverse dynamics problem of neuromuscular control. *Biological Cybernetics, 82,* 133-141.

Hellebrandt, F.A. (1963). Living anatomy. *Quest, 1,* 43-58.

Herbert, R., Moseley, A., Schurr, K., & Wales, A. (1993). Making inferences about muscle forces from clinical observations. *Australian Journal of Physiotherapy, 39,* 195-202.

Herzog, W. (1996a). Muscle function in movement and sports. *American Journal of Sports Medicine, 24,* S14-S19.

Herzog, W. (1996b). Force-sharing among synergistic muscles: Theoretical considerations and experimental approaches. *Exercise and Sport Sciences Reviews, 24,* 173-202.

Herzog W. (2000). Muscle properties and coordination during voluntary movement. *Journal of Sports Sciences, 18,* 141-152.

Herzog, W. (2019). Passive force enhancement in striated muscle. *Journal of Applied Physiology, 126,* 1782-1789.

Hof, A.L. (2001). The force resulting from the action of mono- and biarticular muscle in a limb. *Journal of Biomechanics, 34,* 1085-1089.

Hof, A.L., & Otten, E. (2005). Assessment of two-dimensional induced accelerations from measured kinematic and kinetic data. *Gait & Posture, 22,* 182-188.

Houston, C.S., & Swischuk, L.E. (1980). Varus and valgus—No wonder they are confused. *New England Journal of Medicine, 302*(8), 471-472.

Huijing, P.A. (1999). Muscular force transmission: a unified, dual or multiple system? A review and some explorative experimental results. *Archives of Physiology and Biochemistry, 107,* 292-311.

Huijing, P.A., & Baan, G.C. (2001). Extramuscular myofacial force transmission within the rat anterior tibial compartment: Proximo-distal differences in muscle force. *Acta Physiological Scandinavica, 173*, 297-311.

Hume, P.A., Kerr, D.A., & Ackland, T.R. (2018). *Best practice protocols for physique assessment in sport.* Springer.

Ito, M., Kawakami, Y., Ichinose, Y., Fukashiro, S., & Fukunaga, T. (1998). Nonisometric behavior of fascicles during isometric contractions of a human muscle. *Journal of Applied Physiology, 85*, 1230-1235.

Kachlik, D., Musil, V., & Baca, V. (2015). Terminologia anatomica after 17 years: Inconsistencies, mistakes and new proposals. *Annals of Anatomy, 201*, 8-16.

Kamibayashi, L.K., & Richmond, F.R.J. (1998). Morphometry of human neck muscles. *Spine, 23*, 1314-1323.

Kasprisin, J.E., & Grabiner, M.D. (2000). Joint angle-dependence of elbow flexor activation levels during isometric and isokinetics maximum voluntary contractions. *Clinical Biomechanics, 15*, 743-749.

Kawakami, Y., & Fukunaga, T. (2006). New insights into in vivo human skeletal muscle function. *Exercise and Sport Sciences Reviews, 34*, 16-21.

Kelly, B.T., Kadrmas, W.R., & Speer, K.P. (1996). The manual muscle examination for rotator cuff strength: An electromyographic investigation. *American Journal of Sports Medicine, 24*, 581-588.

Kendall, F.P., McCreary, E.K., Provance, P.G., Rodgers, M.M., & Romani, W.A. (2005). *Muscles: Testing and function* (5th ed.). Lippincott Williams & Wilkins.

Komi, P.V. (1990). Relevance of in vivo force measurements in human biomechanics. *Journal of Biomechanics, 3*(S1), 23-34.

Komi, P.V., Belli, A., Huttunen, V., Bonnefoy, R., Geyssant, A., & Lacour, J.R. (1996). Optic fibre as a transducer of tendomuscular forces. *European Journal of Applied Physiology, 72*, 278-280.

Kornecki, S., Kebel, A., & Siemienski, A. (2001). Muscular co-operation during joint stabilization, as reflected by EMG. *European Journal of Applied Physiology, 84*, 453-461.

Kouzaki, M., Shinohara, M., Masani, K., Kanehisa, H., & Fukunaga, T. (2002). Alternate muscle activity observed between knee extensor synergists during low-level sustained contractions. *Journal of Applied Physiology, 93*, 675-684.

Kubo, K., Kaakami, Y., & Fukunaga, T. (1999). Influence of elastic properties of tendon structures on jump performance in humans. *Journal of Applied Physiology, 87*, 2090-2096.

Kuhman, D.J., & Hurt, C.P. (2019). Lower extremity joints and muscle group in the human locomotor system alter mechanical functions to meet task demand. *Journal of Experimental Biology, 222*, jeb206383.

Kulmala, J-P, Korhonen, M.T., Ruggiero, L., Kuitunen, S., Suominen, H., Heinonen, A., Mikkolo, A., & Avela, J. (2016). Walking and running require greater effort from the ankle than the knee extensor muscles. *Medicine and Science in Sports and Exercise, 48*, 2181-2189.

Lafave, M.R., & Tomkins-Lane, C.C. (2014). Survey of Canadian human anatomy course in kinesiology and physical education. *European Journal of Anatomy, 18*(3), 199-204.

Lieber, R.L., & Friden, J. (2000). Functional and clinical significance of skeletal muscle architecture. *Muscle and Nerve, 23*, 1647-1666.

Loeb, G.E. (2021). Learning to use muscles. *Journal of Human Kinetics, 76*, 9-33.

Maas, H., & Finni, T. (2018). Mechanical coupling between muscle-tendon units reduces peak stresses. *Exercise and Sport Sciences Reviews, 46*, 26-33.

Maas, H., Baan, G.C., & Huijing, P.A. (2004). Muscle force is determined by muscle relative position: Isolated effects. *Journal of Biomechanics, 37*, 99-110.

Maganaris, C.N., & Paul, J.P. (2000). Load-elongation characteristics of the in vivo human tendon and aponeurosis. *Journal of Experimental Biology, 203*, 751-756.

Mirka, G., Kelaher, D., Baker, D., Harrison, H., & Davis, J. (1997). Selective activation of the external oblique musculature during axial torque production. *Clinical Biomechanics, 12*, 172-180.

Morgan, D.L., Whitehead, N.P., Wise, A.K., Gregory, J.E., & Proske, U. (2000). Tension changes in the cat soleus muscle following slow stretch or shortening of the contracting muscle. *Journal of Physiology* (London), *522*, 503-513.

Muraoka, T., Kawakami, Y., Tachi, M., & Fukunaga, T. (2001). Muscle fiber and tendon length changes in the human vastus lateralis during slow pedaling. *Journal of Applied Physiology, 91*, 2035-2040.

Nakazawa, K., Kawakami, Y., Fukunaga, T., Yano, H., & Miyashita, M. (1993). Differences in activation patterns in elbow flexor muscles during isometric, concentric, and eccentric contractions. *European Journal of Applied Physiology, 66*, 214-220.

Norton, K., & Eston, R. (2019). *Kinanthropometry and exercise physiology* (4th ed.). Routledge.

Olewnik, L., Tubbs, R.S., Ruzik, K., Podgórski, M., Aragonés, P., Waśniewska, A., Karauda, P., Szewczyk, B., Sanudo, J.R., & Polguj, M. (2021). Quadriceps or multiceps femoris? Cadaveric study. *Clinical Anatomy, 34,* 71-81.

Pappas, G.P., Asakawa, D.S., Delp, S.L., Zajac, F.E., & Drace, J.E. (2002). Nonuniform shortening in the biceps brachii during elbow flexion. *Journal of Applied Physiology, 92,* 2381-2389.

Patel, T.J., & Lieber, R.L. (1997). Force transmission in skeletal muscle: From actomyosin to external tendons. *Exercise and Sport Sciences Reviews, 25,* 321-364.

Patla, A.E. (1987). Some neuromuscular strategies characterising adaptation process during prolonged activity in humans. *Canadian Journal of Sport Sciences, 12*(Supp.), 33S-44S.

Paton, M.E., & Brown, J.M.N. (1994). An electomyographic analysis of functional differentiation in human pectoralis major muscle. *Journal of Electromyography and Kinesiology, 4,* 161-169.

Paton, M.E., & Brown, J.M.N. (1995). Functional differentiation within latissimus dorsi. *Electromyography and Clinical Neurophysiology, 35,* 301-309.

Pheasant, S. (1990). *Anthropometrics: An introduction* (2nd ed.). British Standards Institution.

Prilutsky, B.I. & Zatsiorsky, V.M. (1994). Tendon actin of two-joint muscles: Transfer of mechanical energy between joints during jumping, landing, and running. *Journal of Biomechanics, 27,* 25-34.

Rabita, G., Perot, C., & Lensel-Corbeil, G. (2000). Differential effect of knee extension isometric training on the different muscles of the quadriceps femoris in humans. *European Journal of Applied Physiology, 83,* 531-538.

Richmond, F.J.R. (1998). Elements of style in neuromuscular architecture. *American Zoologist, 38,* 729-742.

Roberts, T.J., Marsh, R.L., Weyand, P.G., & Taylor, D.R. (1997). Muscular force in running turkeys: The economy of minimizing work. *Science, 275,* 1113-1115.

Rowlands, L.K., Wertsch, J.J., Primack, S.J., Spreitzer, A.M., & Roberts, M.M. (1995). Kinesiology of the empty can test. *American Journal of Physical Medicine and Rehabilitation, 74,* 302-304.

Ruzik, K., Waśniewska, A., Olewnik, L., Tubbs, R.S., Karauda, P., & Polguj, M. (2020). Unusual case report of seven-headed quadriceps femoris muscle. *Surgical and Radiologic Anatomy, 42,* 1225-1229.

Shadwick, R.E., Steffensen, J.F., Katz, S.L., & Knower, T. (1998). Muscle dynamics in fish during steady swimming. *American Zoologist, 38,* 755-770.

Sheard, P.W. (2000). Tension delivery from short fibers in long muscles. *Exercise and Sport Sciences Reviews, 28,* 51-56.

Strzelec, B., Chmielewski, P.P., & Gworys, B. (2017). The terminologia anatomica matters: Examples from didactic, scientific, and clinical practice. *Folia Morphologica, 76*(3), 340-347.

Szucs, K.A., & Borstad, J.D. (2013). Gender differences between muscle activation and onset timing of the four subdivisions of trapezius during humerothoracic elevation. *Human Movement Science, 32,* 1288-1298.

Talbot, J.A., & Morgan, D.L. (1996). Quantitative analysis of sarcomere non-uniformities in active muscle following a stretch. *Journal of Muscle Research and Cell Motility, 17,* 261-268.

van Ingen Schenau, G.J., De Boer, R.W., & De Groot, B. (1989). Biomechanics of speed skating. In C. Vaughan (Ed.), *Biomechanics of sport* (pp. 121-167). CRC Press.

van Ingen Schenau, G.J., Pratt, C.A., & Macpherson, J.M. (1994). Differential use and control of mono- and bi-articular muscles. *Human Movement Science, 13,* 495-517.

van Ingen Schenau, G.J., va Soest, A.J., Gabreels, F.J.M., & Horstink, M. (1995). The control of multi-joint movements relies on detailed internal representations. *Human Movement Science, 14,* 511-538.

Vieira, T.M., & Botter, A. (2021). The accurate assessment of muscle excitation requires the detection of multiple surface electromyograms. *Exercise and Sport Sciences Reviews, 49,* 23-34.

Vitali J., Blackmore, C., Mortazavi, S., & Anderton, R. (2020). Tertiary anatomy and physiology: A barrier for student success. *International Journal of Higher Education, 9,* 289-296.

Watanabe, K., Vieira, T.M., Gallina, A., Kouzak, M., & Moritani, T. (2021). Novel insights into biarticular muscle actions gained from high-density electromyogram. *Exercise and Sport Sciences Reviews, 49*(3), 179-187.

Werkhausen, A., Cronin, N.J., Albracht, K., Bojsen-Møller, J., Seynnes, O.R. (2019). Distinct muscle-tendon interaction during running at different speeds and in different loading conditions. *Journal of Applied Physiology, 127,* 246-253.

Wickham, J.B., & Brown, J.M.M. (1998). Muscles within muscles: The neuromotor control of intramuscular segments. *European Journal of Applied Physiology, 78*, 219-225.

Wickham, J.B., Brown, J.M.M., & McAndrew, D.J. (2004). Muscles within muscles: anatomical and functional segmentation of selected shoulder joint musculature. *Journal of Musculoskeletal Research, 8*, 57-73.

Winter, D.A. (1984). Kinematic and kinetic patterns of human gait: Variability and compensating effects. *Human Movement Science, 3*, 51-76.

Zajac, F.E. (1991). Muscle coordination of movement: A perspective. *Journal of Biomechanics, 26*(S1), 109-124.

Zajac, F.E., & Gordon, M.E. (1989). Determining muscle's force and action in multi-articular movement. *Exercise and Sport Sciences Reviews, 17*, 187-230.

Zernicke, R.F., Broglio, S.P., & Whiting, W.C. (2024). *Biomechanics of injury* (3rd ed.). Human Kinetics.

CHAPTER 3

Measurement and Statistics

Matthew T. Mahar

CHAPTER OBJECTIVES

In this chapter, we will

- explain why measurement is important in every area of exercise science;
- define terms essential to understanding measurement, evaluation, and statistics;
- explain how different types of evidence of validity are appropriate for different research questions;
- discuss the importance of different types of reliability evidence; and
- explain how statistics are used to make comparisons and look at relationships to help researchers answer questions.

Lola wants to become more physically active. She used to run cross country in high school but slowly became less active over the years that she birthed and raised four children. Now her children are grown, and she feels it is time to focus more on herself, including on her health and physical activity. Like many people, Lola has a wrist-worn fitness tracker that tells her how many steps she takes every day. Because she's heard about taking 10,000 steps per day, she sets that as her goal. Is that a smart goal? Does Lola really need to take that many steps a day for her health, or should it be more than that? Measurement and research on our most basic form of ambulation, walking, helps organizations like the Centers for Disease Control and Prevention (CDC) and the World Health Organization (WHO) set evidence-informed guidelines, helps exercise science professionals provide guidance to their clients, and helps people like Lola know how much walking might be too much and how much might be too little.

Measurement is fundamental to professional practice and research in exercise science. Health professionals, including physical therapists, athletic trainers, coaches, and personal fitness trainers, use measurements to diagnose and monitor their clients and to prescribe exercise and rehabilitation programs. Without accurate and consistent measurement of the variables important in exercise science we cannot place any credibility in the results or use them to guide our exercise prescriptions. This holds true for all subdisciplines within exercise science. The measurement and evaluation course that you will take will cover topics of validity and reliability of measurement. These are indispensable constructs; gaining a basic understanding of them at this time will help you to start thinking like an exercise scientist.

Most, although not all, of the subdisciplines and research within exercise science have a quantitative focus, with an interest in measuring variables by assigning numbers to observations according to rules. For example, a researcher might be interested in whether time spent in vigorous physical activity is related to cognitive function. To examine this question, both vigorous physical activity and cognitive function must be measured. Vigorous physical activity might be quantified as the number of minutes spent per day in vigorous physical activity using an instrument known as an accelerometer. Cognitive function might be measured with software that quantifies attention during performance of a task on a computer. To answer the question about whether the variables of vigorous physical activity and cognitive function are related, researchers would apply a selected statistical analysis to the data they measured. Exercise science has a long, rich history of research on developing accurate tests, measurements, and statistical analyses of health and performance variables.

In this chapter, you will learn historical highlights and benefits of measurement research in exercise science. Next, you will be introduced to essential concepts that help exercise science professionals document the quality of measurements and appropriateness of the interpretation of measurements. Basic descriptive statistics and statistical analyses commonly used to examine relationships among variables or differences among groups will also be presented. This chapter will not cover concepts used in analytical and qualitative research. These fields of exercise science use logical argument or systematic, qualitative judgments about observations in their research.

Benefits and History of Measurement in Exercise Science

One of the oldest and most robust subdisciplines of exercise science is measurement, sometimes referred to as *measurement and evaluation*. Since the 1920s, researchers have focused on developing accurate tests and measurements of a wide variety of human movement, health, and performance variables. Many physiological and medical tests commonly used today are based on research and knowledge developed by measurement and evaluation scholars from exercise science. Here are just a few measurement innovations and how they are used in research and professional practice.

Measurement and evaluation researchers have made critically important contributions to the field of youth fitness assessment. In the 1950s, Kraus and Hirschland (1954) published a study that showed that European children scored higher on the Kraus-Weber test of minimum muscular fitness than American children. Although the Kraus-Weber test was not a true test of youth fitness, it was the motivating force behind the development of the first national youth fitness test in the United States, the American Association for Health, Physical Education, and Recreation (AAHPER) Youth Fitness Test, first published in 1958. This test included motor fitness items that measured running, jumping, and throwing—items that encouraged athletic excellence.

A major shift in youth fitness testing away from an athletic emphasis to a health-related emphasis occurred in the 1970s. A group of measurement specialists and exercise physiologists met to discuss the growing medical evidence that supported the role of physical activity and fitness in health. This meeting led to the development of the term *health-related fitness* and the publication of an important paper that influenced the trajectory of youth fitness testing (Jackson et al., 1976). In the United States, a youth fitness test battery called FitnessGram continues to provide the most scientifically validated tests of health-related youth fitness. Health-related fitness components have included aerobic capacity, body composition, muscular strength and endurance, and flexibility. Research evidence that muscle power is associated with bone health, that muscular strength is a good predictor of health and function, and that core endurance is linked to balance and spinal stability led the FitnessGram Advisory Board to include measures of lower-body power via the vertical jump (Mahar et al., 2022), maximal strength via a handgrip test (Saint-Maurice et al., 2018), and core endurance via the plank test (Laurson et al., 2022) in their youth fitness testing battery. As new research evidence accumulates to suggest changes to current practice, exercise scientists need to be willing to embrace new ways of doing things so their practices and decisions remain evidence based.

Regression analysis, a statistical concept addressed later in this chapter, has been used by measurement researchers to produce findings that have greatly influenced research and practice in exercise science. For example, Jackson and Pollock (1978) and Jackson and colleagues (1980) developed generalized regression equations to estimate percentage of body fat from the sum of skinfolds in men and women, respectively. A skinfold is a pinch of skin and the underlying subcutaneous fat and is measured with calipers. These measurement procedures have been used in thousands of subsequent studies to estimate percent fat and by practitioners throughout the world to provide feasible estimates of percent fat for

their athletes, clients, and patients. Over time, other field-based measures of body composition, such as bioelectrical impedance analysis, have been developed to ease measurement for practitioners. These approaches to body composition estimation also use regression analyses to obtain estimates of percent fat and to quantify the accuracy of those estimates. Understanding the science behind the tests and measurements you choose will inform your decisions and communications to your clients and is part of what will make you a quality exercise science professional.

Measurement and evaluation researchers also have been at the forefront of understanding, developing, and using criterion-referenced cut points to interpret health status. After a test is administered and a score is obtained (i.e., measurement), we then need to make a value judgment about the measurement. That value judgment is an evaluation. To evaluate a score—for example, determining whether the value is in the healthy or unhealthy range—the score needs to be compared to something. When we compare the measurement with a predetermined standard, we are using criterion-referenced standards. Setting criterion-referenced standards for health status is a

Professional Issues in Exercise Science

Does It Matter How You Measure Physical Activity?

Physical activity can be measured a number of ways, including by self-report and with objective measurement devices such as pedometers and accelerometers. Self-reported physical activity can be inaccurate because it is difficult for some people to accurately remember the type, time, and intensity of physical activity they did and because of substantial reporting bias. That is, people may tend to report more activity than they actually performed because they think it is socially desirable to be more active. Objective measurement devices, like wearable physical activity trackers, can assess steps taken and the intensity of movement. In a classic study, Troiano and colleagues (2008) demonstrated that it matters—a lot—how physical activity is measured. These researchers used accelerometer data to objectively measure physical activity in a large, nationally representative sample from the National Health and Nutrition Examination Survey (NHANES). The accelerometer data showed that less than 5% of adults obtained the recommended levels of physical activity. This can be contrasted with self-reported measures of physical activity from NHANES, in which approximately 51% of adults met physical activity recommendations. Although error no doubt exists in every measure of physical activity, it is likely that such a large discrepancy in physical activity levels between self-reported and objectively measured physical activity is mainly due to people greatly overestimating their own physical activity when asked via self-report. The lack of confidence in self-reported physical activity may have been one of the factors that led to the great explosion of commercially available wearable activity trackers such as Fitbit, Apple Watch, Garmin Vívoactive, and Polar Grit, among many others.

Citation style: APA

Troiano, R.P., Berrigan, D., Dodd, K.W., Masse, L.C., Tilert, T., & McDowell, M. (2008). Physical activity in the United States measured by accelerometer. *Medicine and Science in Sports and Exercise, 40*(1), 181-188. https://doi.org/0.1249/mss.0b013e31815a5lb3

complex measurement issue that requires collaboration among measurement experts and content experts, like exercise physiologists, athletic trainers, and physical therapists, emphasizing the need to address measurement issues from a multidisciplinary perspective. For youth fitness, scientifically based criterion-referenced standards for aerobic capacity, body composition, muscular strength, muscular endurance, and muscular power have been developed using sophisticated statistical techniques, like lambda-mu-sigma (LMS) growth curves and receiver operating characteristic curve analysis.

With a criterion-referenced framework based on measurements from youth fitness tests, participants are categorized into two or more fitness zones (e.g., Healthy Fitness Zone, Needs Improvement Zone, or Needs Improvement—Health Risk Zone). Setting the appropriate criterion-referenced standards for these zones is an important validity issue. Because validity is about the appropriate use of test scores, a diagnostically accurate criterion-referenced standard helps users of these tests, like physical education teachers or pediatricians, make valid decisions. If, for example, a participant is classified into the Needs Improvement—Health Risk Zone for aerobic capacity, an appropriate interpretation is that they have insufficient aerobic capacity. The exercise science professional might then provide a specific exercise prescription to improve aerobic capacity. Rikli and Jones (2013a, b) published the Senior Fitness Test, a widely used functional fitness test for older adults. They established clinically supportable, criterion-referenced standards for these tests to help test users evaluate the fitness of older adults and to subsequently provide appropriate exercise prescriptions to improve health and fitness of their clients.

Measurement Concepts in Exercise Science

Recall that chapter 1 noted that scientific fields have specific, technical terms and definitions of fundamental concepts. This section summarizes these important concepts, their meaning, and the associated measurement terms common in exercise science.

Definitions

Definitions of these important terms and concepts are presented here to make it easy for you to refer to this list. Exercise science research uses these words with specific meanings that do not always align with more common public use.

- *Test:* An instrument or tool used to take measurements and gather data.
- *Measurement:* The act of assessment to collect numerical information. This is done by assigning numbers to observations according to rules.
- *Evaluation:* A value judgment placed on the measurement. Evaluation involves interpretation of whatever has been measured.

We use tests to take measurements. Evaluation involves a decision about the value of the data obtained from the measurement.

Measurement and statistics are interconnected, but they are not the same. We see statistics nearly every time we read a newspaper, magazine, or website. Generally, these statistics (e.g., batting average, median home price, percent increase in screen time, rankings of best colleges) do not induce anxiety because we understand what they mean. Similarly, you do not need to be intimidated by a measurement or statistics course. You just need to learn what the statistics mean, keep up with the assignments in the course, and realize that you can master this material. As you become more familiar with statistics, any anxiety will fade away. Let's become familiar with definitions related to statistics.

- *Statistic:* A number (datum) or numbers (data) calculated from measured data.
- *Statistics:* Techniques that deal with the collection, organization, analysis, description, interpretation, and presentation of information stated numerically. Not much can be more useful than that.

The use of all numbers is not created equal. Remember that measurement is the assignment of numbers to observations following some rules. These rules help us understand not only what that number means or how it can be interpreted (evaluation), but also what statistics can be calculated with those measurements. Scales of measurement signify how numbers are defined and categorized. The four scales of measurement are defined here and presented in table 3.1 to facilitate understanding.

- *Nominal:* Numbers on a nominal scale represent names and are a set of mutually exclusive categories used to categorize participants. No meaningful order exists to this categorization.
- *Ordinal:* Numbers on an ordinal scale have the characteristic of order, such that one can be classified as higher or lower. No common unit of measurement exists between ordinal

Table 3.1 **Characteristics of the Four Scales of Measurement**

Characteristic	SCALE OF MEASUREMENT			
	Nominal	Ordinal	Interval	Ratio
Categorical or different	X	X	X	X
Order or ranking		X	X	X
Common unit of measure between scores			X	X
Absolute zero point				X

numbers, so ordinal data cannot be averaged (and maintain any meaning).

- *Interval:* Numbers on an interval scale have a meaningful order (like ordinal data) and a common unit of measurement or equal distance between scores. The zero point of interval data is arbitrary.
- *Ratio:* Numbers on a ratio scale have a common unit of measurement (equal distance) between scores and an absolute zero point. This means that a score of zero indicates a lack of or true zero amount of the measurement of interest.

Exercise science professionals use data from all four measurement scales in different ways. A few examples are provided to clarify how data measured on all four scales are necessary and useful. When researchers conduct studies, they assign numbers to observations in various ways. For the nominal variable of sex, one researcher might code females as 0 and males as 1. (Others might code sex in a different way; it really does not matter.) Sex coded as 0 or 1 is on the nominal scale. If participants are coded as 0, they are not coded as 1 (i.e., the categories are mutually exclusive). In addition, no meaningful order exists to the coding. That is, we would not say that sex coded a 0 is lower than that coded a 1. It just means that they are in different categories. Other examples of nominal-level data include race, blood type, group membership, political party, and genotype.

As the name implies, data on the ordinal scale are used to show that order matters relative to distinguishing more than or less than. However, because no common unit of measure exists between scores represented on an ordinal scale, differences between adjacent numbers or categories do not necessarily have the same meaning. Exercise science professionals might code the socioeconomic status of participants as "low income," "middle income," or "high income." We can see that order exists, as one

category is higher or lower than another category, but the distance between categories is not necessarily similar. This limits the appropriate statistics that can be applied to ordinal-level data. In particular, it would be meaningless to average ordinal-level data. A variable measured on the ordinal scale that might be familiar to most of you is the ranking of the top 25 college football teams. Here we can determine that rankings are higher or lower than other rankings, but the distance between the team ranked first and the team ranked second is not necessarily the same as the distance between the team ranked second and the team ranked third.

Whenever you come across an average (arithmetic mean), you know you are dealing with either interval- or ratio-level data. Because interval-level data have a common unit of measure between scores, we can conduct mathematical calculations (such as add, subtract, multiply, and divide) and end up with meaningful results. Temperature in degrees F (or C) is a good example of interval-level data. An average temperature of 70 °F makes sense. To demonstrate that interval-level data have a common unit of measure between scores, we can see that the difference between 20 ° and 50 ° is 30 °. Likewise, the difference between 50 ° and 80 ° is 30 °. Interval-level data, however, do not have a physically meaningful zero point. For example, although 0 °F is cold, it does not represent the absolute lowest value for temperature. (The true absence of heat is 0 ° Kelvin.) Other examples of variables measured on an interval-level scale include IQ, SAT, and GRE scores.

Ratio-level data have the characteristic of a common unit of measure between scores, similar to interval data, and also have the quality of an absolute zero. This means that a score of zero indicates the absence of the trait being measured. Many of the physical measures, like height and weight, that we deal with in exercise science are examples of ratio-level data. With ratio-level variables, the ratio of two measurements has a meaningful interpretation. For example, 50 kilograms is half of 100 kilograms. This is meaningful because mass has an absolute zero point. On the other hand, it is not meaningful to state that 50 °F is half of 100 °F, because temperature in °F is an interval-level variable and has no meaningful zero point. Other examples of ratio-level variables include reaction time, heart rate, income, vertical jump in centimeters, 40-yard (37 m) dash time in seconds, and daily intake of calories.

KEY POINT

Numbers have particular characteristics depending on their level of measurement. It is important to understand the levels of measurement to help determine what statistics are appropriate to use to analyze your data and how the measurements can be interpreted.

The data we obtain from our measurements need to be compared to something to facilitate evaluation and interpretation. That is, the measurements we take need to be referenced against something to help us determine the merit or goodness of the measurement. In order to make an evaluative judgment, our measures are compared to either a norm-referenced standard or to a criterion-referenced standard.

- *Norm referenced:* Measurements are compared to the performance of others (i.e., norms) to judge the quality and make an evaluation. Norms in exercise science can be developed for multiple levels or groups (e.g., international, national, teams, positions, ages, injured and uninjured) of people.

- *Criterion referenced:* Measurements are compared to a predetermined standard or criterion to judge the quality and make an evaluation.

KEY POINT

Depending on the purpose of the evaluation, we compare our measurements to either norm-referenced standards or criterion-referenced standards to help us understand and interpret the measurements.

Validity and reliability are key concepts for measurement and evaluation and can be determined under a norm-referenced framework or under a criterion-referenced framework. For example, evalu-

ation of youth fitness variables has changed over the years from a norm-referenced evaluation framework to a criterion-referenced evaluation framework. In a norm-referenced framework, evaluations often took the form of reporting percentiles. A student was told, for example, that her performance on the vertical jump was at the 70th percentile, meaning she performed better than 70% of the girls her age, based on the norms used. Current youth fitness tests, like FitnessGram, provide criterion-referenced standards for each test. In a criterion-referenced framework, for example, a 12-year-old girl must perform 7 or more push-ups to reach the Healthy Fitness Zone (i.e., pass the test). Determination of criterion-referenced standards is a highly involved and scientific process in which researchers determine the value of the outcomes for each test for each age and gender group that reflects the criterion of interest (in this example, likely association with health).

- *Validity:* Extent to which inferences made from specific measures are appropriate or evidence that a test measures what it is supposed to measure.

- *Reliability:* Consistency of test scores; the ability of a measurement instrument to produce consistent results.

Validity

The concept of validity has become more nuanced over the years. An intuitive and useful way to think about what validity means is to assess whether a test measures what it appears or is intended to measure. It is recognized in measurement, however, that it is really the use of a test outcome that is validated. That is, validity is about whether the inferences made from specific measures are appropriate. For example, one might use the measurement from a validated physician's scale to represent body mass for an individual. This would be considered a valid inference from that measurement. On the other hand, if one were to infer that a person is very smart based on the measurement from a physician's scale, that inference would not possess validity.

That said, the question of validity of measurement inference is not a yes-or-no question. Before using a particular instrument, an exercise science professional should examine whether evidence of validity for their intended use of the measurement data exists. Validity evidence can come from various sources, and the importance of the source depends on the instrument and intended use of the data obtained. The next few paragraphs will provide examples of how validity evidence is obtained for various purposes.

Evidence of Content Validity

Organizations and people who administer written knowledge tests often use expert judgment to determine whether the items on the test represent all of the important content areas the test claims to measure. Organizations such as the American College of Sports Medicine, the American Council on Exercise, the Athletics and Fitness Association of America, and the National Strength and Conditioning Association that offer certification examinations are invested in demonstrating evidence of content validity for their certification exams. Your teachers who administer written knowledge tests also are interested in the content validity evidence for their tests. Specific structured approaches to demonstrating content evidence of validity are available and may be covered in your measurement and evaluation course.

Evidence of Logical Validity

Many people in the field of exercise science administer tests of motor skills (e.g., soccer dribbling skill, volleyball setting skill, jumping ability). For these types of tests, evidence of logical validity is analogous to evidence of content validity for written knowledge tests. Evidence of logical validity would be demonstrated by the extent to which a test is judged to measure the most important components of skill necessary to perform a motor task adequately.

Evidence of Criterion-Related Validity

Techniques to demonstrate the correlation of a test of a construct with a criterion measure of that construct are widely used in exercise science. These techniques are known as *regression techniques*. For this approach, a criterion measure of the construct of interest is identified. Ideally, the criterion measure is considered the best measure available or gold standard measure of the construct of interest. An example should make this clear. Aerobic capacity is an important construct of interest in exercise science because it is highly related to both health and to performance in many sports. Exercise scientists agree that a criterion measure of aerobic capacity is maximal oxygen consumption (denoted as $\dot{V}O_2max$). $\dot{V}O_2max$ stands for the maximal volume (V) of oxygen (O_2) that a person can use during intense exercise. Measurement of $\dot{V}O_2max$ requires expensive analyzers and highly trained personnel (figure 3.1).

However, many practitioners need to estimate $\dot{V}O_2max$ in a way that does not require expensive

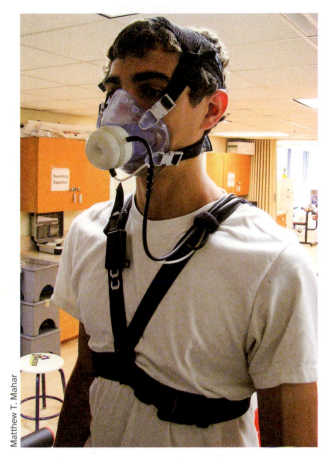

Matthew T. Mahar

Figure 3.1 Measurement of $\dot{V}O_2max$.

equipment. These practitioners look for evidence of criterion-related validity in easier and more practical tests. Two such tests are used extensively in youth fitness testing: the 1-Mile Run/Walk and the PACER. Criterion-related validity evidence of the PACER has been demonstrated (e.g., Mahar et al., 2018) by measuring a large number of people on both the PACER and on $\dot{V}O_2max$. Then, using regression techniques, these measures are correlated and the error (remember, everything we measure has some measurement error) associated in predicting $\dot{V}O_2max$ from PACER performance is estimated. In the example presented here, because the PACER and $\dot{V}O_2max$ were measured at about the same point in time, this evidence of criterion-related validity is known as *concurrent evidence*.

When the criterion is measured sometime in the future, the term *predictive evidence* is often used. When SAT tests are used by colleges in their admissions decisions, the criterion is measured sometime in the future. For example, the criterion might be whether the individual graduated from college or the GPA after the freshman year in college. In some predictive designs in exercise science, we might not be interested in predicting scores on a criterion measure but might instead try to predict health outcomes such as falls, heart attacks, or premature death (Blair et al., 1989; Myers et al., 2002; Sabia et al., 2021). In these designs, we assess the validity of scores on some tests (e.g., amount of physical activity, sleep duration, cholesterol level) as predictors of some later outcome. Specific statistical analyses (e.g., survival analysis or logistic regression) are available for these types of designs.

Other Types of Analyses

Several other types of analyses (e.g., multitrait-multimethod matrix, factor analysis, known difference method, item response theory) are available to demonstrate evidence of validity for different measures and different purposes. These analyses can be fascinating, and if you are lucky, you will learn about some of these methods in advanced measurement courses.

Validity Summary

Validity is so important that a few summary statements are essential.

- The process of validation is continuous and context specific.
- Many different types of validity evidence are possible, and more evidence of validity allows us more confidence in our decisions.
- Validity is focused on the intended use and interpretation of test scores.
- Use of test scores to make decisions has an ethical component, because test bias against certain groups has been identified.
- Although researchers use terms to describe the type of validity evidence presented, in reality, all types of evidence of validity are viewed in terms of how they help us evaluate construct validity. This is because test interpretation requires us to understand the construct we are measuring, and all types of validity evidence help us to do that.

KEY POINT

Researchers might report or demonstrate different types of evidence of validity depending on the purposes of their measurements and analyses.

Reliability

Reliability is about consistency and minimizing measurement error. Reliability of an assessment, similar to validity, is not a yes-or-no question. We look for some evidence of reliability in the measures we choose. Professionals, whether they are researchers, physical therapists, or athletic trainers, judge whether the measurement technique they are considering has strong enough evidence of reliability to allow for some degree of trust in the measure over time. Without adequate evidence of reliability, you cannot have confidence in the measurement method. Inconsistency in measurements can come from the participant, tester, and procedures.

KEY POINT

Reliability is about the consistency or repeatability of a measurement. Reliability indicates the extent to which repeated measurements of the same thing under the same conditions are reproducible.

How much evidence you need depends on how the measurement will be used and what decisions you, as a professional, need to make. For high-stakes decisions, such as whether someone is admitted into a doctor of physical therapy program, high levels of test reliability are essential. One measure of reliability is a statistic called the **intraclass correlation coefficient**. The intraclass correlation coefficient ranges between 0.0 and 1.0 and can be interpreted like a percentage. The closer to 1.0, the more reliable the measurement. Several other measures of reliability exist in relative (like a percentage) and absolute (same units as the measurement) terms.

KEY POINT

Depending on the type of decisions to be made, exercise science professionals generally want reliability measured by intraclass correlation to be 0.8 or higher, with higher values desirable for more high-stakes decisions.

Because reliability is a measure of consistency, the construct that is being measured must be sampled more than one time in order to estimate reliability. Often this means taking two or three trials of a physical measurement before the attribute of interest can change. For example, to estimate the reliability of a vertical jump measurement that might be used to estimate lower-body power, at least two trials of the vertical jump would need to be administered. To estimate reliability of a cognitive measure, such as a multiple-choice test, the test needs to be comprised of more than one item. An important caveat is that accurate estimates of reliability depend on administering the tests to a large sample size of people from a population.

Reliability estimated with an intraclass correlation coefficient normally needs a minimum sample size of at least 30 people. However, for high-stakes tests such as certification exams, sample sizes to estimate test reliability usually consist of 500 participants or more. It is hazardous to calculate any kinds of correlations on small sample sizes (e.g., less than 30 people) because the resulting correlation coefficient is unstable and can be unduly influenced by just one measurement.

One approach to understanding reliability is through classical test theory. The following equation is used to explain classical test theory:

$$X = T + e$$

"X" represents the variable we are measuring. For example, if we are measuring body composition with a handheld bioelectrical impedance analyzer, X stands for the measure of body composition for the person we are measuring. "T" represents true score.

True score is the actual body composition if no measurement error existed. Measurement error always exists, and "e" represents measurement error. Therefore, in the classical test theory equation, the smaller the measurement error, the closer our X measurement is to what it is supposed to be, the true score.

KEY POINT

If measurement error equals 0, then the measurement that was taken (X) is equal to the true score (T). This is, of course, what we desire, but in practice we almost always have some measurement error.

In practice, we should consider how we can minimize measurement error whenever we administer a test. Some considerations to minimize measurement error include the following:

- Assure the instrument is calibrated.
- Assure the tester has been appropriately trained to take the measurement and follows standardized procedures.
- Explicitly describe how the measurement should be taken.
- Prepare participants for the tests so they know what to expect and know how to perform the test.

Research and Evidence-Based Practice in Exercise Science

Can Walking More Keep You Alive Longer?

The respect with which the general public holds exercise science professionals has grown in recent decades because exercise science professionals improve public health by basing their recommendations on research findings. That is, the best advice from exercise science professionals is data informed rather than merely anecdotal. Paluch and associates (2021) conducted an important study with sophisticated statistical analyses to examine the association between step counts and mortality. This is important because the number of steps people take each day is an easy-to-understand metric, and the widespread use of wearable devices that provide the user with a measure of steps per day provides exercise science professionals with an amazing opportunity to monitor their clients' levels of physical activity.

The study followed Black and White middle-aged adults (both women and men) for an average of about 11 years (an impressive amount of time), during which some participants died. The authors analyzed the relationships between step volume and mortality and demonstrated that more steps per day were associated with a lower risk of mortality. The authors statistically controlled for many other variables that could contribute to mortality, making their conclusions stronger and more credible. Because the study included Black and White people of both sexes, the authors were able to determine that similar relationships exist between steps per day and mortality across race and sex subgroups. In general, the study findings demonstrated that middle-aged adults who took more than 7,000 steps per day had significantly lower mortality rates than those who took fewer than 7,000 steps per day. Exercise science professionals can use these findings to make data-informed recommendations for their clients. The data also indicated that encouraging increased walking among the least active segment of the population, perhaps by building accessible walking trails or adding sidewalks and bike lanes to cities, might help to prolong lives.

If you plan to measure something, you should plan to measure it with as little error as possible. This takes some forethought.

No matter how well planned the measurement is, we should always be aware that measurement error exists. Measurement error can be both positive and negative. A positive error exists when the X score we measure is higher than the true score. In the body composition example, that would mean the body composition outcome we measure is greater than the person's true body composition. A negative error exists when the X score we measure is lower than the true score. Any error means that the test score we measure is different than the true score. If that difference is small (as we hope), the impact on our decisions will be minimal.

Multiple types of reliability exist. The evidence of reliability one needs to look for (or establish if it does not yet exist) depends on the variable being assessed. If the variable being assessed is a physical measure (e.g., vertical jump, body composition, plank), stability reliability is usually estimated. Stability reliability is often called *test-retest reliability* because participants are tested with two or more trials or on two or more occasions. These scores are then correlated to estimate reliability. The type of correlation used to estimate reliability is beyond the scope of this chapter but likely will be covered in your measurement and evaluation class.

Internal consistency reliability is the type of reliability calculated when a test with multiple items (like a multiple-choice test or survey instrument) is administered one time. The estimate of internal consistency reliability is calculated from a correlation of items believed to measure the same trait (e.g., athlete burnout) or concept.

KEY POINT

Internal consistency reliability is used to estimate the reliability of multiple-choice tests and some survey instruments.

A final type of reliability that will be explained here is equivalence reliability (also sometimes called *parallel forms reliability*). When two forms of a test designed to measure the same construct exist, administering both forms to the same people and then correlating the scores on the two forms provides an estimate of parallel forms reliability. In exercise science, we sometimes have the situation in which the same construct can be measured with different tests. For example, in youth fitness testing aerobic capacity is measured with the 1-Mile Run/Walk and with the PACER. Both tests are designed to assess aerobic capacity. One could examine equivalence reliability of the two tests by administering both tests to the same people and then correlating the scores on the two tests (e.g., Mahar et al., 1997).

Researchers who publish estimates of the reliability of the tests they use not only provide evidence of the consistency for their use of the test, but also help other researchers who need to decide whether the test has suitable measurement properties for their use. By examining published estimates of reliability, researchers can determine whether the test is appropriate for their intended use.

Validity and Reliability in a Criterion-Referenced Framework

When results of a test are evaluated categorically, such as pass versus fail or achieve standard versus does not achieve standard, a criterion-referenced framework is being used. Thus, a criterion-referenced testing framework is being used when results of tests are presented categorically and determination of a minimum level of proficiency is the desired goal. This is commonly important in occupational fitness testing and is also the case with youth fitness testing in schools.

For example, students might be tested for their level of upper-body strength with a modified pull-up test (Meredith & Welk, 2010). The direct outcome measure from the modified pull-up test is the number of pull-ups completed. In a criterion-referenced framework, the number of pull-ups completed is compared to the criterion-referenced standard that indicates the desired minimum level of proficiency. As an example, let's say the criterion-referenced standard for a 12-year-old boy is seven modified pull-ups completed. The number of modified pull-ups completed by each participant is compared to this criterion-referenced standard to determine if they achieved the standard (i.e., completed seven or more modified pull-ups) or did not achieve the standard (completed fewer than seven modified pull-ups).

A criterion-referenced validity example is one in which categorization from a test, such as the PACER, is compared with categorization from a criterion measure, such as directly measured $\dot{V}O_2$max. In a validity study, participants are often tested with a field-based test and with the gold standard criterion to which researchers would like to compare the field-based test. As noted earlier in this chapter, the PACER is a field-based test of aerobic capacity. The criterion measure of aerobic capacity is $\dot{V}O_2$max measured with a metabolic system during a maximal effort test administered on a treadmill. Estimated $\dot{V}O_2$max from the PACER test is compared to the established criterion-referenced standard. Then measured $\dot{V}O_2$max from the treadmill test is compared to the same criterion-referenced standard.

KEY POINT

Criterion-referenced tests are used to make categorical decisions, like whether a person passed or failed a test or whether a person met or did not meet the standard.

Estimation of validity or reliability in a criterion-referenced framework requires use of different statistics than those used in a norm-referenced framework.

Proportion of Agreement

Proportion of agreement represents the consistency of classification. Reliability in this situation examines whether participants would be classified (e.g., achieved standard versus did not achieve standard) the same way on both testing occasions. To visualize the results of testing the same participants over 2 days, see table 3.2. Note that participants are categorized as passing or failing the test on day 1 and then categorized as passing or failing the test on day 2.

Proportion of agreement is calculated by dividing the number of participants that are consistently classified (the number in cells A + D) by the total number of participants (the number in cells A + B + C + D). Proportion of agreement ranges between 0.00 and 1.00, with higher proportions indicating greater consistency of classification. If proportion of agreement equals 0.80, then 80% of participants were consistently classified. Proportion of agreement is easy to calculate and easy to interpret. However, because some of this agreement could be due to chance, the modified **kappa (K) coefficient**, which corrects the estimate of reliability for chance agreements, is usually presented with proportion of agreement.

Table 3.2 **Contingency Table to Calculate Proportion of Agreement**

		DAY 2 TEST	
		Pass	Fail
DAY 1 TEST	Pass	A	B
	Fail	C	D

Table 3.3 **Contingency Table for Equivalence Reliability of the PACER and 1-Mile Run/Walk Tests**

		1-MILE RUN/WALK TEST	
		Pass	Fail
PACER TEST	Pass	73	11
	Fail	10	32

Proportion of agreement is a versatile statistic. In a criterion-referenced situation in which two tests are designed to measure the same construct (e.g., PACER and 1-Mile Run/Walk are both field tests of aerobic capacity), proportion of agreement can be used to estimate the classification agreement or equivalence reliability of these two tests (e.g., Mahar et al., 1997; Saint Romain & Mahar, 2001). That is, use of the statistics in such a design would examine whether participants would be classified similarly from the PACER and 1-Mile Run/Walk. This could help a teacher or coach determine which test to use.

Data from a study of criterion-referenced equivalence reliability of the PACER and 1-Mile Run/Walk are used here as an example. Table 3.3 demonstrates that 73 participants passed both the PACER and 1-Mile Run/Walk test and that 32 participants failed both tests. These numbers represent participants who were consistently classified. Note that 21 people were classified differently on the PACER and 1-Mile Run/Walk. Ten participants failed the PACER but passed the 1-Mile Run/Walk. Eleven participants passed the PACER but failed the 1-Mile/Run Walk. Calculation of proportion of agreement is demonstrated here:

$$\text{Proportion of Agreement } (Pa)$$
$$= (73 + 32) \div (73 + 11 + 10 + 32) = 105 \div 126 = 0.83$$

This means that 83% of participants were classified consistently on the PACER and 1-Mile Run/Walk tests and that 17% of participants were inconsistently classified.

Modified Kappa Coefficient

Modified kappa provides an estimate of criterion-referenced reliability with correction for chance agreement. Because modified kappa takes chance agreement into account, it provides a more conservative (or lower) estimate of agreement than the proportion of agreement.

$$\text{Modified Kappa } (K) = (Pa - 1/q) \div (1 - 1/q)$$

where Pa is the proportion of agreement and q is the number of classification categories. When categorizing participants as pass or fail, the number of categories is two, although in some situations exercise science professionals might be interested in classifying participants into more than two categories. For the data presented in table 3.3, modified kappa is calculated as follows:

$$\text{Modified Kappa } (K) = (0.83 - 0.50) \div (1 - 0.50)$$
$$= 0.66$$

Thus, a conservative estimate of criterion-referenced equivalence reliability between the PACER and 1-Mile Run/Walk is 0.66. The interpretable range for modified kappa is 0.0 to 1.0, with values closer to 1.0 indicating higher classification consistency. Guidelines to judge the magnitude of kappa have been suggested: 0.81 to 1.0 as almost perfect agreement, 0.61 to 0.80 as substantial agreement, 0.41 to 0.60 as moderate agreement, 0.21 to 0.40 as fair agreement, and 0.0 to 0.20 as slight agreement (Landis & Koch, 1977).

KEY POINT

Different statistics are used to estimate validity and reliability in criterion-referenced frameworks than in norm-referenced frameworks. Typical statistics used to estimate reliability and validity with criterion-referenced testing include proportion of agreement and modified kappa.

Statistics Commonly Used in Exercise Science

Measurement and research rely on statistics. Consequently, exercise science professionals must be knowledgeable about how statistics are used to describe data, calculate associations among variables, and compare groups. This section first defines basic statistical terms and then illustrates their application in exercise science research and professional practice. Statistics is usually divided into two branches:

- *Descriptive statistics:* Statistics used to organize or summarize a set of measurements
- *Inferential statistics:* Statistics used to make inferences from a sample from which they were calculated about the larger population the sample represents

Definitions

- *Central tendency:* A measure that best represents typical or central scores of data. Common measures of central tendency include the mean, median, and mode.
- *Variability:* The dispersion or spread of the scores in a data set. Common measures of variability include the range, standard deviation, and variance.
- *Standard deviation (S):* The average deviation of each score in the data set from the mean of the data set.
- *Correlation coefficient (Pearson's r):* A measure of the strength of the linear relationship between two variables.
- *Coefficient of determination:* Squaring the correlation coefficient (r^2) to document the strength of association between two variables (percentage of variance accounted for in one variable by the other variable).
- *Regression:* A mathematical model in which one outcome variable is predicted from one or more predictor variables.
- *Standard error of estimate:* Standard deviation of prediction errors associated with a given regression equation.
- *Normal curve:* A probability distribution that is perfectly symmetrical and known to have certain properties.
- *Standard score:* A score that has been converted into a standard unit of measurement. Standardization allows comparison when different units of measurement were used. A Z-score is an example of a standard score.
- *Z-score:* A score expressed in standard deviation units. Z-scores have a mean of 0 and a standard deviation of 1.
- *Percentiles:* A measure of the rank of a score that represents the percentage of rank-ordered scores that fall at or below that score.
- *Effect size:* A measure of the magnitude or size of a relationship or effect.
- *Cohen's delta (d):* An effect size that presents the difference between two means in standard deviation units.
- *t-test:* A statistical test used to determine whether the difference between two sample means is significantly different from zero, or whether two means differ significantly.
- *Analysis of variance (ANOVA):* A statistical test used to determine the simultaneous equality of two or more sample means.
- *Statistical significance:* A phrase used in inferential statistics that represents the probability of a result occurring by chance based on the sample data and test assumptions.

Descriptive Statistics

Understanding measurement principles and quantitative research in exercise science requires a basic understanding of descriptive statistics. Do not let that scare you. A basic understanding of all of the descriptive statistics presented in this chapter is easily within reach of students mastering 8th grade arithmetic and algebra.

Descriptive statistics are used to describe important aspects of data. To understand a data set, we need to know something about the central tendency (typical or middle scores) of the scores and something about the variability or the spread of the scores. You are no doubt familiar with the term *average*. The average or typical score (like a heart rate, blood pressure, or run time) can be expressed by many statistics; however, most people use the **mean**. To calculate the arithmetic mean you simply add up all of the scores and divide by the total number of scores. In equation form this looks like:

$$\bar{X} = \Sigma X \div N,$$

where $\bar{X}$ stands for the mean, ΣX indicates the sum of all X scores (the Greek letter sigma is used to represent the sum of whatever follows), and N represents the total number of scores. The mean is usually used to represent the central tendency of a data set when the data set is close to normally distributed. The mean is not the best measure of typical scores when the data are skewed or not normally distributed. We will talk about a normal distribution of data shortly.

KEY POINT

The mean is the average of all the scores in the data set and represents the central tendency or typical score if the data are normally distributed.

The actual middle value of a data set is called the **median**. It is the value above which and below which 50% of the scores fall. When the data set is clearly not normally distributed or skewed, then the median is usually used to represent the central tendency of that data set. The most frequently occurring score in a distribution is called the **mode**. The mode can be useful in certain situations but is not often used in exercise science as the best index of central tendency. Small data sets might have multiple modes.

Let's take a second to calculate the mean and median of two small data sets to make sure you understand this.

Data set A: 10, 20, 30, 40, 50

Data set B: 28, 29, 30, 31, 32

The mean for data set A is 30:

$$\bar{X} = (10 + 20 + 30 + 40 + 50) \div 5 = 150 \div 5 = 30$$

The mean for data set B is 30:

$$\bar{X} = (28 + 29 + 30 + 31 + 32) \div 5 = 150 \div 5 = 30$$

A value of 30 represents both the mean and median (note that 30 is the middle value) for both data sets, but it's clear that these data sets are different. What makes them different? Note that the scores in data set A are much more spread out than the scores in data set B. It is the variability or spread of these data sets that makes them different. This simple example is designed to emphasize that a measure of central tendency is important but is never enough to adequately describe all important attributes of a data set. Both a measure of central tendency and a measure of variability are needed to understand a data set.

The measure of variability that is usually presented with the mean when the data set is close to normally distributed is called the **standard deviation (S)**. This measure of variability is extremely important to understand because it describes the variability in a data set in the same measurement units as the data set and has special properties relative to normally distributed data. These benefits also make the standard deviation essential to understanding more advanced measurement and statistical matters such as standard scores, correlation, regression, and the branch of statistics (inferential statistics) used to make decisions from data sets.

In this class, you will acquire a basic understanding of standard deviation. In more advanced classes, be sure to focus whenever the topic of standard deviation is being considered. The standard deviation can be conceptually understood as the average deviation of each score from the mean. A deviation of a score from the mean is simply that score minus the mean. We will use the letter S to represent standard deviation. In equation form the standard deviation looks like:

$$S = \sqrt{\Sigma(X - \bar{X})^2 \div (N - 1)}$$

where S stands for the standard deviation, $\Sigma(X - \bar{X})^2$ indicates the squared deviation of each score from the mean, and N represents the total number of scores. Note that because the deviation of each score from the mean is squared, the square root is taken at the end to return the standard deviation to the original units of measure. Similar to understanding the mean, it helps to actually calculate the standard deviation to understand it. We will calculate the standard deviation of the two small data sets presented earlier.

Data set A: 10, 20, 30, 40, 50

Data set B: 28, 29, 30, 31, 32

Remember, a deviation score is the score minus the mean. So for data set A, we have the following deviation scores: –20, –10, 0, 10, 20. We cannot just sum deviation scores to represent variability because they always sum to zero. Thus, we square the deviation scores to eliminate the negative signs, resulting in the following squared deviation scores: 400, 100, 0, 100, 400.

The standard deviation for data set A is 15.8:

$$S = \sqrt{(400 + 100 + 0 + 100 + 400) \div (5 - 1)}$$
$$= \sqrt{(1000 \div 4)} = \sqrt{250} = 15.8$$

The standard deviation for data set B is 1.58:

$$S = \sqrt{(4 + 1 + 0 + 1 + 4) \div (5 - 1)}$$
$$= \sqrt{(10 \div 4)} = \sqrt{2.5} = 1.58$$

As you could tell just by looking at the data sets, data set A has far more variability than data set B. The standard deviation allows us to quantify the variability in a data set.

KEY POINT

The standard deviation is used as a measure of the variability in a data set. It is the square root of the variance.

Standard Scores (Z-scores)

Standard scores are scores that have been standardized to a constant mean and standard deviation. This means that scores are converted from the units in which they were measured (e.g., inches or watts) into a standard unit of measurement (e.g., Z-scores), which allows comparison of performance on tests that were originally measured in different units. Standard scores that you might be familiar with include IQ scores, SAT scores, and GRE scores. Standard scores are used by professionals in exercise science for various reasons, and you can use standard scores to determine how well you performed on a test relative to other test takers. One of the most common and flexible standard scores is the Z-score (see example 3 below). Several examples are presented here to demonstrate how informative standard scores can be.

1. Standard scores can be used to help determine your standing on a test compared to other test takers or compared to normative values. For example, if you scored 80% correct on a 100-item test (i.e., you got 80 out of 100 items correct), how did you do compared to others who took the test? It is really not possible to answer this question unless you know both the mean and standard deviation of the test and calculate a standard score. If the mean of the test was 70 and the standard deviation was 10, then you can see that you scored 1 standard deviation above (i.e., 70 + 10 = 80) the mean, which is pretty good. If the mean of the test was 70 and the standard deviation was 5, then you scored 2 standard deviations above the mean (i.e., 70 + 5 + 5 = 80). That is really good.

2. Standard scores can be used by teachers to combine different tests into one overall grade. An instructor teaching a soccer unit might use several tests to evaluate their students—for example, skills tests for passing, shooting, and dribbling; a written knowledge test; and a participation grade. Each of these tests will have different means and standard deviations, so it is not possible just to sum the scores on different tests to determine a final overall grade. In fact, the shooting and dribbling tests are not only measured in different units, but also in different directions. That is, high scores for shooting represent better performance, while low scores for dribbling (assuming it is a timed test) represent better performance. Furthermore, any time tests have different standard deviations they should not be combined until the tests are standardized; otherwise, the test with the larger standard deviation will carry a greater weight in the combined grade. With standard scores, it is also possible for an instructor to assign different weights to different tests. For example, an instructor might determine that the final exam will count toward 30% of the final grade, other tests are weighted just 10% of the final grade, and other assessments also receive some predetermined weight toward the final grade.

3. Standard scores can be used by selective academic programs (e.g., scholarships or law schools) to identify the top specific percentage of performers. For example, the LSAT is probably the most important factor used by law schools for admissions decisions. LSAT scores range from 120 to 180, with an average score of about 152 and a standard deviation of about 10. Using standard scores and the normal curve presented shortly, it is fairly easy to determine the LSAT score that identifies the top 10% of performers. (Using these rounded values, it is an LSAT score of about 165.)

A basic standard score that is used in the field of exercise science is the Z-score. When a set of scores is transformed to Z-scores, the transformed scores have a mean equal to 0 and a standard deviation equal to 1. These basic units make it easy to determine how well someone performed on a test relative to other test takers. If someone has a Z-score greater than 0, that person scored higher than the mean (assuming high scores are better than low scores, such as on a vertical jump test). If someone has a Z-score less than 0 (a negative value), that person scored below the mean. In equation form a Z-score looks like

$$Z = (X - \bar{X}) \div S$$

where Z stands for the Z-score, $(X - \bar{X})$ indicates the deviation of the score from the mean, and S represents the standard deviation. Because the denominator is the standard deviation, the difference between the score (X) and the mean is represented in standard deviation units. That is, a Z-score tells you how many standard deviation units the score is from the mean. If you calculate a Z-score for the score of 40 in data set A, you will find a Z-score equal to 0.68. This means that in data set A, a score of 40 is 0.68 standard deviations above the mean. If you calculate a Z-score for a score of 31 in data set B, you will find a Z-score of 0.68. Thus, the performance relative to the mean for a score of 40 in data set A is the same as the performance of a score of 31 in data set B. Both are 0.68 units above the mean.

Z-scores are used in exercise science research when it is important to standardize scores for understanding or analysis. For example, the Centers for Disease Control and Prevention uses BMI-for-age Z-scores to express the body mass index (BMI) of children for their age and sex relative to a reference population. These standardized values are helpful for interpretation of growth measurements and to define obese and overweight categories (e.g., Freedman et al., 2007). Other researchers use Z-scores to express health-re-

lated and performance-related fitness variables (e.g., Larsen et al., 2015). You will likely find many uses for Z-scores in your career.

KEY POINT

Standardization converts variables into a standard unit of measure. When scores are standardized, we can compare data that were originally measured in different units.

Normal Curve

Many of the variables we deal with in exercise science are normally distributed. A **normal curve** is often referred to as a bell-shaped curve. Figure 3.2 presents a normal curve with Z-scores indicated on the x-axis. Note that a Z-score of 0 falls at the mean, median, and mode on the normal curve, with 50% of scores above and 50% of scores below a Z-score of 0. The distance between any Z-score and the mean can be easily determined by the standard normal distribution table (not presented in this chapter). We also can determine the percent of scores that fall below any point on the normal curve. This is known as the **percentile rank**. The percentile rank for a Z-score of 0 is 50, because 50% of scores fall below that point. The percentile rank for a Z-score of 1 is 84.13 because 84.13% of scores fall below that point (i.e., 50% below a Z-score of 0 and 34.13% between Z-scores of 0 and 1). Likewise, the percentile rank for a Z-score

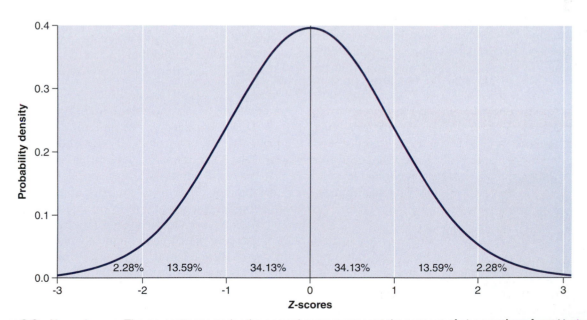

Figure 3.2 Normal curve. The percentages under the normal curve represent the percent of observations found between adjacent Z-scores. The mean Z-score equals 0.

of 2 is 97.72%. Only 2.28% of scores fall above a Z-score of 2. Percentile ranks are typically rounded to whole numbers, so we would say a Z-score of 2 is at the 98th percentile.

KEY POINT

A normal curve is perfectly symmetrical and has known properties that allow us to make probability statements about scores and statistical tests.

Correlation Coefficient

Life is about relationships and so is research. Researchers look for relationships between variables to explain phenomena and predict outcomes. A correlation coefficient is a statistic that describes the strength of the relationship or association between two variables. Several types of correlation coefficients exist, but we will focus on the Pearson product-moment correlation coefficient (so named for the mathematician Karl Pearson), which describes the linear relationship between two variables. If the relationship between two variables is curvilinear, the relationship should not be described with a Pearson correlation because a Pearson correlation would underestimate such a relationship.

We usually denote the Pearson correlation statistic with the letter *r*. The correlation coefficient ranges from –1.00 to +1.00 and is usually reported to two places past the decimal point. Because variables can be negatively correlated or positively correlated, a correlation coefficient can be a negative or positive value. A value of 0 indicates that the two variables of interest are not correlated. A perfect correlation is represented by a value of –1.00 or +1.00. The sign of the correlation denotes the direction of the relationship (either positive or negative) and the value of the correlation coefficient describes the size of the relationship.

When two variables are positively correlated, as the value of one variable increases, the value of the other variable tends to increase. When two variables are negatively correlated, as the value of one variable increases, the value of the other variable tends to decrease. Note that the two previous sentences use the phrase "tends to." This is because a correlation does not indicate causation. Just because two variables are correlated does not mean that one variable causes the other variable. However, the higher the correlation (either negative or positive), the better we can predict one variable from the other variable.

With a little practice, you will be able to examine a scatterplot of the relationship between two variables and estimate the value of the correlation coefficient (*r*). Figure 3.3 presents a scatterplot of a positive correlation between $\dot{V}O_2max$ and PACER laps. The correlation for these data is 0.65. Because this is a positive correlation, it indicates that more PACER laps completed is associated with a higher $\dot{V}O_2max$ and fewer PACER laps completed is associated with a lower $\dot{V}O_2max$. Note that the data points do not all fall exactly on the straight line (called the *trend line*) but instead are spread out above and below the line. In a perfect correlation where *r* = 1.00, all points would fall directly on the trend line. This is a positive moderately high correlation (*r* = 0.65). You can also see that the points tend to increase from left to right, indicating that as the value of one variable increases, the value of the other variable tends to increase. Note also that each point in the scatterplot represents one individual's score on $\dot{V}O_2max$ and PACER. This scatterplot has $\dot{V}O_2max$ and PACER

EXERCISE SCIENCE COLLEAGUES

Matt Soto

Matt Soto is the director of the San Diego State University Adaptive Fitness Clinic. He is a registered kinesiotherapist and holds a bachelor of science degree in kinesiology with an emphasis in kinesiotherapy. He is responsible for making sure the Adaptive Fitness Clinic's goals of serving San Diego's physically disabled community and providing service-learning opportunities for San Diego State University students are achieved. In addition, Matt makes sure the clinic is financially sustainable. A vital part of programming for the clients are regular measurements, which allow the staff to track their clients' improvements in range of motion, strength, and many of their other rehabilitation goals.

Courtesy of Matthew Soto.

scores for 180 people. This was noted in the section on reliability and is worth noting again: it is hazardous to calculate correlations on small samples (i.e., fewer than 30 people is often considered a small sample for correlations) because the resulting value might not be representative of the true relationship in the population, and scores for just one person that might be outliers can have a large impact on the resulting correlation.

Figure 3.4 presents a negative correlation between vertical jump height and percent fat ($r = -0.54$). The negative correlation means that as one variable

increases, the other variable tends to decrease. In this example, as percent fat increases, vertical jump performance tends to decrease. Physiologically, this makes sense because it is more difficult to jump when one has a high level of body fat. Note that the points are spread out around the trend line slightly more than they are in figure 3.3, where the correlation is higher.

One last note about correlations before we move on to regression. The square of the correlation coefficient (r^2) is called the **coefficient of determination**. This statistic is often presented with regression analysis because the interpretation of r^2 is straightforward.

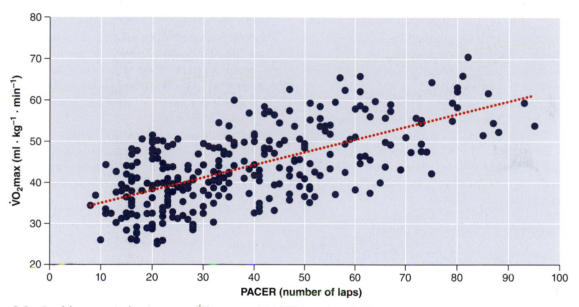

Figure 3.3   Positive correlation between $\dot{V}O_2$max and PACER laps ($r = 0.65$).

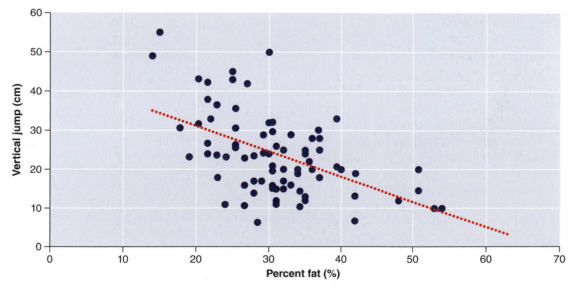

Figure 3.4   Negative correlation between vertical jump and percent fat ($r = -0.54$).

The coefficient of determination is interpreted as the proportion of variance in one variable that can be explained by the other variable.

When using regression analysis, we would like to explain 100% of the variance in the variable we are trying to predict. If we were able to do this, we would have perfect prediction and no error. This never happens for several reasons. In exercise science, variables are not perfectly related (i.e., the correlation between variables is always less than 1.00 or –1.00). In addition, as noted earlier in the section on reliability, everything we measure has error. Whenever error is present, prediction will be less than perfect.

As an example, if one tried to predict $\dot{V}O_2max$ from the number of laps completed in the PACER (see figure 3.3), we could determine the proportion of variance in $\dot{V}O_2max$ explained by PACER laps by squaring the correlation coefficient of 0.65. In this case, $r^2 = 0.65^2 = 0.42$. That is, 42% of the variability in $\dot{V}O_2max$ could be explained by PACER laps. You might be thinking, that is not very much—and you would be right. In order to increase the proportion of variance in the variable we are trying to explain (called the *criterion variable* or *dependent variable*), researchers often develop regression equations with more than one predictor variable, as we will see in the next section.

KEY POINT

A Pearson product-moment correlation is a measure of the strength of the relationship between two variables. It can take on any value between –1.00 and +1.00. The coefficient of determination (r^2) describes the strength of that relationship as a percentage.

Regression Analysis

Measurement researchers often use regression analysis to develop equations to allow prediction of one variable from one or more other variables. The variable that we try to predict is known as the *criterion* or **dependent variable**. The variables that we try to predict from are known as *predictors* or **independent variables**. To develop prediction equations, researchers must measure both the criterion and predictor variables, often in laboratory settings. However, the criterion variable is often difficult or expensive to measure in field-based environments such as schools or sport programs. Once the prediction equation is developed using regression analysis, practitioners can measure the predictor variables and use the prediction equation to estimate the criterion variable.

Because lower-body external muscular power is related to both sport performance and bone strength in youth (Henriques-Neto et al., 2020; Janz et al., 2015), coaches and pediatricians might wish to know the lower-body power of their athletes or patients. However, lower-body power is difficult to measure directly because the necessary equipment (e.g., force plates) are typically only available in laboratory settings. Several research groups have published equations to estimate lower-body power (in watts) from vertical jump height and body mass (Duncan et al., 2013; Gomez-Bruton et al., 2019; Mahar et al., 2022). In field-based settings, vertical jump height and body mass are relatively easy to measure.

Regression equations with one predictor are called *simple linear regression equations*. Most regression equations in exercise science include more than one predictor variable because prediction accuracy is generally better than when only one predictor is used. When regression equations include more than one predictor variable, they are known as *multiple regression equations*.

To demonstrate how regression equations are used, one multiple regression equation to predict lower-body power from vertical jump height and body mass is presented here (Mahar et al., 2022).

$$\text{Lower-body power (watts)}$$
$$= -1354.820 + (35.455 \times \text{vertical jump height [cm]})$$
$$+ (43.942 \times \text{body mass [kg]})$$

For a calculation example, assume an individual jumps 35 centimeters and has a body mass of 50 kilograms. Do not worry that the regression formula looks unusual (large negative constant or steep slope of 35); regression finds the line of best fit of the data and sometimes the mathematics mapping one variable to another in a line does not have an obvious logic. Remember that correlation and regression do not infer causation.

$$\text{Lower-body power (watts)}$$
$$= -1354.820 + (35.455 \times 35) + (43.942 \times 50) = 2083$$

Thus, by measuring vertical jump height and body mass and plugging these values in the multiple regression equation, you can obtain an estimate of lower-body power in watts. Teachers, coaches, or physicians who use prediction equations should remember that no prediction equation is perfect and that a certain amount of prediction error is expected. Whenever a prediction equation is published, the prediction error associated with that equation is also presented. Although, unfortunately, we cannot take a deep dive into statistics that quantify the error of prediction equations, you should at this point realize that these statistics are important to select which equation to use and to determine how much confidence you can have in the predicted values. Statistics like the coefficient of determination and the standard error of estimate are used to quantify prediction accuracy. Recall that the coefficient of determination is the proportion of variance in the criterion explained by the predictor variables. The more variance explained, the more accurate the prediction. The **standard error of estimate** is like the standard deviation discussed earlier in this chapter. Specifically, the standard error of estimate is the standard deviation of the error. The error is a residual score (i.e., actual value of the criterion variable minus predicted value of the criterion variable). A lower standard error of estimate indicates greater prediction accuracy.

KEY POINT

A regression line on a scatterplot represents the regression equation between the criterion variable and predictor variables.

Career Opportunity

Research Physiologist at the Naval Health Research Center

Research physiologists work on environmental physiology teams at the Naval Health Research Center. The team looks for ways to mitigate the physiological impact extreme environmental conditions have on warfighter readiness and how to improve both physical and cognitive performance. They collect and analyze physiological data in austere environments, primarily focusing on heat, cold, and altitude stress. Measurements typically include heart rate, oxygen saturation, skin temperature, core temperature, thermal and shivering sensation, rating of perceived exertion, simple and choice reaction time, and cognitive fatigue.

Research physiologists collaborate with biomechanists, neuroscientists, registered dietitian nutritionists, sleep researchers, and other exercise physiologists when conducting human performance field and laboratory research. Inputting data from various physiological measurements, organizing data into statistical software, running statistical analysis, managing data, and technical and scientific writing for publication is expected.

Effect Size

One of the most basic questions that researchers have is how large the relationship is between variables or how large the difference is between groups. We already discussed the correlation coefficient, which is one statistic that estimates the size of the relationship between variables. In many research studies, the means of groups are compared. For example, males might be compared to females. A sample of older adults might be compared to a sample of young adults. A group that undergoes high-intensity interval training (HIIT) might be compared to a group that undergoes high-volume training. Preintervention scores of a group might be compared to postintervention scores of the same group.

Several ways exist to provide an estimate of effect size. Here we will discuss a useful and intuitive measure of effect size called *Cohen's delta*. Cohen's delta is the size of the difference between means in standard deviation units. This is somewhat similar to Z-scores, except rather than calculating a residual score $(X - \bar{X})$ for the numerator, the numerator is the difference between two means $(\bar{X}_1 - \bar{X}_2)$.

One use of Cohen's delta is to calculate the size of the difference between an experimental group and a control group. This would be interpreted as the size of the experimental effect. The larger the effect size, the greater effect the experimental intervention had on the outcome (termed the *dependent variable*).

$$\text{Effect size (Cohen's delta)} = (\bar{X} \text{ of experimental group} - \bar{X} \text{ of control group}) \div \text{standard deviation}$$

Another use of Cohen's delta is to calculate the size of the change in the dependent variable due to an intervention.

$$\text{Effect size (Cohen's delta)} = (\text{postintervention } \bar{X} - \text{preintervention } \bar{X}) \div \text{standard deviation}$$

Note that an effect size of 0.0 indicates that the means for the two groups did not differ. An effect size of 1.0 indicates that the means of the two groups differed by 1 standard deviation. Interpretation of effect sizes is presented in table 3.4 (Cohen, 1988). A medium effect size of 0.50 means that the size of the difference between means is half of the standard deviation. Because Cohen's delta is standardized by dividing by

Table 3.4 Interpretation of Cohen's Delta Measure of Effect Size

Relative size	Effect size
Small	0.20
Medium	0.50
Large	0.80

the standard deviation, effect size estimates from one study can be compared to effect size estimates from another study. Note that two groups are compared when calculating Cohen's delta. That means that we are dealing with two standard deviations, one for each group. When calculating Cohen's delta, the researchers should justify the standard deviation they use in the denominator. Often, when a true control group exists, the standard deviation of the control group is used. When no control group is involved in the comparison, researchers often use a pooled standard deviation of the two means being compared.

Mahar and colleagues (2006) developed a school-based physical activity intervention called Energizers. These are activities that teachers can use to get their students active in the classroom. In this study, teachers in the intervention group were trained to lead their classes in one 10-minute Energizers activity each day. The teachers in the control group did not use Energizers in their classrooms until after the study was over. To determine if students in the intervention group that performed Energizers were more active over the course of a school day, the total number of steps taken during the school day by these two groups were compared. Table 3.5 presents the sample sizes, means, and standard deviations for the two groups in this study.

$$\text{Effect size} = (5{,}587 - 4{,}805) \div [[(1{,}633 \times 135) + (1{,}543 \times 108)] \div (135 + 108)]$$

$$\text{Effect size} = 782 \div 1{,}593 = 0.49$$

The effect size was 0.49, which is interpreted as a medium effect. This means that the size of the effect of the Energizers intervention on increasing physical activity was medium. That is quite good for a school-based intervention. You probably noticed that calculation of the denominator looks complex. Because the two groups had different sample sizes, the authors decided to pool the standard deviations by weighting them by their sample sizes rather than

using the sample size for either the Energizers group or control group in the denominator.

When other school-based interventions designed to increase school-day physical activity are conducted, the authors can calculate the effect size with Cohen's delta and compare those results with the effect size calculated in this study.

KEY POINT

Cohen's delta is an effect size that represents the standardized magnitude of the difference between two means. Other statistics are also used to represent effect size.

t-tests

t-tests belong to a group of statistics knows as *inferential statistics*. The term *inferential* comes from the idea of trying to infer findings from study samples to larger populations. *t*-tests are used to test a null hypothesis that states no difference exists between two means. If the null hypothesis is not rejected based on the analysis of sample data, the conclusion is that the two means are not significantly different. If the null hypothesis is rejected based on analysis of sample data, the conclusion is that the two means are significantly different. When statistical analysis software is used to conduct *t*-tests, results are reported to allow you to determine whether the means are judged to be significantly different. These statistics, although interesting and important, are beyond the scope of this chapter.

If the people in one group are independent of the people in the other group (e.g., different people randomly assigned into two groups), the *t*-test is called an independent groups *t*-test. For example, a research study comparing sleep duration of college-age females with college-age males could use an independent groups *t*-test. *t*-test results for the Energizers study presented in table 3.5 are shown in table 3.6.

Table 3.5   **Comparison of School-Day Steps of Energizers Group Versus Control Group**

Group	Sample size	Mean	Standard deviation
Energizers	135	5,587	1,633
Control	108	4,805	1,543

Table 3.6   ***t*-test Results for the Energizers Study**

Mean difference	*t*	*df*	Significance (*p*)
782	3.80	241	0.0001

Hot Topic

Setting Criterion-Referenced Standards for Assessment of Musculoskeletal Fitness in Youths

Muscle power (overall external muscular power flow) is a key indicator of youth fitness assessment and surveillance (Institute of Medicine, 2012). Several studies have demonstrated that muscle power is positively associated with bone health (Baptista et al., 2016; Janz et al., 2015). Youth fitness tests are used in schools to provide feedback to students and for national surveillance of the state of youth fitness. A barrier to the adoption of tests of muscle power for youth fitness tests has been the lack of criterion-referenced (healthy levels of bone strength) standards to evaluate fitness based on test performance. Recall that the first 100 years of youth fitness data, including vertical jump, were norm-referenced and noninvasive ways to measure bone density and strength did not exist. Researchers recently developed criterion-referenced standards to provide insight into the levels of muscle power needed for health benefits.

To develop these criterion-referenced standards, researchers measured bone strength (a health outcome), muscle power, and vertical jump (a field-based youth fitness test), among many other variables. Measurement of bone strength is expensive, complex, and time consuming. Bone strength is assessed with dual-energy X-ray absorptiometry. Muscle power is complicated to assess because it involves both speed and strength (i.e., work per unit of time). Researchers use force plates to measure muscle power. Vertical jump performance also must be accurately measured on the same youth. Adding to the complexity of the study is the fact that hundreds of youths must be assessed to represent the age range of interest robustly. Sophisticated statistical analyses are used to develop models to predict muscle power from vertical jump performance and to set criterion-referenced standards that link muscle power to bone strength. This work has been ongoing for several years by members of the FitnessGram Advisory Board, and several papers documenting the process and results were recently published (Janz et al., 2021; Laurson et al., 2022; Mahar et al., 2022; Welk et al., 2022).

The mean difference is simply one mean minus the other mean. The t column is the t-statistic that is calculated from these data. The df column stands for degrees of freedom, which is calculated based on sample size. The significance column is the column that is checked to determine whether the two means are considered significantly different or not. Note that t-tests lead to a binary decision. The difference between means is considered to be significant or not significant. By convention, a significance level of 0.05 (or 0.01) has been used as a decision rule. We will not delve deeply into the understanding of significance levels, but the letter p, which stands for probability, is used for this binary decision. The p-value from these sample data is $p = 0.0001$. If the p-value is less than or equal to 0.05, the null hypothesis is rejected. In this case, 0.0001 is less than 0.05, so the null hypothesis is rejected. Because the null hypothesis is rejected, the conclusion is that the two means are significantly different.

If the people in one group are related in some way to the people in the other group, the t-test is called a dependent groups t-test. Groups can be related in several ways. For example, in a comparison of means from a pretest and posttest, the same individuals are used to calculate both means. In another study, participants in one group might be paired with participants in the other group by matching them on an important variable. If participants are matched for percent body fat before being assigned to one of the groups, for example, the means for the groups would be compared with a dependent groups t-test.

KEY POINT

A t-test is used to test whether the difference between two means is significantly different from zero.

Analysis of Variance (ANOVA)

You probably noted that t-tests are used to compare two (and only two) means. In situations in which more than two means are involved in a study, some type of **analysis of variance (ANOVA)** is usually calculated. ANOVA is a family of statistics that includes, among others, one-way ANOVA, factorial ANOVA, and repeated measures ANOVA.

We will not get into the weeds about this, but ANOVA is just a special case of the linear regression model (presented earlier in the chapter) when the predictor or independent variable is nominal (i.e., a categorical variable of interest in a study with two or more levels). Like the t-test, each ANOVA measures a response or indicator variable on a continuous scale called the *dependent variable*. Furthermore, the *t*-test (presented previously) is just a special case of ANOVA when the nominal variable only has two groups.

One-Way ANOVA

A one-way ANOVA has only one predictor or independent variable, although it can have two or more groups or levels. For example, a researcher might wish to determine if the following three groups differed on their average physical activity level: undergraduate exercise science students, graduate exercise science students, and exercise science professors. The continuous outcome variable in this example is physical activity level. The categorical predictor variable is group (i.e., undergraduate, graduate, or professor). A one-way ANOVA would be used to compare the average physical activity levels of these three exercise science knowledge groups simultaneously. If the test was statistically significant, researchers usually use post hoc tests to see exactly which specific means were different from each other.

Factorial ANOVA

The term *factorial* is used to indicate a situation in which the effects of more than one categorical predictor variable on one continuous outcome are being studied. A two-way ANOVA is a factorial design with two predictor variables, and a three-way ANOVA is a factorial design with three predictor variables. We rarely see factorial designs with more than three predictor variables, but theoretically any number of predictor variables could be involved in the design. If we added a variable, such as gender, to the study described in the previous paragraph, we would change the one-way ANOVA design to a two-way ANOVA design. We then would be able to determine (1) whether the average of the three exercise science groups differed, (2) whether the average for males and females differed, and (3) whether an interaction between group and gender exists. Interactions are often the main interest of researchers who choose this type of design. An interaction examines whether the pattern of effect in one independent variable depends on the level of the other independent variable. In this example, a significant interaction might be a situation where females in the undergraduate groups were more active than males in the undergraduate group, but no difference between genders existed for the graduate student group or the professor group.

Repeated Measures ANOVA

Many studies in exercise science involve measuring participants on the same outcome variable on several occasions over time. For example, Stylianou and associates (2016) measured on-task behavior of third and fourth grade students on days students attended a before-school running program and on days they did not attend. Thus, the outcome variable of on-task behavior was assessed under two different conditions. (By the way, these authors found that students had higher levels of on-task behavior on days they participated in the running program compared to days they did not participate. Although this study was not conducted on college students, if you have trouble paying attention in class, it might be worth trying to get a workout in during the day before your class.)

Repeated measures ANOVA studies can involve only the outcome variable measured twice, similar to the dependent groups *t*-test mentioned earlier in this chapter, or several other variables. A useful repeated measures design involves the outcome variable measured more than once and a categorical variable. This is often referred to as a *two-way factorial ANOVA with repeated measures on one factor*. Repeated measures designs are powerful and versatile, and give the ability to incorporate multiple categorical variables and assess the change in the outcome variable across time.

> **KEY POINT**
>
> Analyses of variance (ANOVA) are statistical tests use to determine whether two or more group means differ. A variety of ANOVAs exist to match different experimental designs.

Wrap-Up

Many health-related and educational decisions rely to a great extent on the processes of measurement and evaluation. Over your educational and professional careers, you will have to make decisions for yourself and for your clients. As exercise science professionals, these decisions should be informed by data. Understanding how data are collected and the evidence, or lack thereof, of validity and reliability of the measurements will help you make better decisions. Statistics are tools to help us organize, analyze, report, and understand our data.

Validity is the most important characteristic of a measure. Several types of evidence of validity exist, and researchers present that evidence to try to determine whether the test measures what it was intended to measure. The type of evidence of validity presented depends on the proposed use of the test outcome. Reliability is another important characteristic of a test and represents the consistency of the measurement. Once data are collected, evaluation of the measurements can be made in a norm-referenced or criterion-referenced context. For norm-referenced evaluation, performance of the participant is compared to others who were tested or to previously established norms. For criterion-referenced evaluation, performance is compared to a predetermined standard that has been linked to a health outcome or behavior.

Statistics are used to summarize data (called *descriptive statistics*) and to explore relationships and make comparisons (called *inferential statistics*). The most common measure of the central tendency of a distribution of data is the mean (i.e., arithmetic average), and the associated measure of the spread of the distribution is the standard deviation. The mean and standard deviation of a data set can be used to standardize scores and to understand where a partic-

ular score falls within the distribution. Standardized scores (e.g., Z-scores) are scores with a constant mean and standard deviation. They can be used to determine how a score compares with other scores, to combine tests that were measured with different units, and to select top performers.

Some valuable basic statistics that will help prepare you to understand and produce research include the correlation coefficient, regression analysis, effect size, *t*-tests, and analysis of variance. A basic understanding of these statistics will provide you with a solid foundation for learning some of the many other statistics that are available. The great statistician Karl Pearson noted, "Statistics is the grammar of science." In order to communicate and understand quantitative science, it is important to develop an appreciation for, if not a love of, statistics.

More Information on Measurement and Statistics

Organizations
American College of Sports Medicine

American Educational Research Association

National Council on Measurement in Education

National Strength and Conditioning Association

SHAPE America

Journals
BMC Medical Research Methodology

British Journal of Sports Medicine

European Journal of Sport Science

International Journal of Behavioral Nutrition and Physical Activity

Journal for the Measurement of Physical Behaviour

Journal of Health Education

Journal of Physical Activity and Health

Measurement in Physical Education and Exercise Science

Open Journal of Statistics

Physiological Measurement

Public Health Nutrition

Research Quarterly for Exercise and Sport

Sports Medicine

Statistical Methods in Medical Research

Go to *HKPropel* to complete the activities for this chapter.

Review Questions

1. Define the various types of evidence of validity. Provide an example in exercise science relative to when evidence of criterion-related validity is important.

2. What is the basic definition of reliability? Explain the different types of reliability.

3. What do central tendency and variability mean? What statistics are used to describe central tendency and variability?

4. How is a Z-score calculated? Explain what a Z-score tells us and why Z-scores might be used.

5. Explain a negative correlation and a positive correlation. Provide examples of each using variables from exercise science.

6. Describe the statistical tests that can be used to compare two means and two or more means simultaneously.

Suggested Readings

Bassett, D.R., Toth, L.P., LaMunion, S.R., & Crouter, S.E. (2017). Step counting: A review of measurement considerations and health-related applications. *Sports Medicine, 47,* 1303-1315.

Bassett, D.R., Mahar, M.T., Rowe, D.A., & Morrow, J.R., Jr. (2008). Walking and measurement. *Medicine and Science in Sports and Exercise, 40*(7 Suppl), S529-S536.

Cohen, J. (1994). The earth is round (*p* < .05). *American Psychologist, 49,* 997-1003.

Institute of Medicine. (2012). *Fitness measures and health outcomes in youth.* National Academies Press.

Mahar, M.T., & Rowe, D.A. (2014). A brief exploration of measurement and evaluation in kinesiology. *Kinesiology Review, 3,* 80-91.

Mahar, M.T., & Rowe, D.A. (2008). Practical guidelines for valid and reliable youth fitness testing. *Measurement in Physical Education and Exercise Science, 12,* 126-145.

Schmidt, F.L. (1996). Statistical significance testing and cumulative knowledge in psychology: Implications for training of researchers. *Psychological Methods, 1,* 115-129.

References

Baptista, F., Mil-Homens, P., Carita, A.I., Janz, K.F., & Sardinha, L.B. (2016). Peak vertical jump power as a marker of bone health in children. *International Journal of Sports Medicine, 37,* 653-658.

Blair, S.N., Kohl, H.W., Paffenbarger, R.S., Clark, D.G., Cooper, K.H., & Gibbons, L.W. (1989). Physical fitness and all-cause mortality: A prospective study of healthy men and women. *JAMA, 262,* 2395-2401.

Cohen, J. (1988). *Statistical power analysis for the behavioral sciences* (2nd ed.). Lawrence Erlbaum Associates.

Duncan, M.J., Hankey, J., & Nevill, A.M. (2013). Peak-power estimation equations in 12- to 16-year-old children: Comparing linear with allometric models. *Pediatric Exercise Science, 25,* 385-393.

Freedman, D.S., Kahn, H.S., Mei, Z., Grummer-Strawn, L.M., Dietz, W.H., Srinivasan, S.R., & Berenson, G.S. (2007). Relation of body mass index and waist-to-height ratio to cardiovascular disease risk factors in children and adolescents: The Bogalusa Heart Study. *American Journal of Clinical Nutrition, 86,* 33-40.

Gomez-Bruton, A., Gabel, L., Nettlefold, L., MacDonald, H., Race, D., & McKay, H. (2019). Estimation of peak muscle power from a countermovement vertical jump in children and adolescents. *Journal of Strength and Conditioning Research, 33,* 390-398.

Henriques-Neto, D., Magalhães, J.P., Hetherington-Rauth, M., Santos, D.A., Baptista, F., & Sardinha, L.B. (2020). Physical fitness and bone health in young athletes and nonathletes. *Sports Health, 12,* 441-448.

Institute of Medicine. (2012). *Fitness measures and health outcomes in youth.* National Academies Press.

Jackson, A.S., Franks, B.D., Katch, F.I., Katch, V.L., Plowman, S.A., & Safrit, M.J. (1976). *A position paper on physical fitness. Position paper of a joint committee representing the Measurement and Evaluation, Physical Fitness, and Research Councils of AAHPER.* American Association for Health, Physical Education, and Recreation.

Jackson, A.S., & Pollock, M.L. (1978). Generalized equations for predicting body density of men. *British Journal of Nutrition, 40,* 497-504.

Jackson, A.S., Pollock, M.L., & Ward, A. (1980). Generalized equations for predicting body density of women. *Medicine & Science in Sports & Exercise, 12,* 175-181.

Janz, K.F., Laurson, K.R., Baptista, F., Mahar, M.T., & Welk, G.J. (2021). Vertical jump power is associated with healthy bone outcomes in youth: ROC analyses and diagnostic performance. *Measurement in Physical Education and Exercise Science.* https://doi.org/10.1080/1091367X.2021.2013230

Janz, K.F., Letuchy, E.M., Burns, T.L., Francis, S.L., & Levy, S.M. (2015). Muscle power predicts adolescent bone strength: Iowa bone development study. *Medicine & Science in Sports & Exercise, 47,* 2201-2206.

Kraus, H., & Hirschland, R.P. (1954). Minimum muscular fitness test in school children. *Research Quarterly, 25,* 177-188.

Landis, J.R. & Koch, G.G. (1977). The measurement of observer agreement for categorical data. *Biometrics, 33,* 159-174.

Larsen, L.R., Kristensen, P.L., Junge, T., Rexen, C.T., & Wedderkopp, N. (2015). Motor performance as predictor of physical activity in children: The CHAMPS Study-DK. *Medicine & Science in Sports & Exercise, 47,* 1849-1856.

Laurson, K.R., Baptista, F., Mahar, M.T., Welk, G.J., & Janz, K.F. (2022). Designing health-referenced standards for the plank test of core muscular endurance. *Measurement in Physical Education and Exercise Science.* https://doi.org/10.1080/1091367X.2021.2016409

Mahar, M.T., Welk, G.J., & Rowe, D.A. (2018). Estimation of aerobic fitness from PACER performance with and without body mass index. *Measurement in Physical Education and Exercise Science, 22,* 239-249.

Mahar, M.T., Murphy, S.K., Rowe, D.A., Golden, J., Shields, A.T., & Raedeke, T.D. (2006). Effects of a class-room-based program on physical activity and on-task behavior. *Medicine & Science in Sports & Exercise, 38,* 2086-2094.

Mahar, M.T., Rowe, D.A., Parker, C.R., Mahar, F.J., Dawson, D.M., & Holt, J.E. (1997). Criterion-referenced and norm-referenced agreement between the one mile run/walk and PACER. *Measurement in Physical Education and Exercise Science, 1,* 245-258.

Mahar, M.T., Welk, G.J., Janz, K.F., Laurson, K., Zhu, W., & Baptista, F. (2022). Estimation of lower body muscle power from vertical jump in youth. *Measurement in Physical Education and Exercise Science.* https://doi.org/10.1080/1091367X.2022.2041420

Meredith, M.D. & Welk, G.J. (2010). *FitnessGram and ActivityGram test administration manual* (4th ed.). Human Kinetics.

Myers, J., Prakash, M., Froelicher, V., Do, D., Partington, S., & Atwood, J.E. (2002). Exercise capacity and mortality among men referred for exercise testing. *New England Journal of Medicine, 346,* 793-801.

Paluch, A.E., Pettee Gabriel, K., Fulton, J.E., Leis, C.E., Schreiner, P.J., Sternfeld, B., Sidney, S., Siddique, J., Whitaker, K.M., & Carnethon, M.R. (2021). Steps per day and all-cause mortality in middle-aged adults in the Coronary Artery Risk Development in Young Adults study. *JAMA Network Open, 4*(9), e2124516.

Rikli, R.E., & Jones, C.J. (2013a). Development and validation of criterion-referenced clinically relevant fitness standards for maintaining physical independence in later years. *The Gerontologist, 53,* 255-267.

Rikli, R.E., & Jones, C.J. (2013b). *Senior fitness test manual.* Human Kinetics.

Sabia, S., Fayosse, A., Dumurgier, J., van Hees, V.T., Paquet, C., Sommerlad, A., Kivimäki, M., Dugravot, A., & Singh-Manoux, A. (2021). Association of sleep duration in middle and old age with incidence of dementia. *Nature Communications, 12.* https://doi.org/10.1038/s41467-021-22354-2

Saint-Maurice, P.F., Laurson, K., Welk, G.J., Eisenmann, J., Gracia-Marco, L., Artero, E.G., Ortega, F., Ruiz, J.R., Morena, L.A., Vicente-Rodriguez, G., & Janz, K.F. (2018). Grip strength cutpoints for youth based on a clinically relevant bone health outcome. *Archives of Osteoporosis, 13,* 92.

Saint Romain, B., & Mahar, M.T. (2001). Norm-referenced and criterion-referenced reliability of the push-up and modified pull-up. *Measurement in Physical Education and Exercise Science, 5,* 67-80.

Stylianou, M., Kulinna, P.H., van der Mars, H., Mahar, M.T., Adams, M.A., & Amazeen, E. (2016). Before-school running/walking club: Effects on student on-task behavior. *Preventive Medicine Reports, 3,* 196-202.

Welk, G., Janz, K., Laurson, K., Mahar, M., Zhu, W., & Pavlovic, A. (2022). Development of criterion-referenced standards for musculoskeletal fitness in youth: Considerations and approaches by the FitnessGram Scientific Advisory Board. *Measurement in Physical Education and Exercise Science.* https://doi.org/10.1080/1091367X.2021.2014331

PART II
Major Subdisciplines of Exercise Science

Exercise science professionals are defined by their mastery of a complex body of theoretical knowledge about exercise and its application in evidence-based practice. Some people who have not formally studied exercise science might think they know a good deal about working out or staying fit, but their information is often incomplete, out of date, or simply untrue. As an exercise science major, you—more than the typical person on the street, celebrity trainer, or your friends in other disciplines—need to possess a solid base of knowledge from many subdisciplines of exercise science.

To that end, part II of the book introduces you to the scholarly study of exercise in five major subdisciplines of exercise science. Although the field also includes other subdisciplines as well as various professional areas of research, these five subdisciplines have gained recognition as the major categories of knowledge.

- Chapter 4 introduces you to the fascinating world of biomechanics of human movement.
- Chapter 5 addresses basic physiological aspects of exercise, including the effects of exercise on muscular and cardiovascular systems and the physiological basis for training.
- Chapter 6 directs your attention to the performance of skilled movement and factors affecting it, including those related to human development.
- Chapter 7 covers the psychological aspects of sport and exercise.
- Chapter 8 explores the disease-prevention power of exercise and population-level issues of exercise on public health.

Even though these chapters are jam packed with exercise science knowledge, they provide only a preview of these scholarly subdisciplines. Thus, when you finish reading and studying them, you will not have mastered the content or the holistic body of knowledge that constitutes exercise science. However, you will have taken a glimpse at what lies ahead of you in your program of studies, in which you will delve into each of these subdisciplines in greater depth.

CHAPTER 4

Biomechanics

Kathy Simpson

CHAPTER OBJECTIVES

In this chapter, we will

- describe what biomechanics is and what it encompasses,
- explain how biomechanics is useful to you and in careers in exercise sciences,
- address what biomechanists and related specialists do,
- explain how biomechanics emerged within the field of physical activity, and
- introduce biomechanical concepts and the processes by which biomechanists and professionals in exercise science answer questions of interest in professional settings.

Katsumi is a physically active, older person with mild degenerative knee osteoarthritis (OA, which wears away joint cartilage) that is now causing stiffness, pain during certain movements, and decreased mobility. They want to remain active at a moderate intensity level. Therefore, they want to know what they can do to prevent further cartilage degeneration and reduce the pain while staying active and fit. To help Katsumi, imagine that you are serving in one of the following professional roles:

- Physical or occupational therapist, athletic trainer, or other rehabilitation specialist
- Personal trainer
- Medical personnel (e.g., orthopedist or nurse practitioner)
- Scientist in the research and development department at a company that manufactures knee braces
- Science or health journalist
- Another relevant role of interest to you

Think of at least five pieces of information you would want to know that would help you answer Katsumi's question. How might you obtain this knowledge?

In your role as a professional in the situation just described, which of the following did you believe important to answering Katsumi's question: movement technique; potential anatomical problems (e.g., knock-knee); prior injury, training, and physical activity history; body weight; current footwear; other information? If you were to observe or measure Katsumi's movements, how would you know if they posed a concern? How would you know whether information obtained from other sources was valid?

Exercise science professionals use biomechanics to answer questions such as Katsumi's by incorporating assessments of the client's movements and other information with research-based biomechanical knowledge of the factors affecting knee osteoarthritis. We will consider Katsumi's situation throughout this chapter. Your opinions might shift as you learn about biomechanics and, later, knee OA.

Goals of Biomechanics

- To understand how the basic laws of mechanical physics and engineering affect and shape the structure and function of the human body
- To apply this understanding to (1) improve the outcomes of our movements (i.e., performance effectiveness) or (2) increase or maintain the safety and health of our tissues

Benefits of Biomechanics Knowledge

Biomechanics applies mechanical principles to understand how forces affect the structure (i.e., physical body or its components) and functions (i.e., specific purposes) of a biological entity. Mechanics is the branch of physics that describes motion (kinematics) and the causes of motion (kinetics). Within that branch, the field of biomechanics encompasses the study of any living organism, be it bacteria, fungi, plant, animal, or human. Biomechanics can be applied to any part of an organism or to the entire organism. Because we focus here on biomechanics of human movement, we use the term *body* to refer either to the entire human body or to any part of it (e.g., heart, muscle, leg, bone cell).

Let's explore this definition of biomechanics and why biomechanics is useful. When Katsumi exercises, one function that their heart must accomplish is pumping blood throughout their body. Meanwhile, their skeletal muscles function to move their body. For example, functions of their tibia (lower leg bone) include holding up their body and serving as a struc-

ture for muscles to pull against in order to create leg movement such as running. The tibia's structure consists of its anatomy, including its compositional elements (e.g., bone cells, minerals, proteins, water, marrow, connective tissue) and the way in which those elements are put together and shaped. The tibia is shaped like a slightly bowed rod, with dense, compact bone on the outer part for strength and porous bone inside, which is arranged to help keep the marrow in place and reduce bone weight. This combination of structural elements allows the tibia to be lightweight enough to move rapidly yet also to resist considerable force when muscles pull on it or when a performer hits the ground hard when landing.

Our structures help us to function—for example, to breathe, run, text, and eat. Conversely, our movement (i.e., our functioning) affects our structure. For example, one function of our tendons (structures that attach muscles to bone) is to help us move effectively. If we engage in heavy physical activity for months, our tendon structure changes—it gets thicker and stronger.

Forces are needed to accomplish any task involving movement; they also help us maintain the health of our body structures. A force consists of a push or a pull on an object or body. To kick a ball, for instance, you must apply force to the ball with your foot. For hip replacement patients to stand up, they must be able to create muscle forces to push themselves upward. Forces also are used by microscopic elements (e.g., proteins), cells, muscles, bones, connective tissues, and organs to accomplish their functions. For example, heart muscle creates force to pump blood. If we engage in enough physical activity—whereby forces are applied to bones, connective tissues, and muscles—these tissues become stronger and better able to carry out their functions. The converse is also true—that is, disuse weakens body tissues. In addition, excessive or repetitive forces with inadequate rest can injure tissues, as in the brain cell damage that can be caused by a head impact.

The forces commonly involved in physical activity include those that you likely know about, such as air resistance, gravity, and friction (see figure 4.1). You have also experienced forces applied by another object (e.g., the ground) or person (e.g., friend helps you up from the ground), as well as forces that you applied to another object (e.g., lifting a box). Finally, you experience forces generated inside of your body (e.g., muscle forces pulling on bone, bones pushing or rubbing against one another at a joint).

Mechanical laws of physics and engineering are principles that explain how force is generated or manipulated or how force affects a system's structure or

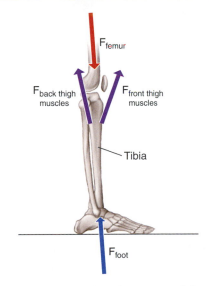

Figure 4.1 Some typical forces that influence body structure and function (including movement): *(a)* forces applied to us by another object, *(b)* forces that we apply to another object in order to manipulate its motion, and *(c)* forces acting on a bone (the tibia).

function (e.g., produces movement). Newton's law of acceleration tells us that a system's acceleration is greater when more total force acts on the system or when the system has less mass. Thus, at the start of a master's sprint race, when high acceleration is needed, Katsumi will want to generate as much muscle force as possible in the right direction and carry only minimal excess fat mass. Another principle holds that if the amount of force applied to a structure exceeds a certain limit, permanent damage occurs. This principle explains how certain injuries occur in bone, muscle, and connective tissue—for example, mild and moderate ankle sprains.

> **KEY POINT**
>
> Force is needed to (1) produce and control our movements and (2) maintain or improve the health of our tissues. Excessive amounts or repetitions of force, however, can injure tissues.

Biomechanics in exercise science is often called **human movement biomechanics**. This is only one of several specialized areas in the field of biomechanics. Here are examples of other specializations (Sharma & Khurana, 2018):

- Nanobiomechanics: forces and movements at the molecular level
- Continuum biomechanics: forces and movements produced from molecular to organ level
- Computational biomechanics: use of engineering computation techniques to investigate forces and movements
- Biotribology: wear, friction, lubrication of joints
- Animal, dental, plant, and forensic biomechanics

Human movement biomechanics can include these specializations and overlap with other areas such as sport and dance, orthopedic (muscle, connective, and bone tissue), and clinical biomechanics, which focuses on issues related to persons with medical concerns, movement disorders, or injury problems (Innocenti & Galbusera, 2022).

Goals of Human Movement Biomechanics

Human movement biomechanics has two main goals: theoretical discovery and application. The first goal is to discover the fundamental, universal theories that explain how forces affect our movements and our bodies' structures. Here are some examples of fundamental theoretical questions:

- How do we use our muscles to control our balance?
- How are the structure and flexibility of connective tissues affected by mechanical loading factors such as the speed and duration of stretching?
- What trunk movements apply the most and least force to different spinal tissues?

The second goal is to use discovered theories and applied biomechanical research to find answers to real-life questions and determine whether the theories work in real-life conditions or whether new theories are needed. Here are some examples of practical research questions asked by biomechanists in exercise science:

- Why, compared to younger adults, do some older adults use different muscle-activity strategies to control their balance?
- What weight-training exercises, equipment, and loads are best for improving my client's sport performance?
- How do I know when it is safe for a young person who had a moderately serious knee ligament injury to return to physical activity or competitive sport?
- Should my patient, whose arm trembles when eating, use a heavier spoon to reduce some of the trembling effects and produce smoother hand movements?
- What can my client and I do to reduce the risk of another back injury occurring at work?

If you become an exercise science professional, biomechanical knowledge will help you answer questions about improving performance effectiveness, increasing the health of tissues, preventing and treating injuries, selecting or modifying assistive devices or joint or tissue replacements, and modifying sport or fitness equipment or rules in order to ensure safety.

KEY POINT

Biomechanical knowledge will help you in a career related to exercise science, rehabilitation, or medicine to improve clients' performance effectiveness; improve their tissue health; prevent or treat injury; and select or modify assistive devices, equipment, and gear related to sport and fitness.

The performance of any person can be improved in almost any situation that involves physical movement or even no movement because forces always

act on the body—for example, repetitive work tasks, activities of daily living, exercise, occupational or sport training, sport performance, music playing, dance, and even lying in bed. If we understand how living organisms can best exploit the mechanical laws that govern how motion is controlled, we can intelligently select the best **movement techniques** for a given performer to use in completing a certain movement task. For example, an athletic trainer can share scientific evidence with a coach demonstrating that a new sport technique greatly increases tissue loading, thereby likely increasing risk of injury to their athletes. An occupational therapist can apply biomechanics in order to determine how to help a client with a neuromuscular disorder adapt their movements or use an assistive device (e.g., modified spoon) to perform daily movements.

We can also improve the health and safety of a person's tissues. For instance, rehabilitation and training exercises are most effective if we have theoretical and practical knowledge of how tissues are affected by conditions related to the application of force (e.g., speed of movement and amount of force). This knowledge also helps us understand how to reduce the risk of movement-related tissue injuries that can occur during work, daily life, and other physical activity. Consider this injury prevention question: Does stretching before we engage in strenuous physical activity reduce the risk of injury? See the Research and Evidence-Based Practice in Exercise Science sidebar on stretching for an answer.

Performance and tissue health also can be affected by the equipment and flooring used during human movement. Of course, the equipment can vary widely, ranging from bats to assistive devices (e.g., wheelchairs or braces) to footwear and headgear. Biomechanical researchers can establish optimal characteristics for such equipment, including size, weight, racket shape, and characteristics of prosthetic limbs. In terms of flooring, we can use spring-loaded floors and mats in gyms and dance studios to reduce impact forces on the body during landings.

KEY POINT

Using biomechanics to evaluate and design equipment is important because the materials and design of these devices interact with performers' biological tissues, affecting the forces applied to the body and by the body. Biomechanical assessment of equipment, rehabilitation devices, and flooring by exercise science professionals and rehabilitation specialists also ensures that performance is either maintained or improved for clients.

Research and Evidence-Based Practice in Exercise Science

Does Stretching Before Activity Prevent Injury?

You might have seen people get ready for physical activity by stretching and doing nothing else. They hold a stretch for a bit, then go on to the next static stretch. This type of stretching does not reduce one's risk of muscular or connective tissue injury during subsequent activity. In fact, if a warm-up consists only of static stretches, it neither reduces the risk of injury nor enhances performance (Behm et al., 2016; Junior et al., 2017).

Behm and associates (2016) rigorously reviewed the best scientific research investigating the effects of pre-exercise stretching on injury and performance. They evaluated many types of flexibility exercise protocols. The outcomes showed mixed answers and depended on several factors such as the type of stretch and sequence of movements. So, what to do? The authors deduced that performing an appropriate warm-up prior to dynamic stretching—that is, doing several general movements, such as light jogging, then stretching with slow movements—might reduce injury risk in muscle connective tissue and increase joint range of motion without reducing muscle force production during subsequent physical activity.

What Do I Need to Know to Use Biomechanics Professionally?

Most exercise science degree program specializations require at least one biomechanics course. In it, you will learn to apply the mechanical principles for manipulating movement as well as other principles that can affect tissue health. You also will learn how to use practical methods and tools to answer human movement questions related to assessing or improving performance effectiveness or tissue health, including injury prevention. You might also learn how to perform or interpret qualitative walking or gait analysis that is useful in athletic training, physical and occupational therapy, and prosthetics and orthotics.

Practitioners and biomechanists often apply biomechanics knowledge in conjunction with knowledge from other disciplines and subdisciplines of exercise science (Knudson, 2021). For instance, anatomy, exercise physiology, and motor behavior are crucial to understanding how our body structures produce movement. An exercise science professional recognizes that our movements are also affected by psychological, sociological, and cultural factors. Consider the question of why athletes who self-identify as females have a higher rate of anterior cruciate ligament (ACL) knee injury than do equally skilled self-identified male athletes (Hughes, 2014). Does this difference relate to sex differences in anatomy (possibly some contribution) that cause self-identified females to land in a more knock-kneed position? Or does it relate to effects of female hormone levels on ligaments (some contribution), muscular function, or unidentified cultural influences that affect how certain females perform movements or generate muscle force? What is known (nothing at present) about knee injury causation or risk of injury for individuals who do not identify as male or female? We hope to better prevent ACL injury when this question is pursued not just from a biomechanical perspective, but when an interdisciplinary approach is used involving researchers from many subdisciplines of exercise science. Moreover, involving researchers from other disciplines within and beyond kinesiology and the athletes themselves is needed if the biomechanics community is to include all important perspectives and generate useful and innovative solutions. Drawing on the wisdom and perspectives of members of all areas of society will enable us to better answer biomechanical problems such as this, which will then serve all members of our society.

KEY POINT

Biomechanics is a helpful tool when answering a professional question of interest in exercise science and health- and medical-related fields. Moreover, incorporating knowledge from other exercise science areas with an understanding of the client or performer provides better answers to a biomechanical question.

What Do Biomechanists Do?

Career opportunities in biomechanics include such positions as researcher, clinical biomechanist, sport performance specialist, ergonomist, forensic biomechanist, and university professor. Related positions include certified orthotist and certified prosthetist.

Biomechanics researchers work in biomechanics laboratories, where they perform experiments that address problems of interest to various industries or assist with product development. For example, to design better footwear, a biomechanist working at a footwear corporation would collaborate with design engineers to understand the interaction between people's anatomy, the way they move, and the forces that act on them. A clinical biomechanist in a biomechanics laboratory at a hospital might work with physicians and therapists to understand how best to help patients regain normal walking patterns with medical treatment. The clinical biomechanist also might be expected to collaborate on research with physicians.

A performance enhancement biomechanist might work with collegiate athletes, elite athletes, or professional teams and their coaching staffs in order to improve athletes' performance. For example, a biomechanics company or sports medicine clinic could operate a facility in which biomechanists analyze athletes' techniques to assess performance effectiveness or detect injury-related errors. Several biomechanists have founded their own companies, which offer access to equipment, software, testing, and consultation in sport biomechanics. This type of biomechanist might be hired, for example, by national-level athletic teams and dance companies to work with their performers as a consultant.

Biomechanists might also work in occupational settings as physical ergonomists or human factors engineers. The International Labour Union and the International Ergonomics Association (International Ergonomics Association, n.d.) define physical ergonomics as the study, evaluation, and reduction of the physical demands of performing work on the human body, focusing on reducing discomfort, pain, and injury. IEA (n.d.) notes that relevant topics include "working postures, materials handling, repetitive movements, work-related musculoskeletal disorders, workplace layout, physical safety and health."

Thus, some ergonomists work in research and development departments as part of a team of people who design equipment. Others might work for specialized ergonomics-focused corporations that perform job-site analyses. These analyses involve evaluating how and why employees perform their work tasks. The ergonomist generates data and then recommends appropriate modifications in tasks, equipment, employee training, or incentives to encourage employees to modify their behaviors in ways that improve their safety or efficiency.

Forensic injury biomechanists use biomechanical knowledge and principles to answer questions related to civil and criminal lawsuits (Franck & Franck, 2016). For example, a forensic biomechanist might be hired as an expert witness in a civil lawsuit to testify about whether a biomechanical basis exists for an alleged work-related injury. Suppose that an employee develops a back injury and sues the employer. The employee alleges that the injury was caused by having to lift excessively heavy boxes of materials. The forensic biomechanist would try to determine whether the back-muscle forces required when performing the lifting task correctly exceed the maximums allowed by government standards and are high enough to injure this employee. The biomechanist also would have to determine whether it is likely that other factors caused or contributed to the injury, such as failing to use safe lifting techniques taught to workers.

Clinical biomechanists who work in medical-related settings, such as research hospitals that include biomechanics laboratories, perform biomechanical analyses of patients or research participants so that physical therapists and physicians can determine treatment and scientifically study treatments. For example, several Shriners hospitals have a biomechanical gait analysis laboratory. In this setting, the surgeon, physical therapist, and biomechanist work together to assess whether a child with a disorder (e.g., cerebral palsy) requires treatment. If the child does receive treatment (e.g., surgery or physical therapy), the treatment team works together to determine whether the treatment sufficiently improved the child's ability to walk. The biomechanist's responsibility is to perform and interpret the gait analysis and report the findings to the treatment team.

Although prosthetists and orthotists are not biomechanists, they use many biomechanical concepts and methods. Prosthetists and other professionals

Career Opportunity

Research Scientist in Big Tech

Did you know that Apple has a Biomechanics Research Center? Or that FitBit's parent company, Google, has the Human Research Lab? Some large sports equipment companies also have biomechanics laboratories. With the potential of wearable technology to provide data and feedback on human movement, companies now are hiring a few human movement biomechanists. These biomechanists are required to be creative, innovative, and highly skilled in biomechanics research and computer methodology. More rehabilitation centers also are hiring skilled human movement biomechanists to provide movement analyses to their clients. However, because of the scarcity of these positions, they are very competitive. Talk to biomechanists if you are interested in such a career.

(e.g., physical therapists and podiatrists) sometimes also become certified orthotists. A prosthetist's main focus is to help a client obtain and use a replacement body part such as an artificial hand. Orthotists, on the other hand, assess a client's body structure and function to understand what might be preventing a body part from completing its tasks effectively. For instance, injury or pain can result from anatomical deviations, which in children can result from abnormal bone growth, improper muscle functioning, neural defect, or some combination of these factors. The orthotist then fits the patient with an orthosis, a limb-supporting device such as a molded plastic brace that holds a child's foot in the proper position.

Prosthetists and orthotists also reassess their clients' movements, watch for potential problems (e.g., injury, skin issues, ineffective movements), and make adjustments as needed.

Many biomechanists work in college or university departments of exercise science, kinesiology, engineering, medicine, or other disciplines. A professor in an exercise science department teaches biomechanics to students with interests in a variety of fields related to physical activity, rehabilitation, or medicine. The professor also conducts research in an area of biomechanics. Some biomechanists might also be clinicians and integrate patient care with their research and teaching.

EXERCISE SCIENCE COLLEAGUES

Marika Walker

Courtesy of Marika Walker.

Marika Walker became interested and involved in biomechanics by an unusual route. During high school, after visiting the headquarters and research labs of a major sports apparel company, she thought, "Being a researcher in a sport company like this one—that's perfect. I would get to blend my love of sport, science, technology, engineering, and math in this type of work." Her career interests and path changed over time. She conducted research as an engineering intern at Nike and, as a master's student researcher, in a textile research center. During her doctoral degree work at the University of Georgia, Marika combined her interest in sport product development and human movement biomechanics by investigating how different rib protector designs influenced the throwing performance of American football quarterbacks. She also investigated whether a stiffer protector would cause the quarterbacks to feel less able to throw effectively. Dr. Walker is now a senior associate in biomechanics at EXponent, an engineering and scientific consulting group. Dr. Walker also formed a company, ADEPT Movement Academy, whose mission is to improve athletic performance and reduce injury prevalence. Her goal is to educate coaches and performers on foundational biomechanical principles that underlie optimal technique and learn how to apply those principles when analyzing and correcting movements.

Does a Career in Biomechanics Interest You?

To determine whether you are interested in a career in biomechanics, answer the following questions:

- Do you like observing and analyzing how people move?
- Do you find yourself trying to figure out a better way to perform a task?
- Do you enjoy trying to solve puzzles?
- Do you like the idea of applying the biology, physics, and math that you have learned to movement situations of interest?
- Do you enjoy using your mathematical skills?
- Are you interested in investigating the internal structures of the body, such as tendons, to determine how they act and are affected by factors such as exercise?

If you answered yes to some of these questions, consult the website of the International Society of Biomechanics for information about the field of biomechanics, job opportunities, graduate programs, and biomechanists available for contact. You can also talk with biomechanics professors to help you identify programs aligned with your interests.

History of Biomechanics

Throughout history—both for practical purposes such as survival and to satisfy our innate need to know and understand—we humans have been applying natural laws (often unknowingly) to investigate the structure and function of living organisms. The field of biomechanics, however, is relatively young.

Setting the Stage

Interest in learning how the body moved and its structure began thousands of years ago and was influenced by the science, math, and medical contributions of Middle Eastern, African, Near Eastern, Indian, Asian, and Greco-Roman civilizations (Imhausen et al., 2007; Tipton, 2008; Unschuld, 1985). Within these civilizations, sophisticated theories emerged of how the body functions. In Europe, this knowledge was built upon during the scientific Renaissance. You might be familiar with the name René Descartes (1596-1650). Although he was a mathematician, he strongly hoped to understand the human body in order to cure disease and slow down aging. He found no cures for disease or aging, but he did conclude that the physical body was simply a machine that could maintain itself without assistance from a soul or essential being. Another Renaissance example is Giovanni Alfonso Borelli (1608-1679), who is considered by some to be a founder of biomechanics. He understood correctly how muscles act and use leverage to move the body, and he discovered how to calculate forces acting on or produced by organs (Pope, 2005).

These two individuals are mentioned because this view of body structure and function established the European-based philosophy that our bodies act simply as machines, a notion that still influences Western biomechanical thinking today. For instance, we might view the heart as a pump or consider how we might improve the leverage of the outside hip muscles in order to help improve hip function of patients with a hip replacement.

The early philosophies of health, movement, and medicine developed within some of the world civilizations mentioned above continue to have an influence in some parts of the world today. Practitioners who use these philosophies and practices in their work with clients consider a person's health from a more whole-person approach. As biomechanists, we, too, should consider that movement and other functions are influenced by more than peoples' machine-like structures.

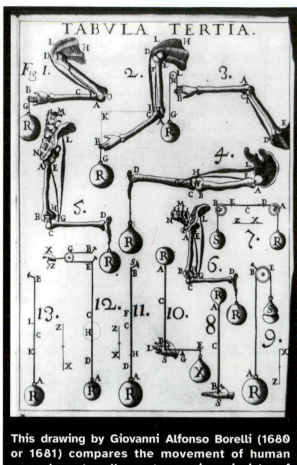

This drawing by Giovanni Alfonso Borelli (1680 or 1681) compares the movement of human appendages to pulley systems using principles of mechanics and statics.

Courtesy of the Library of Congress, LC-USZ62-95253.

Beginnings of the Subdiscipline of Biomechanics

By the late 1800s and early 1900s, rapid progress in both knowledge and technology had made the time right for kinesiological sciences, including biomechanics, to flourish. Concurrently, the Efficiency Movement (late 1890s to early 1930s), popular in European and North American industrial countries, was thought to reduce waste and improve the quality of life (and increase profit for business owners). Frederick Winslow Taylor, an American engineer and a leader of the movement, scientifically studied the motions and time requirements of work tasks and generated principles for increasing workers' efficiency in order to improve productivity—for example, using certain movement techniques to reduce muscle fatigue (Taylor, 1911). Taylor-style efficiency analyses might have been some of the first assessments of human movement based on quantitative movement data. Taylorism also laid the foundation for the field of ergonomics (occupational biomechanics) or human factors engineering, and efficiency analyses began to be applied to sport movements in the 1920s. Efficiency movement proponents viewed workers as machines.

In Russia, Taylor's ideas influenced Vsevolod Meyerhold, who was perhaps the first to use the term *biomechanics*, although its meaning then differed somewhat from today's definition. Meyerhold was an innovative but controversial theater director who viewed movement as the most powerful mode of theatrical expression (Cash, 2015). He developed his version of biomechanics between 1913 and 1922 as a method of acting and theater production that integrated laws of biomechanics (from mechanical physics) with other components such as movement efficiency, balance, coordination, and rhythm (Pitches, 2004).

The development of biomechanics and other kinesiology and exercise science subdisciplines in physical education, sport, and dance applications began in the very late 1800s and early 1900s. In the United States, two Swedish men (Posse, 1890; Skarstrom, 1909) interested in gymnastics were among the first to apply the term *kinesiology* (term initially used in North America instead of *biomechanics*) to the analysis of muscles and movements in an education setting (Atwater, 1980). (See chapter 1 for more on the term *kinesiology.*)

Ruth Glasgow was one of several early biomechanists who profoundly affected the field. Her groundbreaking work helped lay the foundation for the biomechanical understanding of human movements.

University of Wisconsin—Madison Archives

KEY POINT

Human movement biomechanics was initially called *kinesiology* (primarily in the United States) in the late 1800s and early 1900s. The term *biomechanics* later became the accepted name for the entire field of biomechanics with its current definition.

Mechanical analyses of basic movements were not emphasized originally as part of the formal training of physical and dance educators until 1923 (and before exercise science was its own discipline). At that time, Ruth Glasgow and her students at the University of Wisconsin began three decades of work to classify activities into categories such as locomotion, throwing, striking, and balance. They also quantified movements (e.g., joint motions) using Glasgow's pioneering high-speed film analysis technology, and they applied fundamental principles of mechanics

to understand the skills in each category (Atwater, 1980). Glasgow used this knowledge to develop learning objectives for teaching physical education long before this practice became common in education (Sloan, 1987).

Many research leaders in university kinesiology programs in the 1930s and 1940s came from areas outside of kinesiology (e.g., medicine). Those who researched questions that were best answered by biomechanics, such as Glasgow, had a profound effect on human movement biomechanics. They were the first modern scientists to answer such questions, which required them to develop the needed research methodology and tools. They also educated the first specialized biomechanists and were among the first professors to teach exercise science and physical and dance education preprofessionals how to use **biomechanical principles**.

Era of Contemporary Biomechanics

In the 1960s biomechanics began to get established as a scholarly area and a recognized subdiscipline of university physical and dance education, kinesiology, and exercise science programs. Newly created scientific societies and journals for biomechanics produced the first generation of an international community of biomechanics researchers. In addition, biomechanics courses and graduate-level university programs (Atwater, 1980) created pathways in the field of exercise science for learning biomechanics or becoming a biomechanist. Since then, the world has seen a rapid expansion in the number and scope of national and international professional organizations and university programs in biomechanics (see the International Society of Biomechanics' website). Moreover, many areas of biomechanics are now represented in these organizations and can be studied not only in human movement–related disciplines such as exercise science but also in other disciplines, especially engineering and orthopedics.

Many of the most successful biomechanics researchers and faculty came primarily from the dominant group of their country (e.g., in the United States, white males). While white women also were involved in biomechanics from the beginning, few were able to achieve the same recognition or access to opportunities as their white male counterparts. This trend has continued to affect all biomechanics areas. However, some biomechanics organizations have begun to address the lack of diversity. Thus, although some progress has been made, it will take a committed, continued effort by more of the biomechanics community and its leaders striving for equity until we can bring about true equality and inclusion.

KEY POINT

The era of contemporary biomechanics began in the late 1960s. Many biomechanists from kinesiology and exercise science contributed to the development of biomechanics as a new area of scholarly study.

Era of Low-Cost Computer Technology

Biomechanical research and applications have flourished, with rapid advances in low-cost computing and technology beginning in the 1980s. Biomechanics research was painstakingly slow prior to this time. Much work had to be done by hand, with very limited access to computers, no Internet, and bulky electronic equipment that could not store large amounts of biomechanical data. Imagine having to track by hand the location of every major joint in the body for every frame of high-speed (at least 100 images per second) film or video of a performer's movement from a single camera. This resulted in more simplistic and less accurate analyses of complex movements compared to later studies using multiple cameras and technology that could track a performer's movements more automatically.

Now we are beginning to understand better the true motions of bones, limbs, and the spine with a level of accuracy not possible before (Hamill et al., 2021). We also can simulate movements of an entire organism or the small molecular-level interactions inside a cellular structure (see Computer Modeling and Simulations in this chapter). This also allows us to predict loads on tissues such as ligaments and cellular structures such as muscle fibers (Seth et al., 2018). Simulations allow a practitioner or researcher to ask questions that are difficult to answer using actual performers. Any number of questions about why and how the body moves as it does, how tissues respond to exercise and activity, and how to improve movement or surgical outcomes are being investigated.

Research and Evidence-Based Practice in Exercise Science

Do Knee Injury Prevention Programs Work?

The knee joint is one of the most injured body regions in sport and exercise (Owoeye et al., 2020; Ángel Rodríguez et al., 2022). Injury often occurs during a landing or some type of stop-and-go movement. Knee ligament injuries can require expensive surgery and substantial recovery time, and they often lead to osteoarthritis within 10 years after injury (see Nessler et al., 2017, for review). It appears that appropriate movement technique, combined with the ability to generate knee muscle forces quickly to stabilize the knee joint, can reduce forces on the knee ligaments and excessive stretching that cause ligament tears.

Based on this body of biomechanical research, the National Athletic Training Association (NATA) recommends incorporating at least three of the following types of exercises into a knee ligament injury prevention program: strength, plyometrics (fast, large movements), agility, balance, and flexibility, along with feedback on proper exercise technique (Padua et al., 2018). But do these programs work?

Yes! Huang and associates (2020) reviewed evidence from only the highest-quality studies that met the NATA recommendations and found that most of the programs incorporated plyometrics, strengthening, and agility exercises along with feedback on proper landing technique. Encouragingly, they found that these programs reduced knee ligament injury rates by 53 percent.

Research Methods for Biomechanics

What tools do biomechanists use to accomplish the goals of the field? For example, how do they perform gait analyses, research the cushioning properties of newly designed shoes, analyze the performance of Olympic athletes, and perform research to answer fundamental questions about how we move? In the pursuit of answers to such questions, biomechanics research has benefited greatly from technological advances.

What tools could you use to measure movements made or forces generated by a performer? Examples include timer and camera devices and apps for measuring time and motion, barbells and free weights for measuring lifting force, and laser devices or tape measures for measuring distances or lengths. In addition, a variety of sports equipment (e.g., bats, basketballs, footballs, golf clubs) can be equipped with sensors and software to indicate the equipment's speed and perhaps its accuracy. High-speed video, still images, and simple phone apps also can measure the body's

speed and direction of travel and the distance that a limb traveled, quantify body movements and positions such as knee-joint angles during running, and assess trunk flexibility. Biomechanical researchers have used many tools, ranging from the ones listed here to more sophisticated systems. Figure 4.2 shows parts of instrumentation systems typically used in a biomechanics laboratory for research and for clinical, ergonomic, performance, or other types of analysis.

Motion Measurement Devices

Much of the animation technology used in video game and movie production began with biomechanical technology and is still being used today. The use of cameras and other motion-detection technology is called *motion capture* (or mocap for short). Digital high-speed cameras are used to trace the motion of reflective markers placed at selected points on a human body to reconstruct the motion of the various body segments. Mocap without sensors relies on software to track the contours of a performer's body to reconstruct movement. MEMS sensors, like those in a smartphone, cars, and some sports equipment,

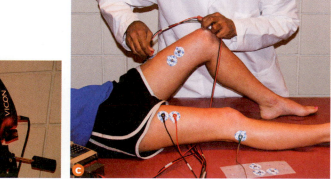

Figure 4.2 Biomechanical instrumentation. Components for motion measurement include *(a)* a digital, high-speed camera that tracks the location of reflective markers identical to those used in video game and movie animations; *(b)* microelectromechanical systems (MEMS) sensors that track body movements via miniature gyroscopes, accelerometers, and magnetic field detectors; *(c)* electrodes used in electromyography to measure electrical activation of muscle; and *(d)* force platforms used to measure ground reaction forces.

Photos a, c, and d © Kathy J. Simpson; photo b courtesy of Xsens Technologies B.V.

combine miniature mechanical sensors (e.g., gyroscopes) with microcircuitry on a tiny chip. MEMS sensors used for mocap detect some combination of the following factors: acceleration, gravity, magnetic fields, and gyroscopic data. When one MEMS unit is attached to each body segment of interest (e.g., head or arm), the movements of those segments can be calculated. Wearable sensor technology, combined with complex artificial intelligence mathematics and biomechanical knowledge of gait and running, enables human movement practitioners and rehabilitation specialists to monitor training and rehabilitation progress and potentially to predict and prevent injury (Benson et al., 2020; Preatoni et al., 2022).

For research purposes, mocap output can be used to understand movements or to provide feedback to performers about their movements in many activities such as sport, dance, music, and physical rehabilitation. This technology also can be used to help practitioners analyze movements observed in occupational and clinical settings in order to improve people's functioning or prevent injury.

Force Measurement Devices

When we need to know what is affecting a performer's movement or contributing to injury, we can measure quantities related to force. To measure the force placed on a joint, ligament, or object, tiny force-measuring devices (force transducers) can be surgically attached to or inserted into living tissues, artificial joints, or nonliving tissues. Transducers can also be used to measure how much force a patient can exert against a strength-measuring device. For instance, force platforms can be used to measure **ground reaction forces** (GRF) when a performer applies force to the ground; this approach can help diagnose postural and balance disorders. GRF values are also used in calculating internal force loading on bones and cartilage during weight-bearing movements by means of various biomechanical models and laws of mechanics.

Pressure is determined by the amount of force applied to a given amount of surface area. Areas of high pressure applied to human tissue can cause health problems. For example, in a person with diabetes who cannot sense foot pain, too much pressure repeatedly placed on one area of the foot while walking can cause the skin in that area to deteriorate and become infected. Therefore, identifying high-pressure areas on the feet of a patient with diabetes helps the clinician select appropriate footwear or make an individualized shoe insert to spread the ground reaction forces more evenly across the foot and thus prevent further tissue damage (Igiri et al., 2019). Pressure devices can also be placed between a device (e.g., wheelchair seat or lower-limb prosthesis) and areas of the body that come into contact with the device. Such contact areas could be subjected to high pressures if not monitored and the device altered.

Because of the difficulty of directly measuring muscle forces, motion and force platform data are

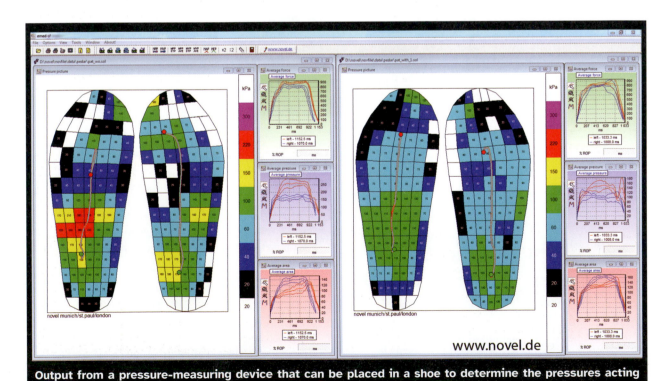

Output from a pressure-measuring device that can be placed in a shoe to determine the pressures acting on a client's foot.

Photo courtesy of Novel Electronics, Inc.

used to calculate estimates of muscle forces using biomechanical models. Another method of estimating muscle forces involves measuring the electrical activity of the muscle cells (fibers) when nerve cells stimulate them to contract. This method, called *electromyography* (EMG), uses electrodes placed on or in muscles to record their electrical activity. The amount of electrical activity generated is a measure of muscle activation and can be converted into an estimation of muscle force when calculated using a mathematical model of human anatomy and muscle biomechanical properties.

EMG is used for other purposes such as identifying the muscles that are active during a particular movement. We can then know what muscles to train for a particular movement such as standing up or throwing a baseball pitch, understand what muscles are being used during various exercises such as abdominal crunches, investigate muscle injury causation, or identify why muscles are not working correctly for individuals with a pathological condition. EMG also can be used by neurologists to assess whether muscles and nerves are functioning incorrectly. Understanding muscle activation (timing and intensity) also has allowed researchers to build a wearable device that stimulates the appropriate muscles to contract at the right time, thus enabling a person with paralysis or with powered prostheses to walk.

How Do You Measure Up?

What instrumentation would you like to use to help answer a biomechanical question that you have? Returning to the chapter-opening scenario, what instrumentation could be used to assess the movements of Katsumi's leg with OA and the healthy leg?

Computer Modeling and Simulations

All of the instrumentation described in the previous paragraphs measures one quantity or another. However, many internal biomechanical quantities (e.g., joint angles or muscle forces) cannot be directly or easily measured; therefore, they must be derived through mathematical computations. Often, these techniques also require mathematical representations or models of the body's anatomy and other structures such as muscle attachment locations and bone lengths (see photo that shows a model of the lower body). An interrelated approach is to use computer simulations to compute forces or movements of a virtual human or its parts. Biomechanical models are used in simulations to help us understand basic mechanics at the cellular, molecular, tissue, and whole-body levels. They also allow us to try out

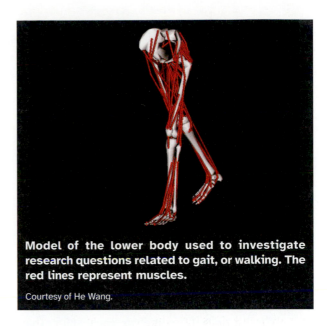

Model of the lower body used to investigate research questions related to gait, or walking. The red lines represent muscles.

Courtesy of He Wang.

hypothetical scenarios such as the effects of different placements of muscle attachments during surgery, estimating the loads on brain tissues during a concussion event, or varying the techniques of a gymnastics movement. In these examples, it is more insightful or safer to use simulations.

For example, simulated motion of the knee joint can be used to model the anatomy and forces involved in various structures (e.g., ligament, tendon, cartilage, bone, muscle), thus helping us to understand typical motion in the knee joint. We can then change the model's knee joint alignment to understand how poor knee alignment contributes to degeneration of knees with OA. For instance, a sport biomechanist could change the aerial technique of a virtual gymnast and watch the resulting performance to assess safety before having real gymnasts try the technique.

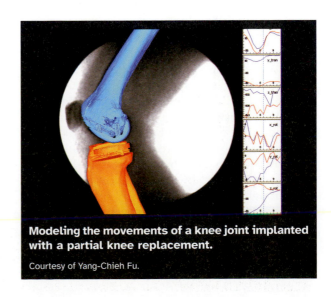

Modeling the movements of a knee joint implanted with a partial knee replacement.

Courtesy of Yang-Chieh Fu.

Overview of Knowledge in Biomechanics

Much of biomechanics consists of learning why and how to apply the mechanical laws of physics and engineering in order to answer questions about movement or tissue health. Contemplate this: While watching players using wheelchairs at a national basketball tournament, I noticed that some players shot the ball with two hands, starting the ball from behind the head or from chest height; others used a variety of one-hand form, including a technique typically used by stand-up players. The rules dictate that players with differing levels of upper-body muscle function must be in the game at all times. Thus, some players had only partial use of their arms, while others had complete use of trunk and arm muscles. So, what movement technique is the most effective? Are any of these techniques more likely to cause injury than others? These situations are examples of two major questions to be explored: Why do we move in certain ways and not others? Is there such a thing as the perfect movement technique, or is there a range of optimal technique given certain conditions?

Moving Masterfully

"Why do people move the way they do?" "Is one technique best for performing a given movement or exercise?" These are two of the biggest questions in human movement biomechanics, and practitioners must consider them carefully, regardless of whether they are working with clients or designing new equipment or assistive devices.

Movement technique involves using particular body positions and movements in a prescribed manner with proper timing. For example, for a squat exercise, you might position your feet shoulder-width apart and then bend at your hips, knees, and ankles without arching your back. Even before thinking about technique, however, we must always remember that how and why we move as we do is shaped by specific characteristics in the following five categories: task, performer goal, performer characteristics, environment, and laws of nature (including biomechanics). In addition, these factors interact with each other. Let's consider an example.

A shooting task for our basketball players shapes how they move because it reduces the possible movements to those that can be done from a wheelchair. The *performer's goal* in this situation is to get the ball through the basket. *Performer characteristics* can include anything related to the performer: available musculature, arm length, emotions, motivation, skill level, strength, and culture. For example, the player's arms can move only in certain directions because bone and muscle structures constrain possible movements. Another factor is the *environment*, anything outside of the performer; in this case, the wheelchair, other players, rules, and the cheering crowd. Relevant *laws of nature* include physical principles, such as Newton's laws, that are related to how forces, such as air resistance, gravity, and muscle forces, are produced and negotiated, as well as biomechanical principles of how tissues respond to forces applied to them. Here, the players cannot lean too far back in their sport chairs or the force of gravity will tip the chair backward. Sufficient muscle force is needed to project the ball. *Interaction* of factors might influence the technique—for example, interactions between the distance between the basket (environment) and player, the muscle force required to accelerate the ball (law of nature), and the amount of musculature available to produce the force (performer characteristics). Biomechanists often are most interested in the effects of interactions between mechanical laws of nature and the anatomical and neuromuscular factors.

Next, contemplate the notion of a correct or best technique in relation to the goals of the task and the performer. In competitive movements in which technique is judged (e.g., dance, diving, gymnastics), movement technique is everything. However, what about other competitive activities? Ask yourself and others whether performing a given movement (e.g., high jump, running, rising from a chair) with the right technique is important for performance success and why (or why not). How do we know whether the current technique or a new one provided by a coach or clinician is the right or best one? Referring back to our basketball players, what is the right technique? Or for the chapter-opening scenario, how is running technique relevant to Katsumi's situation? Now, consider whether there is a right technique for movements with other purposes such as work tasks (e.g., lifting objects or people) and daily activities (e.g., standing and sitting). Also consider characteristics that might affect best technique—for example, the performer characteristics of age, gender, skill level, and medical condition.

For competitive activities, some often assume that the correct technique is that of the world's best performers. After all, they must be doing it right in order to be the best. Not necessarily, given they might use only good technique that is supplemented by exceptional strength, balance, or other neuromuscular factors. What if a movement technique works well only for whole-body coordination from a standing position, not sitting in a sport chair? Thus, sport biomechanists compare the biomechanics of elite performers to many other kinds of athletes in a variety of settings and conditions. They then try to identify biomechanical factors in the different conditions that might be related to high-level performance given unique characteristics such as age, skill, sex, and other factors.

For factory work tasks, the employer's goals of high productivity, safety, and minimal production costs can influence what is considered correct technique. Potentially, if workers are paid by the number of tasks completed, an employee's goal might be to complete the maximum number of tasks. In this case, the employee might use techniques that are the fastest but not necessarily the safest. In such a case, a responsible employer might use an ergonomist (occupational biomechanist) to determine what techniques will accomplish production goals while maintaining worker safety. Even then, it is not always possible to determine a single best movement technique, particularly for tasks that are unusual or involve awkward work spaces.

Lastly, for activities of daily living, what technique is correct for people with movement disorders? The goals of a client with weak leg muscles might be to stand up or sit down safely during daily life without assistance and without feeling self-conscious. Such a client will demonstrate a different pattern of movements than those who have adequate muscle function. However, though we often assume that the movement techniques displayed by those without movement disorders set the standard to achieve, they might not be appropriate for a given client. Even so, they can be useful in assessing a client's movement.

Here are some deductions about best movement technique.

1. How and why we move as we do are influenced by the characteristics of the five main categories discussed earlier (task, performer goal, performer characteristics, environment, and laws of nature including biomechanics) and their interactions.

   • Most humans display relatively comparable basic movement techniques for some fundamental movements (e.g., running and hopping) due to their similarities in the five main categories (e.g., basic skeletal structures and being subject to laws of nature such as gravity).

   • Variations of those characteristics between individuals also mean that the best technique for one person might not be the best for another. Certainly, each of our basketball players must have determined the technique that allows them to use the muscles available to generate the force and accuracy needed to shoot baskets. Also, for a given pitching style, baseball pitchers all throw somewhat differently from each other, due in part to variations in shoulder joint and body structures.

Research and Evidence-Based Practice in Exercise Science

Back Up

Appropriate movement technique for occupational tasks such as lifting objects is critical to get the job done quickly but without producing injury. Low-back pain is a common occupational problem created or exacerbated by movement technique that produces excessive loads to back tissues or excessive repetition. Low-back pain can be debilitating. Beaucage-Gauvreau and colleagues (2020) asked whether using one hand on the thigh to brace the trunk ("bracing") while lifting a 2-kilogram (4.4 lbs) or 10-kilogram (22 lbs) object would reduce the loading on two of the low-back vertebrae compared to not using the bracing technique. Twenty participants with low-back pain and 20 healthy individuals of similar age and gender lifted the objects from the floor to a standing position using or not using the bracing technique. Force applied by the bracing hand to the thigh was measured via a force transducer and video mocap, and perceived back pain was analyzed. Low-back vertebral forces were estimated using simulation software. The bracing technique lowered compression forces pushing the vertebrae together and forwards-backwards vertebral forces but produced greater side-bending and twisting vertebral movements and loads. Pain scores varied little between techniques or groups. The authors concluded that the bracing technique can be useful to reduce back loads when healthy and back-pain individuals lift similar weights. However, side bending, asymmetric, and twisting low-back loads and movements are associated with back injury potential (Rohlmann et al., 2014). Therefore, further research is needed before practitioners have their low-back pain clients use the one-hand bracing technique.

2. A successful movement technique is one that, regardless of how the movement is performed, enables the performer to accomplish movement goals (i.e., performance effectiveness) and health and safety goals at the highest level that the person can or wants to achieve.

- Better choices for movement technique are available when the performer takes advantage of biomechanical principles that best enable the performer to generate and control forces and movements. Given Newton's law of acceleration, for example, when shooting a basketball, a player who does not have the use of trunk muscles to produce force to accelerate the ball sufficiently can create more total force by shooting with two hands rather than one hand.
- The best movement technique might be optimal for achieving some goals but not others. This dilemma can be seen in both factory work and sport, where performance goals can be achieved at the cost of safety goals.

Each person has a unique set of potentially conflicting goals and interactions among characteristics of the five categories. Therefore, it can be difficult for biomechanists to determine one-size-fits-all movement techniques that are common to most performers. These individual and context issues make it challenging for professionals to determine what the best technique is for a given client.

Balancing Performance Effectiveness and Safety

Let's explore how biomechanical principles and analysis can help someone achieve both performance and tissue safety goals. Specifically, we return to Katsumi's scenario from the chapter opening. Katsumi's performance goal is to win in their age group during a national 10K running road race; their tissue safety goals are to minimize progression of knee OA and remain injury free. We will see the conflict between the performance goal that requires force generation to produce high maintainable speed and the tissue safety goal due to the knee joint forces involved.

Performance Effectiveness

To ensure that Katsumi can run at the fastest pace that they can maintain, we can look at the biomechanics of the entire body during various intervals of time within a single running cycle. Here, we only examine one small interval—the braking phase, or roughly the first half of the time period when the foot is in contact with the ground.

First, we describe some of the motions and body positioning. A runner's foot typically strikes the ground slightly in front of the body, and then the body rotates over the foot. During the braking phase, the runner loses a small amount of speed and forward momentum; of course, we want to minimize this loss of momentum.

Second, because forces affect Katsumi's motion, let's identify the forces acting on the body during this phase (see figure 4.3). Of course, the force of gravity is present—that is, the body weight pulling the runner downward. We also consider whether air resistance and wind forces are acting (assume that they are negligible). Next, we search our biomechanical toolbox for principles to help us identify forces that could be generated between the foot and the ground. According to Newton's third law (action–reaction), for every force applied by the foot to the ground, the ground will apply an equal and opposite-acting force back to the foot. Because Katsumi's foot is in front of the body at contact, Katsumi's momentum causes the landing foot to push forward against the ground (action force); in reaction, the ground applies a backward force (GRF) to the body, thus opposing Katsumi's motion and slowing them down. Simultaneously, Katsumi creates another set of action–reaction forces in the vertical direction by pushing down against the ground; the reaction force here is upward.

Third, we determine the biomechanical principles that could help Katsumi minimize loss of momentum. Here, we want principles that explain how forces affect momentum. We identify the principle stating that the amount of momentum gained or lost by a system depends not only on the amount of force but

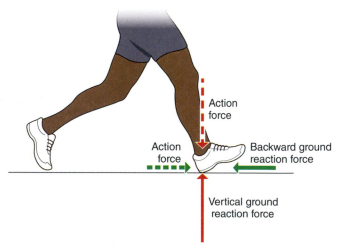

Figure 4.3   Action–reaction forces between the foot and ground during the braking phase. The ground reaction forces acting on Katsumi affect the body's vertical and horizontal momentum.

also on the length of time that the force is applied. Therefore, the higher the backward GRF acting on Katsumi and the longer it acts on the body, the more momentum is lost.

Fourth, we seek to determine how to manipulate the crucial quantities, and this is where we consider Katsumi's movement technique. Specifically, how can we reduce the amount of backward-directed GRF or the time spent in the braking phase? If Katsumi is placing the landing foot too far in front of the body or landing on the heel instead of on the forefoot, that will likely create too much backward-directed GRF that will act for too long, causing too much loss of momentum.

This is also where we find the first potential conflict between performance and tissue safety: If Katsumi changed their footstrike to a forefoot landing in order to shorten the time acted on it by the GRF, this might also increase the loading force on calf muscles or foot connective tissues that could eventually contribute to injury of the foot tissues and knee cartilage and to OA progression.

Hot Topic

Risk Factors for Knee Osteoarthritis

Worldwide, approximately 86.7 million people had knee OA in 2020 (Cui et al., 2020). OA has no cure and can become severe, causing excruciating pain and disability, due primarily to continued wear of the joint cartilage and development of bone abnormality (National Institute of Arthritis and Musculoskeletal and Skin Diseases, 2019) (figure 4.4). Therefore, OA causation exemplifies how forces can affect tissue structure and health.

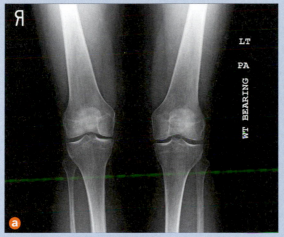 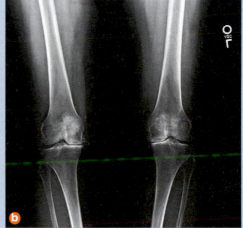

Figure 4.4 Radiographs of *(a)* a healthy knee and *(b)* an osteoarthritic (OA) knee. The space between the OA bones is narrowed because the cartilage (which isn't visible on a radiograph) is worn down, particularly on one side of the knee joint.

© Kathy J. Simpson

What might be your highest risk factors? Previous knee joint injury is the predominant factor causing young-adult knee OA (Emery & Pasanen, 2019; Vannini et al., 2016). Your risk of developing OA is up to six times higher after having certain knee injuries (Snoeker et al., 2020). Otherwise, obesity is the highest risk factor for adults in developed countries (Vina & Kwoh, 2018) because it causes a continuing cycle of damage. Specifically, obesity creates abnormal inflammatory processes and high tissue loading due to high body weight, which leads to chronic, atypical structural changes in bone and weaker, damaged cartilage (Nedunchezhiya et al., 2022). Another risk factor is the length of time spent during one's life performing high-impact, quick-stop activities, as such activities apply high loads to the knee cartilage, ranging from two to four times body weight during level walking to as much as 20 times body weight during landings in certain sports (Vannini et al., 2016). Individuals who perform tasks that put high loads on knee joint tissues during work such as heavy lifting also are at risk (Schram et al., 2020). Individuals who are not of the dominant racial or ethnic group in their country or who live in poor socioeconomic conditions also have greater risk in part because a greater percentage of these individuals perform manual labor, particularly if living in areas with poor occupational safety and because they often have poorer medical care (Bowden et al., 2022).

Tissue Health and Safety

In our ongoing example involving Katsumi, the goals for tissue health and safety are to prevent injury and slow down the knee OA progression. Here again, we examine the braking phase of running, but our interest now focuses on the biomechanics of the tibia. We also limit our analysis here to the vertical direction because it involves the highest bone forces at the knee joint, which could damage joint cartilage.

First, for simplicity, we identify only some of the vertical forces acting on the tibia. The downward force that the femur applies to the tibia's articular surface is created, in part, by Katsumi's downward momentum when the body hits the ground, which is a function of body mass and the speed at which Katsumi is moving. The rest of the femur force is created as a reaction to the tibia being pulled and pushed upward against the femur by knee joint muscles and vertical GRF transmitted to the tibia, respectively.

Next, we seek mechanical principles that explain how tissue health and safety are compromised. We know that we can help keep bone, connective tissue, and cartilage healthy by applying mechanical load (force) in safe amounts over an appropriate number of repetitions. However, another principle holds that high total force combined with too many repetitions (i.e., number of steps) over a long period of time can weaken these tissues, which means that less force is required to create permanent damage when loaded.

Thus, high volumes of running in training coupled with high tibial forces could result in overuse injuries in the tibia (e.g., stress fracture) or connective tissue (e.g., joint cartilage) that, for Katsumi, contributes to progression of knee OA. Conversely, OA progression can be reduced with more moderate force loading because it stimulates joint lubrication to reduce joint friction that wears down the joint surface (Castrogiovanni et al., 2019).

Now, we analyze how we might reduce the forces and the number of times that they are applied to Katsumi's tissues. In terms of movement technique, running more slowly could decrease all of the vertical forces that affect tibial loading, but that is unrealistic in this case because Katsumi wants to compete. Another option is to strike the ground using the heel rather than the forefoot. That would create less knee-joint force, perhaps due to a reduction in the vertical GRF (Futrell et al., 2018; Zimmerman & Bakker, 2019) and muscle forces transmitted to the tibia. Here again, we see the conflict between performance and safety, because a forefoot strike might be better for performance but might also increase the risk of stress fracture or OA progression. In terms of environmental considerations, Katsumi could run on relatively softer, stable surfaces (rubberized track) to reduce vertical GRF. Also, the training volume should be no higher than needed to avoid excessive repetitions of tibial loading.

Professional Issues in Exercise Science

Brain Matters

Worldwide, traumatic brain injury (TBI) contributes to more deaths and disability than any other traumatic injury (Dewan et al., 2019). Certain groups have greater risk of TBI or poorer care of TBI, such as people of color, military members, incarcerated individuals, survivors of domestic abuse (Daugherty et al., 2019) and athletes (Tierney, 2021).

In Tierney's (2021) TBI literature review, they note that permanent damage can affect all areas of a person's life. Damage can occur due to one contact event or due to cumulative damage from multiple impacts over a long period of time. Understanding how brain injury occurs is hard to ascertain. Why? First, the brain, more than any other organ, undergoes the most complex injury mechanisms, because the brain is suspended in fluid in the skull and does not act as a solid object. The brain sloshes and rotates and stops moving very rapidly, excessively stretching brain tissue and blood vessels. Second, damage and subsequent healing to nerve cells are difficult to detect medically. Third, we cannot reproduce injury well in biomechanical simulations when we cannot directly measure the movements and forces of the brain. Consequently, to assess TBI, practitioners have to rely on indirect measurement tools, such as balance or cognitive task tests, or on questionnaires, whose scores are difficult to match to brain injury severity and healing.

Tierney notes that we have much to do to understand, prevent, and treat TBI more effectively. This also means reducing the disparities that cause some groups of people to have greater injury rates and poorer outcomes.

Citation style: APA

Tierney, G. (2021). Concussion biomechanics, head acceleration exposure and brain injury criteria in sport: A review. *Sports Biomechanics*. https://doi.org/10.1080/14763141.2021.2016929

Wrap-Up

The physical laws of nature shape our movements and our body tissues. Biomechanists and exercise science professionals can apply mechanical principles and laws to enhance performance effectiveness; maintain or increase the health of tissues and prevent injury; and select or modify assistive devices, joint or tissue replacements, and exercise equipment. When we understand how mechanical principles influence our body functioning, movement, and structure, we can apply that knowledge to work, leisure activities, sport, exercise, dance, daily tasks, rehabilitation—indeed, to any action involving movement or forces acting on or within the body.

More Information on Biomechanics

Organizations

American Society of Biomechanics

Gait and Clinical Movement Analysis Society

International Society of Biomechanics

International Society of Biomechanics in Sports

International Society of Electrophysiology and Kinesiology

International Society for Posture and Gait Research

Journals

Applied Bionics and Biomechanics

Clinical Biomechanics

Frontiers in Bioengineering and Biotechnology: Biomechanics

Gait & Posture

Human Movement Science

International Biomechanics

International Journal of Experimental and Computational Biomechanics

Journal of Applied Biomaterials and Biomechanics

Journal of Applied Biomechanics

Journal of Biomechanical Engineering

Journal of Biomechanics

Journal of Dance Medicine & Science

Journal of Electromyography and Kinesiology

Journal of Forensic Biomechanics

Journal of Prosthetics and Orthotics

Sports Biomechanics

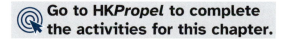 **Go to HK*Propel* to complete the activities for this chapter.**

Review Questions

1. What is similar and what is different between the meanings of the terms *biomechanics*, *human movement biomechanics*, and *kinesiology*?

2. What is meant by "structure" and "function"? Give two new examples for each term and determine how structure and function affect each other for each example.

3. What are the goals of the study of biomechanics?

4. What are the primary application areas of biomechanics in exercise science?

5. In what types of settings do biomechanists typically work?

6. What major instrumentation systems and methods are used in biomechanics research?

Suggested Readings

Hunt-Broersma, J. (2021, August). Everything you need to know about running with a running blade. *Runner's World*. www.runnersworld.com/training/a37209643/running-with-a-prosthetic-leg/

Kuhl, E., & Humphrey, J.D. (2020). Editorial overview: Biomechanics and mechanobiology of tissue growth and remodeling: Current opinions. *Current Opinion in Biomedical Engineering, 15*, A1-A2.

Talman, L.S., & Hiller, A.L. (2021). Approach to posture and gait in Huntington's disease. *Frontiers in Bioengineering and Biotechnology, 9*, 668699.

References

Ángel Rodríguez, M., García-Calleja, P., Terrados, N., Crespo, I., Del Valle, M., & Olmedillas, H. (2022). Injury in CrossFit: A systematic review of epidemiology and risk factors. *The Physician and Sportsmedicine, 50*(1), 3-10.

Atwater, A.E. (1980). Kinesiology/biomechanics: Perspectives and trends. *Research Quarterly for Exercise and Sport, 51*, 193-218.

Beaucage-Gauvreau, E., Brandon, S., Robertson, W., Fraser, R., Freeman, B., Graham, R.B., Thewlis, D., & Jones, C.F. (2020). A braced arm-to-thigh (BATT) lifting technique reduces lumbar spine loads in healthy and low back pain participants. *Journal of Biomechanics, 100*, 109584.

Behm, D.G., Blazevich, A.J., Kay, A.D., & McHugh, M. (2016). Acute effects of muscle stretching on physical performance, range of motion, and injury incidence in healthy, active individuals: A systematic review. *Applied Physiology, Nutrition, and Metabolism, 41*, 1-11.

Benson, L.C., Clermont, C.A., & Ferber, R. (2020). New considerations for collecting biomechanical data using wearable sensors: The effect of different running environments. *Frontiers in Bioengineering and Biotechnology, 8*, 86.

Bowden, J.L., Callahan, L.F., Eyles, J.P., Kent, J.L., & Briggs, A.M. (2022). Realizing health and well-being outcomes for people with osteoarthritis beyond health service delivery. *Clinics in Geriatric Medicine, 38*(2), 433-448.

Cash, J. (2015, July 1). *Meyerhold's biomechanics for theatre*. The Drama Teacher. www.thedramateacher.com/meyerholds-biomechanics-for-theatre/.

Castrogiovanni, P., Di Rosa, M., Ravalli, S., Castorina, A., Guglielmino, C., Imbesi, R., Vecchio, M., Drago, F., Szychlinska, MA., & Musumeci, G. (2019). Moderate physical activity as a prevention method for knee osteoarthritis and the role of synoviocytes as biological key. *International Journal of Molecular Sciences, 20*(3), 511.

Cui, A., Li, H., Wang, D., Zhong, J. Chen, Y., Huading, L. (2020). Global, regional prevalence, incidence and risk factors of knee osteoarthritis in population-based studies. *EClinicalMedicine, 29-30*, 100587.

Daugherty, J., Waltzman, D., Sarmiento, K., & Xu, L. (2019). Traumatic brain injury–related deaths by race/ethnicity, sex, intent, and mechanism of injury—United States, 2000–2017. *MMWR Morbidity and Mortality Weekly Report, 68*(46): 1050-1056.

Emery, C.A., & Pasanen, K. (2019). Current trends in sport injury prevention. *Best Practice & Research Clinical Rheumatology, 33*(1), 3-15.

Franck, H., & Franck, D. (2016). *Forensic biomechanics and human injury: Criminal and Civil Applications—An engineering approach*. CRC Press.

Futrell, E.E., Jamison, S.T., Tenforde, A.S., & Davis, I.S. (2018). Relationships between habitual cadence, foot-strike, and vertical load rates in runners. *Medicine & Science in Sports & Exercise, 50*(9), 1837-1841.

Hamill, J., Knutzen, K.M., & Derrick, T.R. (2021). Biomechanics: 40 years on. *Kinesiology Review, 10*(3), 228-237.

Huang, Y.-L., Jung, J., Mulligan, C.M.S., Oh, J., & Norcross, M.F. (2020). A majority of anterior cruciate ligament injuries can be prevented by injury prevention programs: A systematic review of randomized controlled trials and cluster-randomized controlled trials with meta-analysis. *American Journal of Sports Medicine, 48*(6), 1505-1515.

Hughes, G. (2014). A review of recent perspectives on biomechanical risk factors associated with anterior cruciate ligament injury. *Research in Sports Medicine, 22*, 193-212.

Igiri, B.E., Tagang, J.I., Okoduwa, S.I.R., Adeyi, A.O., & Okeh, A. (2019). An integrative review of therapeutic footwear for neuropathic foot due to diabetes mellitus. *Diabetes & Metabolic Syndrome: Clinical Research & Reviews, 13*(2), 913-923.

Imhausen, A., Robson, E., Dauben, J.W., Plofker, K., & Berggren, J.L. (2007). In V.J. Katz (Ed.), *The mathematics of Egypt, Mesopotamia, China, India, and Islam: A sourcebook*. Princeton University Press.

Innocenti, B., & Galbusera, F. (Eds.). (2022). *Human orthopaedic biomechanics: Fundamentals, devices and applications*. Elsevier Science & Technology.

International Ergonomics Association. (n.d.). *What is ergonomics?* Retrieved July 26, 2022, from https://iea.cc/what-is-ergonomics.

International Labour Union and International Ergonomics Association (ISEA). (2021). *Principles and guidelines for human factors/ergonomics (HFE) design and management of work systems.* International Labour Union.

Junior, R.M., Berton, R., de Souza, T.M.F., Chacon-Mikahil, M.P.T., & Cavaglieri, C.R. (2017). Effect of the flexibility training performed immediately before resistance training on muscle hypertrophy, maximum strength and flexibility. *European Journal of Applied Physiology, 117*(4), 767-774.

Knudson, D. (2021). *Fundamentals of biomechanics* (3rd ed.). Springer.

National Institute of Arthritis and Musculoskeletal and Skin Diseases (NIAMS). (2019). *Osteoarthritis.* NIH Publication No. 15-4617. www.niams.nih.gov/health_info/osteoarthritis

Nedunchezhiyan, U., Varughese, I., Sun, A.R., Wu, X., Crawford, R., & Prasadam, I. (2022). Obesity, inflammation, and immune system in osteoarthritis. *Frontiers in Immunology, 13*, 907750.

Nessler, T., Denney, L., & Sampley, J. (2017). ACL injury prevention: What does research tell us? *Current Reviews in Musculoskeletal Medicine, 10*(3), 281-288.

Owoeye, O.B.A., Ghali, B., Befus, K., Stilling, C., Hogg, A., Choi, J., Palacios-Derflingher, L., Pasanen, K., & Emery, C.A. (2020). Epidemiology of all-complaint injuries in youth basketball. *Scandinavian Journal of Medicine & Science in Sports, 30*(12), 2466-2476.

Padua, D.A., DiStefano, L.J., Hewett, T.E., Garrett, W.E., Marshall, S.W., Golden, G.M., Shultz, S.J., & Sigward, S.M. (2018). National Athletic Trainers' Association position statement: Prevention of anterior cruciate ligament injury. *Journal of Athletic Training, 53*(1), 5-19.

Pitches, J. (2004). *Vsevolod Meyerhold* (2nd ed.). Routledge.

Pope, M.H. (2005). Giovanni Alfonso Borelli the father of biomechanics. *Spine, 15*(30), 2350-2355.

Posse, N. (1890). *The special kinesiology of educational gymnastics.* Lothrop, Lee and Shepard.

Preatoni, E., Bergamini, E., Fantozzi, S., Giraud, L.I., Bustos, A.S.O., Vannozzi, G., & Camomilla, V. (2022). The use of wearable sensors for preventing, assessing, and informing recovery from sport-related musculoskeletal injuries: A systematic scoping review. *Sensors, 22*(9), 3225.

Rohlmann, A., Pohl, D., Bender, A., Graichen, F., Dymke, J., Schmidt, H., & Bergmann, G. (2014). Activities of everyday life with high spinal loads. *PloS One, 9*(5), e98510.

Schram, B., Orr, R., Pope, R., Canetti, E. (2020). Risk factors for development of lower limb osteoarthritis in physically demanding occupations: A narrative umbrella review. *Journal of Occupational Health, 52*, e12103.

Seth, A., Hicks, J.L., Uchida, T.K., Habib, A., Dembia, C.L., Dunne, J.J., Ong, C.F., DeMers, M.S., Rajagopal, A., Millard, M., Hamner, S.R., Arnold, E.M., Yong, J.R., Lakshmikanth, S.K., Sherman, M.A., Ku, J.P., & Delp, S.L. (2018). OpenSim: Simulating musculoskeletal dynamics and neuromuscular control to study human and animal movement. *PLoS Computational Biology, 14*(7), e1006223.

Sharma, M., Khurana, S.M. (2018). Biomedical engineering: The recent trends. In D. Barh & Azevedo (Eds.), *Omics Technologies and Bio-Engineering* (Vol. 2, pp. 323-336). Academic Press.

Skarstrom, W. (1909). *Gymnastic kinesiology.* Bassette.

Sloan, M.R. (1987). *Ruth B. Glassow: The cutting edge.* Academy Papers No. 20, National Academy of Kinesiology, 120-128.

Snoeker, B.A.M., Turkiewicz, A., Magnusson, K., Frobell, R., Yu, D.H., Peat, G., & Englund, M. (2020). Risk of knee osteoarthritis after different types of knee injuries in young adults: A population-based cohort study. *British Journal of Sports Medicine, 54*(12), 725-730.

Taylor, F.W. (1911). *The principles of scientific management.* Harper & Brothers.

Tierney, G. (2021). Concussion biomechanics, head acceleration exposure and brain injury criteria in sport: A review. *Sports biomechanics.* https://doi.org/10.1080/14763141.2021.201692

Tipton, Charles. (2008). Susruta of India, an unrecognized contributor to the history of exercise physiology. *Journal of Applied Physiology, 104*(6), 1553-1556.

Unschuld, P.U. (1985). *Medicine in China: A history of ideas.* University of California Press.

Vannini, F., Spalding, T., Andriolo, L., Berruto, M., Denti, M., Espregueira-Mendes, J., Menetrey, J., Peretti, G., Seil, R., & Filardo, G. (2016). Sport and early osteoarthritis: The role of sport in aetiology, progression and treatment of knee osteoarthritis. *Knee Surgery, Sports Traumatology, Arthroscopy, 24*(6), 1786-1796.

Vina, E.R., & Kwoh, C.K. (2018). Epidemiology of osteoarthritis: Literature update. *Current Opinions in Rheumatology, 30*(2), 160-167.

Zimmermann, W.O., & Bakker, E.W.P. (2019). Reducing vertical ground reaction forces: The relative importance of three gait retraining cues. *Clinical Biomechanics, 69*, 16-20.

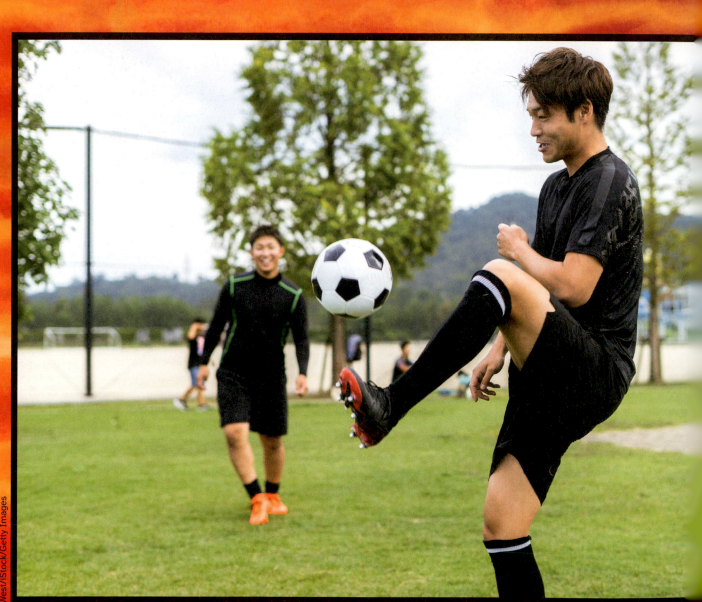

CHAPTER 5

Exercise Physiology

Jennifer L. Caputo

CHAPTER OBJECTIVES

In this chapter, we will

- summarize the knowledge of exercise physiologists,
- provide an overview of the history of the discipline,
- highlight current advances in the field of exercise physiology,
- describe assessment and research tools used by exercise physiologists, and
- identify how exercise physiology research informs practice.

For many people entering the field of exercise physiology, their interest began with personal experiences in sport or athletics. James began playing soccer at 5 years old in a youth recreation program in his community. He learned basic ball skills and had fun playing with his friends. As he continued to improve, he joined an academy program at age 9 to further develop his technical skills. James continued to advance, and when he was 12 years old, he tried out for a competitive travel team.

In high school, James volunteered for a soccer outreach program for children with special needs and became a "soccer buddy" to a child with cerebral palsy. As a volunteer, he learned how to modify activities to help his buddy develop soccer skills in addition to making the activities fun and safe. The experience of sharing the sport he loved with children with special needs helped James understand that everyone can learn and enjoy participating in physical activities with appropriate modifications.

James continued to play competitive soccer through high school. In a game during his junior year, James was hit by an opposing player and tore his anterior cruciate ligament (ACL). Following surgery, he began four months of physical therapy to help achieve his goal of returning to play for his senior year. While in rehabilitation, James learned how to use exercises to help restore his strength and range of motion. After a successful senior season, James began thinking about what he would study in college. His passion for soccer and sport coupled with his experiences helped guide him to exercise science.

As this scenario illustrates, some exercise science majors are interested in the response of the body to exercise training programs. The subdiscipline of exercise physiology provides the scientific basis for the training recommendations to help people in a variety of settings and situations.

If you are curious about how your body and your health status are altered by physical activity—and more specifically, exercise or a lack thereof—the subdiscipline of exercise physiology will be of great interest to you. Exercise physiologists apply principles of biology and chemistry to understand the acute and chronic responses of the body to physical activity. With this knowledge, you can help athletes achieve peak performance, help people participate safely in physical activity, conduct research on preventing and treating disease by means of physical activity and exercise, and help maximize the capacity of people who work in extreme conditions of heat, cold, altitude, underwater settings, and even the microgravity of space.

You regularly experience some of the changes that occur in the human body in response to exercise. For example, if you go for a jog, your heart rate and breathing rate increase and you might begin to sweat. These are responses of the body to meet the increased demand for metabolic energy and cooling the body while jogging. If you continued an exercise program for several months, your resting heart rate and blood pressure likely would decrease, and you might experience changes in your body composition. Exercise scientists seek to understand both how the body responds to meet the immediate demands of being physically active and how it adapts to repeated bouts of exercise. In doing so, exercise physiologists generate knowledge that supports evidence-based recommendations to improve health and physical performance.

Goals of Exercise Physiology

- To understand how to maximize physical performance
- To understand how to improve physical function in altered environments such as those characterized by high temperature or high altitude
- To understand how physical activity and exercise improve health and fitness
- To understand how exercise can be used in treating and preventing disease and alleviating symptoms of disease
- To understand adaptations in physiology and pathophysiology in response to physical activity and exercise

Benefits of Exercise Physiology Knowledge

In chapter 1, you learned exercise is a form of physical activity used to improve or regain performance, health, or physical appearance. **Exercise physiology**

primarily focuses on the exercise components of physical activity, including training, detraining, and how participating in physical activity can improve health and decrease the risk of all-cause mortality. In this section, some of the benefits of understanding the subdiscipline of exercise physiology are highlighted for you.

Enhancing Sport Performance and Training

Applying physiological techniques to understand and improve human exercise performance has been a key feature of exercise physiology since its beginning. This area of exercise physiology, often called *sport physiology*, involves applying "the concepts of exercise physiology to enhancing sport performance and optimally training athletes" (Kenney et al., 2020, p. 3). Principles of exercise physiology form the foundation for developing training programs for athletes. While athletes playing a team sport are working toward a common goal, the physiological demands of different positions on the team require coaches to adapt training programs to target performance outcomes by position. For example, on a soccer team, drills and training programs need to be designed to meet the unique physical and movement demands of goalies, full backs, midfielders, center backs, and forwards. Programs can even be individualized to the fitness characteristics of individual athletes.

Research conducted at the Harvard Fatigue Laboratory (Dill et al., 1930; Margaria et al., 1933) from 1927 to 1947 provided an initial understanding of the responses to exercise that contribute to fatigue, cardiovascular and oxygen uptake responses to exercise, and how changing altitude and temperature alter the physiological response to exercise. This early research provided an understanding of the physiological responses to strenuous exercise. Today, research by sport physiologists covers a range of topics such as

- fluid and food consumption recommendations to support exercise,
- the role of genetics in exercise and physical activity performance,
- the effectiveness of ergogenic aids in improving performance, and
- methods to maximize recovery for athletes.

Sport physiologists also use knowledge and techniques from other disciplines and subdisciplines to study ways of enhancing performance. For example, they use knowledge gleaned from nutrition research in working with sport dietitians and nutritionists to prepare athletes for performance, help athletes maintain performance, and enhance recovery following performance. Just as training programs differ based on the demands of the sport, so do nutrition needs for athletes. Nutritional guidance is individualized based on the sport, the health of the athlete, the external environment, and even practical issues such as whether there are breaks in activity to allow for consumption of food or beverage and whether the athlete must carry their supplies. Maintaining glycogen (stored glucose) and blood glucose levels, hydration strategies, timing of protein consumption, and supplements are important areas where exercise physiologists and sport nutritionists work together.

Sport physiologists are also concerned with the effects of the environment on sport performance. Data from sport physiology studies have been used to develop guidelines to avoid heat illness (heat exhaustion and heat stroke) in sport and to prevent health problems in scuba divers brought on by prolonged immersion in deep water. Before the 1968 Summer Olympics in Mexico City—which sits 7,218 feet (2,200 m) above sea level—leading exercise physiologists researched the acute effects of, and acclimatization to, high altitude. Sport governing bodies used the knowledge gained from these studies to better prepare athletes for the Olympics. Often, physiological knowledge is focused on maximizing performance to increase the chance of winning a medal, but it has also focused on dealing with adverse conditions like environmental pollutants at some Olympic venues.

Improving Physical Fitness

Studying exercise physiology also helps us understand the physiological determinants of physical fitness and the ways training programs can improve fitness. In the 1950s, interest in improving physical fitness was sparked by two events in the United States—President Dwight D. Eisenhower's heart attack and the publication of a study reporting that American children were less fit than European children (Berryman, 1995). These events, combined with concerns about young men's poor fitness for military service, contributed to interest in and development of exercise physiology research on components of physical fitness.

Research conducted in exercise physiology and other areas of exercise science has resulted in recommendations for optimal intensity, frequency, and duration of training programs to develop various components of physical fitness. Just as no distinguishing athletic personality has been discovered, early exercise science researchers concluded that no uniform quality of athletic ability exists. There are, instead, identifiable components of health- and skill-related physical fitness. People who work in fitness centers need to understand this knowledge as well as how to adapt fitness programs to make them safe and effective for different clients—young and old, sedentary and trained, and those with special conditions (e.g., pregnancy or diabetes). Some exercise scientists specialize in studying and working with children and adolescents, and others focus on gerontology (the study of aging). In addition, the influence of heredity on physical fitness components and trainability has attracted considerable research interest among exercise physiologists. Much of this interest has been stimulated by research on identical twins conducted by Claude Bouchard and colleagues (1986) and is discussed later in this chapter.

Health- and Skill-Related Components of Physical Fitness

Health- and skill-related components of fitness can be altered by physical activity and exercise. Components of health-related fitness include body composition, cardiovascular endurance, flexibility, muscular strength, and muscular endurance. Improving health-related fitness assists in performance of daily activities without fatigue, reduces risk of chronic disease, and improves well-being. Skill-related components of fitness do not directly affect health. Instead, altering the components of speed, agility, coordination, balance, power, and reaction time assists in sport performance and skill development.

Promoting Health and Treating Disease With Exercise

Exercise physiology serves as a foundation for understanding how physical activity and exercise help reduce disease risk and help treat disease, particularly hypokinetic diseases. In 1996, the first report from a surgeon general addressing physical activity and health was released. This report emphasized "significant health benefits could be obtained by including a moderate amount of physical activity on most, if not all, days of the week" (U.S. Department of Health and Human Services [USDHHS], 1996). Many researchers have examined the benefits of physical activity and physical fitness in preventing and even reversing coronary heart disease (Lawler et al., 2011; Thompson et al., 2003). Researchers have also documented the role of physical activity in preventing diseases such as breast, lung, and colon cancers (McTiernan et al., 2019) and non-insulin-dependent diabetes (Conners et al., 2014). Researchers have developed exercise tests to screen for obstructive sleep apnea and examined the role of exercise in decreasing the risk of dementia and Alzheimer's disease.

Some exercise physiologists also focus on the big picture—how exercise and physical activity are related to disease in large populations. In these cases, exercise physiologists might collaborate with specialists in epidemiology and public health in conducting two types of studies:

- *Longitudinal:* following the same group of people across time
- *Cross-sectional:* comparing different groups of people at one time

Obesity is one disease that can be studied using these approaches. In a longitudinal example, Di Pietro and colleagues (2004) followed 2,501 healthy men for 5 years and showed that daily physical activity was negatively associated with weight gain across time. The men who reduced their activity gained weight, while those who increased their activity lost weight. In a cross-sectional study, Chau and associates (2012) found that Australian workers with a job that required sitting had a higher overweight and obesity risk than workers who stood at their jobs. Both studies provide knowledge for understanding how physical activity is associated with long-term health, one describing how body weight changes over time relative to physical activity and the other providing a snapshot of how obesity varies across sitting and standing groups.

Physiologists who work in clinical settings prescribing and monitoring exercise in disease management and rehabilitation are known as *clinical exercise physiologists.* Often working under the guidance of a physician, clinical exercise physiologists might work with people who have diabetes, cancer, cardiac diseases, or pulmonary diseases. Knowledge of the pathophysiology of these medical conditions is needed in addition to an understanding of how these conditions are diagnosed and treated. Organizations such as the American College of Sports Medicine (ACSM) provide guidelines to establish safe exercise practices within specific clinical populations.

Understanding Physiological Changes Due to Exercise

Early work in exercise physiology focused on the ways in which exercise affects the functioning of organs and body systems. Exercise physiologists have examined how the cardiovascular, respiratory, muscular, immune, and endocrine systems respond functionally and adapt structurally to different types of acute and chronic exercise.

One branch of exercise physiology is closely linked to **biochemistry**—that is, the chemistry of living things. Research techniques developed in biochemistry (and physical chemistry) have been used by investigators to examine genetic variants to training adaptations. One of the most important advances in exercise physiology resulted from the use of muscle biopsy technique to sample muscle tissue and examine **muscle glycogen** (stored carbohydrate) concentration during exercise (Hultman, 1967). Biochemical techniques also have been used to develop our understanding of lactate production and the use of energy stores during exercise. For individuals with type 2 diabetes, biochemical techniques have provided an understanding of how exercise enhances insulin sensitivity and improves glycemic control.

Research from molecular biology has enhanced our understanding of how cells function. Exercise physiologists use techniques of molecular biology to study how muscle protein synthesis is turned on and off by changing amounts of and kinds of exercise (Atherton & Smith, 2012). This molecular knowledge is important for understanding how muscles increase and decrease in size in response to changes in exercise; it also contributes to our understanding of how muscles are damaged and recover from injury.

Career Opportunity

Working in Cardiac Rehabilitation

In cardiac rehabilitation, exercise physiologists work with patients who have cardiovascular conditions such as high blood pressure (hypertension) or who have had a heart attack (myocardial infarction) in addition to those recovering from bypass or open-heart surgery. These individuals apply guidelines established by the American Association of Cardiovascular and Pulmonary Rehabilitation (AACVPR) and ACSM to develop exercise programs to meet the individual needs of patients. Exercise physiologists use their assessment skills to measure heart rate and blood pressure; monitor ECGs (electrocardiogram) for changes in heart rhythms during exercise; and assess program outcomes such as muscular strength, flexibility, and cardiovascular endurance. Effectively communicating with patients and medical doctors, assisting patients with psychosocial adaptations following a cardiac event, being a motivator, and knowing how to educate patients regarding lifestyle modifications are important skills of exercise physiologists in the cardiac rehabilitation setting.

Increases and Decreases in Muscle Size

Gain and loss of skeletal muscle mass involves complex molecular and cellular events. Muscle damage from a resistance training program can trigger these events, resulting in an increase in muscle size. This hypertrophic response generally occurs one to two months following the start of a training program. Conversely, immobilization, such as following surgery, and aging (**sarcopenia**) trigger cellular changes that lead to a loss of muscle mass. Physical therapists use exercise to help a person rehabilitate following surgery, and resistance training can be used to help reverse the effect of aging on human skeletal muscle.

From this section, you can appreciate that exercise physiology knowledge of how the body responds to exercise expands our understanding of human physiology and can be applied to many situations and environments.

KEY POINT

With knowledge of exercise physiology you can enhance sport performance and training, improve physical fitness, promote health, treat and prevent disease, and understand physiological adaptations to exercise.

What Do Exercise Physiologists Do?

Scholars who study exercise physiology typically work as university faculty members, although employment opportunities also exist in clinics, hospitals, and research centers. Researchers and professors usually hold a doctoral (PhD) degree in exercise physiology. University faculty members teach courses such as exercise physiology and exercise prescription, and they conduct research on topics such as the effects of conditioning programs on sport performance and the effects of physical activity on reducing the risk of chronic disease. Other faculty focus on basic research, using animal models to understand mechanisms and establish core theories of physiology and pathophysiology. To support their research, faculty members often write research grant proposals to federal agencies such as the National Institutes of Health and to foundations such as the Robert Wood Johnson Foundation. Research funded by the U.S. military and by the National Aeronautics and Space Administration (NASA) is also conducted by exercise physiology scholars in university, government, and medical laboratories. Research contracts sometimes support physiological research when commercial applications are of interest.

Graduates who hold bachelor's or master's degrees from exercise science programs might have careers in clinical exercise physiology, often with additional

EXERCISE SCIENCE COLLEAGUES

Liz Ackley

Courtesy of Richard Boyd, Boyd Pearman Photography.

Dr. Liz Ackley is an exercise science faculty member and founder of the Center for Community Health Innovation at Roanoke College, where she strives to reduce community-level health inequities. Ackley uses her passion for and knowledge of exercise physiology to position youth fitness metrics as catalysts for change in infrastructure, programs, and policy initiatives in her community. Dr. Ackley remarked, "Exercise scientists are uniquely positioned to guide community-level change to address factors which promote or detract the public from engaging in health-promoting levels of exercise and physical activity" (L. Ackley, personal communication, June 8, 2021). In her work, Ackley maps findings from an annual health surveillance system on youth health outcomes and family-reported barriers and facilitators to physical activity using geographic information systems (GIS) technology. This method establishes evidence of the need for intervention in specific geographic areas. She then collaborates with multisector partners across city government and nonprofit and private sectors to address the community's needs in culturally appropriate ways. Her translational approach to increasing physical activity has been funded by the Robert Wood Johnson Foundation and ChangeLab Solutions. Ultimately, Ackley's work promotes health equity through the creation of new community programs, safety-promoting strategies, and improved access to neighborhood resources supporting exercise and physical activity, such as parks, playgrounds, crosswalks, and sport facilities.

Career Opportunity

Performance and Sport Scientists

An emerging field allows exercise scientists to use evidence-based training practices to optimize athletic performance. Performance and sport scientists use technology and their knowledge of human performance and sport to gather data on athletes so positive and negative changes in performance can be evaluated. Information on biometrics such as body mass, hydration, hours of sleep, training and recovery heart rates, and training load and intensity are analyzed to better inform coaches' and members of the athletic performance staff on decisions related to athlete training, performance, and recovery. The demand for people with knowledge of exercise sciences, technology to assess and monitor performance, and familiarity with statistics is increasing in high-level sport organizations and international sporting federations to gain a competitive edge. See chapter 10 for more on this area of exercise science.

training and certifications. These individuals provide clinical exercise testing and prescription in affiliation with medical clinics or hospitals to help patients with cardiac and pulmonary rehabilitation. Opportunities also exist to work with people who have diabetes, who are recovering from cancer, or who are part of a weight management program. Those interested in physical rehabilitation from injury or surgery have opportunities to be employed as physical therapy technicians.

History of Exercise Physiology

Exercise physiology evolved from physiology in the 18th century after Antoine Lavoisier and Pierre de Laplace developed the methodology to measure oxygen consumption and carbon dioxide production during respiration in animals (USDHHS, 1996). During the following century, these techniques were further developed and applied to studying how humans responded to exercise and daily tasks such as lifting loads. There has been great interest in exercise physiology research because of the dramatic and immediate physiological responses to exercise and many positive, long-term benefits.

Early Beginnings

Our earliest understanding of what is today recognized as exercise physiology is rooted in ancient times from Hippocrates (460-370 BC), who promoted exercise to maintain and improve health (Berryman, 1992). Exercise physiology has evolved into the study of "the functions and adaptations of living organisms, their bodily parts, and their organic and chemical processes as a result of increased energy demand due to muscle contraction" (Ivy, 2007, p. 34). Our foundational knowledge of exercise physiology is credited to several seminal laboratories and early

researchers from all over the world who not only conducted research, but also trained other scientists.

An early contributor to our understanding of exercise physiology is August Krogh of the University of Copenhagen. Krogh developed one of the first cycle ergometers (exercise bikes) to study physiological responses to exercise. Krogh received the Nobel Prize in Physiology in 1920 for his research on the regulation of microcirculation, and he is known as the father of exercise physiology in Scandinavia (Åstrand, 1991). Krogh initially worked in the laboratory of Christian Bohr, who discovered the effect of carbon dioxide on the dissociation curve of hemoglobin (Åstrand, 1991). The Bohr effect is a process you will learn about in your exercise physiology course.

Archibald Vivian (A.V.) Hill of University College, London earned the Nobel Prize for his work on energy metabolism in 1921. In a foundational paper, Hill and Lupton (1923) presented many of the basic concepts of exercise physiology related to oxygen consumption, lactate production, and **oxygen debt** (i.e., excess oxygen consumption following exercise). In addition to being an outstanding scientist and mentor, Hill assisted European minority refugee scientists who were fleeing the Nazis during the 1930s and 1940s (Rall, 2017).

In 1927 Dr. Peter V. Karpovich established an exercise physiology laboratory at Springfield College in Massachusetts (Kroll, 1982). Karpovich is well known for his research on the effects of ergogenic aids on physical performance and served as one of the founders of ACSM in 1954. The Harvard Fatigue Laboratory, directed by David Bruce Dill, PhD, was also established in 1927. More than 300 peer-reviewed papers stemmed from investigations conducted in this laboratory across 20 years (Ivy, 2007). The Harvard Fatigue Laboratory served as a training site for researchers from around the world. The Physical Fitness Research Laboratory was established in 1944 by Thomas K. Cureton Jr. at the University of Illinois.

Harvard Fatigue Laboratory

The Harvard Fatigue Laboratory, founded in 1927, was the brainchild of L.J. Henderson, a physical chemist. Directed by David Bruce Dill, the laboratory included a room containing a treadmill borrowed from the Carnegie Nutrition Laboratory, as well as a large gasometer, a room for basal metabolism studies, an animal room, and a climatic room (Dill, 1967). Research undertaken in the laboratory included environmental studies conducted at high altitude, in the desert, and in steel mills.

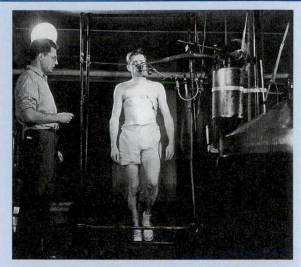

Reprinted by permission from W.L. Kenney, J. Wilmore, and D. Costill, *Physiology of Sport and Exercise*, 7th ed. (Champaign, IL: Human Kinetics, 2020), 7.

Although the laboratory existed only until 1947, it profoundly affected research on exercise physiology in the United States and Europe during the second half of the 20th century. Many young investigators received their formative training at the laboratory as postdoctoral or doctoral students, including Ancel Keys, R.E. Johnson, Sid Robinson, and Steve Horvath. The many international scientists who spent time at the laboratory included Lucien Brouha of Belgium; Rodolfo Margaria of Italy; and E.H. Christensen, Erling Asmussen, and Marius Nielsen of Denmark (Dill, 1967). All these investigators went on to establish their own laboratories and were responsible for training many of the leading scientists in exercise physiology in the early decades of exercise science.

The pioneering work directed by Dr. Cureton in this laboratory established the science behind physical fitness testing and athletic performance. Many leading investigators of the physiology of exercise also trained with Dr. Cureton.

Significant Events, 1950 to 2000

One of the most important events that stimulated research in exercise physiology after 1950 occurred in England. In 1953, Jeremy Morris and colleagues published a study on coronary heart disease and physical activity in which they found that drivers of double-decker buses in London who sat during their shifts had significantly higher disease risk than conductors, who moved around the buses (Morris et al., 1953). This study stimulated interest in epidemiological research on physical activity, physical fitness, and chronic disease that continues to this day. Morris' research also inspired additional studies of the effects of fitness and endurance training on risk factors for coronary heart disease, including serum cholesterol and blood pressure.

Based on the results of these studies and many subsequent studies, U.S. government agencies released two official statements in the 1990s about the role of physical activity in preventing chronic diseases. The first statement followed the NIH Consensus Development Conference on Physical Activity and Cardiovascular Health held in 1995. Among other things, the consensus development panel concluded that "physical *inactivity* is a major risk factor for cardiovascular disease" and that "moderate levels of regular physical activity confer significant health benefits" (NIH Consensus Development Panel, 1996, p. 245; italics added). The other statement came in the form of the surgeon general's first report on physical activity and health (USDHHS, 1996). As noted, it was concluded in this report that regular physical activity can not only reduce the risk of heart disease but can also reduce the risk of diabetes, hypertension (high blood pressure), and colon cancer and can help control body weight (USDHHS, 1996).

Recent Events and Advances in Exercise Physiology

Recently, physiology researchers have continued to highlight the impact of physical activity and exercise not only in mitigating degenerative diseases such as osteoporosis and diabetes, but in prolonging life. The Human Genome Project completed in 2003 provided an opportunity to study how physical activity prevents diseases and can be used to treat disease by influencing the expression of genes. Specifically, this epigenetic research supports the role of physical activity in the prevention of cancer, diabetes, cardiovascular, and neurodegenerative diseases (Grazioli et

Professional Issues in Exercise Science

Why Physical Activity Levels of the Nation Matter to U.S. Government Agencies

Physical activity guidelines released by government agencies serve to minimize the personal and societal consequences of physical inactivity. Only half of adults in the United States perform enough physical activity to reduce or prevent chronic disease, and 50% of adults have at least one chronic comorbidity. If the U.S. population met published physical activity recommendations, 1 in 10 premature deaths could be prevented. In addition to physical activity preserving human life, the yearly cost of treating disease linked to inadequate physical activity is $117 billion. Physical activity ensures a healthier workforce for the nation and limits the number of sick days by employees. The readiness of our military is also affected by physical inactivity, because about 25% of young adults are too heavy to serve. Overall, government agencies promote physical activity to help protect and improve the quality of life of citizens while preserving economic resources.

Citation style: APA

Division of Nutrition, Physical Activity, and Obesity, National Center for Chronic Disease Prevention and Health Promotion. (2020, May 13). *Why it matters*. Centers for Disease Control and Prevention. www.cdc.gov/physicalactivity/about-physical-activity/why-it-matters.html

al., 2017). Additionally, researchers have explored the existence of fitness phenotypes and how genes influence muscular and cardiovascular adaptations to physical activity and exercise (Bray et al., 2009).

Strides in understanding muscle metabolism and AMPK, the enzyme that regulates glucose and fat uptake, have led to better interventions for obesity and type 2 diabetes (Kjøbsted et al., 2018). Greater focus also has been placed on the benefits and efficacy of low-volume, high-intensity interval training (HIIT). This method of training incorporates short, alternating periods of high-intensity activity and rest periods and has been shown to improve maximal oxygen consumption, insulin sensitivity, and endothelial function (Kilpatrick et al., 2014). Researchers also have focused on the benefits of lower-intensity activity on longevity. Older women taking as few as 4,400 steps per day had a lower risk of mortality over a 4.3-year period than those taking 2,700 steps per day. Regardless of stepping intensity, the all-cause mortality risk continued to decrease until the number of daily steps reached 7,500 (Lee et al., 2019).

KEY POINT

Exercise physiology research has contributed to the discovery that both high- and low-intensity physical activity or exercise can lead to improved health and longevity.

The use of devices for measuring step count as well as tracking heart rate and distance covered has been a top fitness trend identified by the ACSM since 2016. Wearable technology includes smart watches, heart rate monitors, and GPS (global positioning system) devices and is estimated to be a $95 billion industry (Thompson, 2019). In addition to movement, these devices can monitor sitting or sedentary time. This is important because evidence from inactivity physiology has uncovered negative cardiovascular and metabolic outcomes of being sedentary (Hamilton et al., 2008).

Sedentary behavior is characterized by the expenditure of 1.5 or less metabolic equivalents (**METs**; units of resting metabolic energy expenditure) while sitting, reclining, or lying (Tremblay et al., 2017). The relationship between sedentary behavior and health is a new area of emphasis in exercise science; most of the research documenting this relationship has been published since 2010. Studies have documented strong evidence for the positive relationship between sedentary behavior and all-cause mortality, cardiovascular disease mortality, incident cardiovascular disease, and type 2 diabetes (Katzmarzyk et al., 2019). As assessed through accelerometer data, children and adults are sedentary 7.7 hours each day (approximately 55% of waking hours; Matthews et al., 2008), emphasizing the potential adverse impact on health. For those who perform low amounts of moderate to vigorous physical activity, the negative impact of sedentary behavior on all-cause and cardiovascular disease mortality is even stronger (Katzmarzyk et al., 2019). With this new knowledge, exercise scientists now face the additional challenge of concurrently decreasing sedentary behaviors while increasing physical activity. Reflecting this evidence, for the first time, the Physical Activity Guidelines for Americans (USDHHS, 2018) includes recommendations to reduce sedentary time in addition to accumulating moderate-intensity activity or exercise to improve health.

Researchers are studying methods to decrease sedentary behaviors in the home, in schools, and in the workplace. As with physical activity interventions, sedentary interventions should be population and site specific. Smartphone, watch, and computer-based applications prompt people to get up and move on a regular basis. Companies are also marketing active workstations such as standing desks, treadmill desks, and cycle desks to help minimize sedentary behavior while at a work desk. The practice of implementing workplace interventions, and the potential of employers covering costs, relies on the accumulation of research evidence demonstrating sustainability of such programs along with potential reductions in absenteeism and medical care costs (Lutz et al., 2020). You can begin to reduce your own sedentary behavior by making small changes such as getting up on a regular schedule while reading, working on the computer, or watching television; walking while on the phone; and standing as often as possible. You could even schedule your next meeting or study session as a walking meeting.

Business closures during the COVID-19 (coronavirus disease 2019) pandemic have led to an emerging trend of online training in the health and fitness industry. In a 2021 survey of fitness professionals around the world of their views on trends in the health and fitness industry, online training, for the first time, was the number one trend identified (Thompson, 2021). This form of training allows for a person to train at home in individual or group sessions and can be livestreamed or prerecorded.

Online training provides benefits such as privacy, reduced travel, and convenience.

In 2020 one out of five members of a gym, health club, or fitness studio also paid for an online fitness service, and 68% of those who began using an online service during the COVID-19 pandemic planned to continue using the service in the future (International Health, Racquet, & Sportsclub Association, 2021). The demand for continued use of online training in the fitness industry generates a need for educating and certifying personal trainers to administer and implement assessments and training sessions through online platforms.

Research can guide the use of online exercise prescription and training in people with musculoskeletal conditions (Bennell et al., 2019), pulmonary diseases (Hansen et al., 2020), rheumatoid arthritis (Williamson et al., 2020), and elderly populations (Hong et al., 2018). However, research on the healthy general population is limited. Work on best practices for effective online exercise programming for the general population should expand with the increased demand for this kind of exercise.

KEY POINT

To optimize health, the Physical Activity Guidelines for Americans (USDHHS, 2018) include recommendations to reduce sedentary time in addition to accumulating moderate-intensity activity.

Research and Evidence-Based Practice in Exercise Science

Can Physical Activity Alter the Association Between Sitting and Mortality?

With the body of literature on the association of sitting time and mortality risk, Ekelund and colleagues (2016) investigated how physical activity time and intensity mediates this relationship. Their findings helped shape the physical activity guidelines for Americans. Underscoring the benefits of physical activity, Ekelund and associates showed performing 60 to 75 minutes of daily moderate-intensity activity eliminates the mortality risk of high sitting time such as at a desk or in a motor vehicle. As society becomes increasingly dependent on technology, more occupations require extended sitting. Coupled with the lack of people meeting physical activity guidelines (approximately 50%; National Center for Health Statistics, 2021), exercise professionals now face two challenges. A two-pronged approach including strategies to increase moderate-intensity physical activity and strategies to decrease sitting time provides the best practice in decreasing mortality risk (figure 5.1).

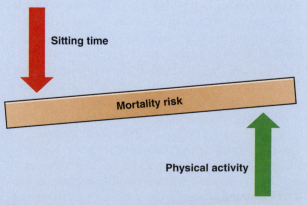

Figure 5.1 A two-pronged approach to decreasing mortality risk.

Research Methods in Exercise Physiology

Exercise physiologists use many methods to measure and assess how the human body responds and adapts to physical activity. This work is conducted both within and outside of the laboratory.

Laboratory Work

Exercise physiology laboratories contain many pieces of equipment used to monitor and evaluate physiological responses. Working in a laboratory space allows control of factors that can affect responses to exercise, such as variations in temperature and humidity. As in other scientific areas, exercise physiologists use standard protocols, techniques, and calibrated equipment to allow work from different laboratories to be compared.

There are many techniques of assessing the components of health-related physical fitness and the traits associated with a lower risk of hypokinetic diseases. For example, cardiovascular endurance can be assessed either by direct measurement, which requires measuring the volume of oxygen and carbon dioxide in expired (exhaled) air or by estimating oxygen consumption. The most widely accepted method of measuring cardiovascular fitness is to determine **maximal oxygen uptake**, or $\dot{V}O_2$**max**. This can be done by having a person exercise while the intensity is increased progressively (i.e., using a graded exercise test) until the person can no longer maintain the required exercise intensity.

Early investigators collected expired air in Douglas bags, which are large inflatable bags, while people were physically active. The concentrations of the gases in the expired air were then determined using chemical analyzers, and the volume of gas in the bag was measured with a spirometer or volume meter. This method could be used while people walked or ran on a treadmill or pedaled on a stationary bicycle in the laboratory. Creative methods also were used to allow gas collection outside using Douglas bags and tubing while people skied (bags were strapped to the person's back), ran outside (a person riding a bike alongside held the bags), or even swam (a person walked alongside the pool holding the bags). Today, investigators measure gas concentrations with electronic analyzers and gas volumes with flow meters. When interfaced with computers, these devices provide nearly instantaneous and continuous information about oxygen uptake as people rest or exercise. In addition to these stationary metabolic carts, some portable units are small enough for people to wear on their body to allow for the measurement of respiratory variables outside of the laboratory in almost unlimited modes of physical activity or exercise.

By measuring oxygen uptake, researchers can obtain information about how the muscles use oxygen and how much metabolic energy is expended during physical activity. They also can make comparisons between elite and recreationally active individuals or before and after a training intervention to assess training adaptations. When a $\dot{V}O_2$max test is not possible, whether due to lack of equipment, trained personnel, or characteristics of the person being assessed, physiologists also use submaximal exercise tests to estimate cardiovascular fitness. These tests do not require people to exercise at maximal level; instead, prediction equations using heart rate responses to submaximal exercise are used to estimate maximal oxygen uptake.

Ergometers are used during exercise tests to measure the external work performed by muscles. Common ergometers used in exercise science include motorized treadmills and leg and arm cycles. These

Blood pressure is easily monitored as a person rides a leg cycle ergometer.

© Jennifer L. Caputo

devices enable physiologists to monitor factors such as heart rate, blood pressure, breathing rate, oxygen consumption, and carbon dioxide production while a person is active. Physiologists also might draw blood samples to monitor changes in lactic acid and glucose.

Because body weight is supported during cycling, cycle ergometers are useful for comparing people of different body weights. Leg cycle ergometers also enable researchers to measure blood pressure and obtain blood samples during exercise more easily because the arms are stationary. Some disadvantages of the cycle ergometer are that oxygen uptake is generally lower and local fatigue occurs earlier because the rider uses only the leg muscles. Motorized treadmills allow people to be tested while they walk or run at different speeds; in addition, the slope (or grade) of the treadmill can be raised to increase exercise intensity at a constant speed. To study upper-body exercise, exercise physiologists use arm ergometers. Arm ergometers are valuable when working with athletes who are rowers and athletes who use wheelchairs. A swimming flume can be used to study physiological responses to swimming.

For studies of moderate to vigorous exercise, treadmills are preferable to cycle ergometers because running is more strenuous than cycling for most people. When using treadmills, however, researchers must account for body weight differences. When walking or running at the same speed and grade, heavier people work harder than lighter people do because body weight is lifted with every step. Treadmills also make it more difficult to measure blood pressure and sample blood, especially when a person is running, because the arms are moving.

Another health-related component of physical fitness commonly assessed by exercise physiologists is body composition. Body composition assessments determine a person's percentage of body fat or ratio of lean tissue to adipose tissue. The traditional gold standard for determining body composition in humans involves hydrostatic weighing, also known as **underwater weighing**, in which, as the name indicates, a person is weighed while submerged in water. This technique makes use of Archimedes' principle, which states that the buoyant force acting on a submerged object is equal to the weight of the water displaced (density = mass ÷ volume). Once body density has been determined, equations are used to estimate percentage of body fat. The BOD POD is an alternative method that uses air displacement instead of water displacement to calculate body composition. Other techniques for estimating body density and body composition include measuring total body water using isotopes (e.g., deuterium), measuring the thickness of subcutaneous fat with skinfold calipers, determining tissue impedance through bioelectrical impedance analysis (BIA), and measuring body fat with dual-energy X-ray absorptiometry (DEXA).

Biochemical Methods

Exercise physiologists use biochemical methods to examine changes at the tissue and cellular levels during and after exercise. These more invasive techniques include blood sampling and muscle biopsy. Bergstrom (1962) is credited with the development of the muscle biopsy technique. Blood samples are obtained from either venipuncture of superficial arm veins or finger pricks. Monitoring changes in the concentration of blood constituents can help physiologists determine the use of substrate (stored carbohydrate and fat), acid–base balance, hydration status, immune function, and endocrine responses. One commonly used biochemical technique in exer-

Research and Evidence-Based Practice in Exercise Science

Rating of Perceived Exertion (RPE)

If you have ever been asked to rate how hard you are exercising using a scale from 6 (no effort) to 20 (maximal effort), you have used the Borg scale, also called a rating of perceived exertion (RPE). Dr. Gunnar A. V. Borg (1927-2020) published this subjective scale in 1982 (Borg, 1982), and it might be the most cited exercise science paper (Khatra et al., 2021). The RPE scale is used to subjectively assess the overall intensity of an activity based on heart rate, respiration, sweating, and fatigue. The heart rate during an activity can be estimated by multiplying the rating by 10. The scale is also the preferred measure of intensity when individuals are taking medications that affect heart rate. Ratings of 12 to 14 suggest a moderate intensity, meeting the Physical Activity Guidelines for Americans (USDHHS, 2018). The RPE scale is used by the general population, practitioners, and researchers, and is posted in most settings where exercise scientists work.

Body fat can be assessed using underwater weighing.

David Madison/Getty Images

cise physiology is to measure lactate from a blood sample taken with a finger prick during exercise as an indicator of the use of **anaerobic** (without oxygen) energy-producing systems.

Physiologists use muscle biopsies obtained before and after exercise to examine changes in muscle glycogen (stored carbohydrate), lactate production, and enzyme activity. Examination of muscle tissue samples under a microscope helps determine structural changes following different types of training as well as structural damage after exercise.

KEY POINT

Muscle biopsies are instrumental in helping exercise physiologists determine the mechanisms through which muscles adapt to training interventions.

Animal Models

The effects of physical activity on organs such as the brain, heart, and liver cannot easily be studied in humans. In some instances, researchers use a variety of animals (e.g., frogs, mice, rabbits) to examine both the functional and the structural changes that occur in response to single and repeated bouts of exercise on animal treadmills and running wheels. Research using animals is often essential to ground-breaking, Nobel Prize–winning basic science. Krogh (1919) earned the Nobel Prize in 1920 by demonstrating the variability in frog skeletal muscle blood flow during exercise and at rest. Also with frogs, Hill (1926) won the Nobel Prize in 1922 for his characterization of the chemical and mechanical features of muscular contraction. In 1938 Krebs and colleagues first described the citric acid cycle (or the Krebs cycle). This research was conducted with pigeons and was recognized with a Nobel Prize in 1953. This biochemistry study is foundational in our understanding of metabolism and the process of **aerobic** energy production (with oxygen) during physical activities lasting longer than two minutes.

Some of the advantages of using animals in physiological research are decreased variability and tighter environmental control than when working with human volunteers. This allows for smaller sample sizes in animal studies. Also, because the life span of some animals is short (rodents live 2 to 3 years), it is possible to study age-related disease using longitudinal designs in a shorter time frame. Animals with genetic abnormalities are helpful in examining the effects of exercise on certain clinical disorders such as obesity and hypertension.

On the other hand, animal research suffers from less ecological validity because not all physiological and chemical mechanisms observed in animal species are identical to those observed in humans. For example, growth hormone produced in humans is not identical to that of any other species. In addition, tissue changes due to training also can differ between species. Therefore, investigators must be careful to select the animal, usually a mammal, that most closely reflects the human responses to physical activity in which they are interested and interpret their results with these limitations in mind.

KEY POINT

Advances in exercise physiology are derived from research with humans and animals in laboratories and in field settings.

Field Work

Whether working with individuals or conducting research, exercise physiologists often go beyond the laboratory. A trade-off always exists between the higher experimental control permitted in lab-based research and the ecological validity of field-based research. For example, the easily obtained field measure of body mass index (BMI) is a good screening tool for risk of obesity-related diseases but is not a measure of body composition. Working in the field can present difficulties with

- monitoring physiological responses,
- controlling exercise intensity, and
- controlling environmental conditions.

Benefits of field research include being able to work with larger groups, to recruit participants who might not be able to travel to you, and to study physical activity interventions in specific environments.

Some field research requires portable instruments and recording equipment. However, some estimates of cardiovascular fitness, for example, can be conducted that do not require transportation of heavy or expensive equipment.

- Researchers can determine how quickly students can run or walk a mile (1.6 km) marked off on a school field and then use students' completion times to estimate maximal oxygen consumption.
- The PACER (Progressive Aerobic Cardiovascular Endurance Run) test is a multistage fitness test adapted from the 20-meter (22 yards) shuttle run, which can be performed in school gymnasiums.
- Bleachers in school gymnasiums can be used to help measure cardiovascular fitness. The Queen's College step test (McArdle et al., 1972) requires a step that is 16.25 inches (41.28 cm) high, the height of most bleachers.

In addition to accommodating multiple students at once, these tests require little equipment and offer flexibility when the amount of space available differs from one place to another.

Alternative means also exist for measuring other physiological responses while working in the field. For instance, whereas heart rate can be monitored in the laboratory with an ECG machine, it also can be measured outside of the laboratory with a low-cost, battery-operated heart rate monitor. Some heart rate monitors allow information to be collected over several days through internal memory of the monitor or an app and then downloaded to a computer. These monitors also are useful in providing information about exercise intensity in the field.

Physical activity can be monitored in the field by motion sensors, which provide an alternative to potentially unreliable self-report measures. For instance, pedometers and accelerometers (worn at the hip, ankle, or arm) measure the quantity or intensity (or both) of physical activity or inactivity and estimate caloric expenditure. These instruments have become increasingly popular due to their low cost, small size, and ease of use. They also allow assessment of activity throughout a person's day in all locations to which they travel. Whereas pedometers generally measure only step count, accelerometers can detect the intensity of movement using an acceleration-versus-time curve. Some pedometers and accelerometers have memory capacity, which allows for monitoring of weekly activity and progress toward caloric expenditure goals. Wearable sensors that use GPS to measure a person's movements will soon permit more precise data that are useful in small spaces like basketball and volleyball courts. Many monitors are commercially available and are used by people to monitor their own physical activity.

Monitoring Your Health and Physical Activity

Think about what tools you use to monitor your physical activity. Have you ever been reminded to get up and move by a notification from your watch? Perhaps you have a watch or application on your phone that tracks your steps. Some watches and phones also contain GPS to track the distance you cover while hiking, skiing, or swimming. These commercially available tools can also assist in monitoring your heart rate and sleep and keeping track of the food you eat.

Overview of Knowledge in Exercise Physiology

Over the past century, exercise physiologists have conducted extensive studies of responses to single and repeated bouts of exercise. In this overview, we examine how physiological systems respond acutely and adapt over time to physical activity; review factors that influence the physiological responses to physical activity; and then consider the relationships among fitness, physical activity, and health. As a student in exercise physiology classes, you will learn and practice how to conduct health and fitness assessments and to apply this knowledge to improve the health and performance of people. How you apply this information will depend on the characteristics of the people you are working with including such factors as their goals, health and injury status, and age.

Skeletal Muscles

The contraction of skeletal muscles, through a process called the *sliding filament theory*, produces joint movement. This theory of the basis of muscular contraction was established by Huxley and Hanson in 1954. Advances in light microscopes allowed these scientists to visualize the contractile proteins actin and myosin sliding over one another as activated and relaxed muscles changed in length.

Muscles are composed of many cells called *muscle fibers*. Muscle fibers contain myofibrils that run the length of the fiber. These **myofibrils** are packed with the contractile elements (actin and myosin) that slide over one another as muscles shorten to generate force and move your bones during physical activity. Humans have three primary types of muscle fibers in skeletal muscles.

Muscle Fiber Types

Have you ever wondered why some people excel at running fast for short distances, such as a 100-yard (91 m) dash, whereas other people are better at running marathons? Varying distributions of fiber types in the leg muscles affect how well people are able to perform these events. Muscle fiber types are broadly classified according to the speed at which they contract and other properties: fast-twitch (FT) or slow-twitch (ST). Fast-twitch fibers are further subdivided according to how energy is generated. Fast-twitch fibers that use almost exclusively anaerobic energy systems are called *fast glycolytic* (FG) fibers, whereas fast-twitch fibers that use both aerobic and anaerobic energy systems are called *fast oxidative glycolytic* (FOG) fibers. In slow-twitch fibers the aerobic energy systems dominate.

During light- to moderate-intensity physical activity, such as walking to class, walking on a golf course, or mowing your lawn, slow-twitch fibers are recruited first. These fibers produce force at a slower rate but are resistant to fatigue and allow this type of activity to be sustained for prolonged periods. As the intensity of activity increases, FOG fibers are recruited followed by FG fibers, at the highest intensities (e.g., for an all-out sprint or lifting heavy weights). Although FG fibers produce high force and power at a fast rate, they fatigue quickly. Thus, the highest intensity physical activity can be sustained only for a short time or a few repetitions.

Resistance Training

Muscular **strength** is the maximum force or moment of force exerted by a muscle or a group of muscles. Muscular **power**, in contrast, is the product of the muscular force generated and the velocity of shortening or lengthening. So, muscle power can only be expressed when there is motion, and it is influenced by the force and velocity of muscular contraction (Winter et al., 2016). A muscle's ability to exert force repeatedly or over a prolonged period is known as **muscular endurance**. Resistance or weight training programs designed to improve muscular strength might use exercises that are

- **isometric** (producing tension without changing overall muscle length),
- **isotonic** (producing changes in muscle length without changing external gravitational resistance), or
- **isokinetic** (producing changes in muscle length with joint movement at approximately a constant angular velocity).

Isokinetic machines are often used in research, clinical testing, and rehabilitation to measure muscular strength and other muscular performance variables.

Isometric contractions occur, for example, if you do a wall squat or a plank. Isotonic contractions occur when you lift a free weight such as a barbell. An isokinetic **dynamometer**, such as a Biodex, can measure joint **torque** throughout the range of motion for angular velocities of joint rotation between 500 (**concentric**) degrees/second and –300 (**eccentric**) degrees/second. Muscular strength is most often measured with isokinetic machines at 0 degrees/second (isometric conditions) because it is the strength of a muscle group at a specific angle, not influenced by the widely varying strength values with changes in muscle length and velocity.

Whether your goal is to increase your muscular strength, power, or muscular endurance, a muscle or muscle group must be overloaded beyond the usual level of activity. This is the training principle of **progressive overload**, which indicates stress must be gradually increased on a physiological system or tissue for improvements to occur. In resistance training, an overload can be created through increases in weight, repetitions, or sets. Likewise, when you stop overloading a system, the training adaptations are lost. This is the training principle of **reversibility**.

KEY POINT

Overloading muscles by increasing the intensity, duration, or frequency of an activity will increase muscular strength, power, or endurance. Removal of this overload will lead to a loss of these muscular adaptations.

Training is also subject to the **principle of specificity**. This principle holds that if you want to improve your performance in a given sport, you must select exercises that match, as closely as possible, the techniques, muscles, and energy systems used in the chosen sport. Strength gains will be greatest across the range of movement of the resistance exercises performed. In some physical activities, the ability to develop force rapidly is critical to performance. For example, a volleyball player must rapidly extend their legs in jumping to block a spike at the net. If your sport requires you to move at high speed, higher-velocity isokinetic or isotonic training will be more beneficial than low-velocity or isometric training.

Increases in muscular strength result from two predominant factors: neural adaptations and changes in muscle size. Neural adaptations occur prior to changes in muscle size and might include recruitment of additional muscle fibers (motor units), better synchronization of muscle fiber contraction, and reduction in neural inhibition. Second, individual muscle fibers increase in size following resistance training (**hypertrophy**). High-intensity resistance training leads to increases in muscle cross-sectional area of all muscle fiber types.

Energy-Producing Systems

Movement requires metabolic energy. The chemical form of energy used in the body is **adenosine triphosphate** (ATP). Only enough ATP to sustain maximal, whole-body muscular activity for several seconds is stored in the body. Therefore, depending on metabolic demand, several mechanisms can be used to produce ATP. The systems differ based on the need for oxygen, the rate at which ATP is produced, the amount of ATP produced, and the substrate required. The energy-producing systems operate on a continuum where, depending on the duration and the intensity of a physical activity and the fitness of the person performing the physical activity, specific systems will predominate. An overview of the primary ATP-producing systems follows.

Assessing Your Maximum Bench Press

The bench press exercise is a traditional measure of overall upper-body, forward-pushing strength. The maximum amount of weight a person can push one time (1RM) is a field measure of muscular strength. Intuitively, it would seem like you could go into the weight room and, with a spotter, find your 1RM by continually adding weight to the bar until you reached a weight at which you could only do one repetition. The drawback to this approach is that potential fatigue from your trial-and-error process would lead to an inaccurate assessment. To help produce a valid measure of strength, organizations such as the NSCA and ACSM provide 1RM assessment guidelines. The protocols have recommended starting intensities based on a person's perceived maximal lift and parameters for increasing the load on successive attempts. Using standardized assessment protocols allows comparison to test norms to see results relative to peers of the same age and sex and for reliable pre- and posttest assessments. Regression equations can be used to estimate 1RM from submaximal lifts that might be relevant for novices or other people with higher risk of injury in establishing maximal muscular strength.

KEY POINT

An individual's fitness level and the intensity and duration of an activity determine which metabolic energy system dominates during exercise.

The immediate need for increased ATP at the start of exercise is met with an anaerobic system that uses creatine phosphate to help generate ATP. The **ATP-PC system** (or phosphagen system) dominates during high-intensity, short-duration activities and when quick bursts of energy are needed such as a breakaway in basketball, a hill climb while cycling, a pole vault, or lifting weights.

The anaerobic component of the glycolytic pathway or fast glycolysis dominates in activities lasting 30 to 90 seconds. This system helps in providing energy for the performance of a 400-meter (437 yards) run or a period in wrestling. This pathway uses muscle glycogen (stored glucose) and blood glucose to rapidly produce ATP, although in limited quantity. As activity exceeds 2 minutes, the aerobic ATP-generating systems will dominate in producing a larger amount of ATP, although at a slower rate, helping activities in this time frame be sustained for a longer duration. Long-distance running, swimming, and cycling are supported by aerobic ATP production. Carbohydrate and fat are the primary energy nutrients in aerobic ATP production. The aerobic pathway includes glycolysis, the Krebs cycle, and the electron transport system (oxidative phosphorylation). In a separate aerobic process, beta-oxidation, fatty acids are used to generate ATP. Proteins, composed of amino acids, serve as the building blocks of the body while playing a small role in ATP production. It is estimated that amino acids contribute 2% to 4% of the total energy substrate (Brooks, 2012). Amino acids can be converted to glucose in the liver or to metabolic intermediates, which are compounds that directly participate in bioenergetics.

Exercise scientists apply their knowledge of the energetic pathways used in producing energy for physical activity to design sport-specific training programs. A training program that emphasizes activities powered by the ATP-PC system will not be beneficial in improving performance of longer-duration activities in which aerobic ATP production dominates. Coaches, personal trainers, and strength and conditioning specialists need to evaluate the metabolic demands of physical activities to identify the dominant energy-producing system to design effective training programs (see figure 5.2).

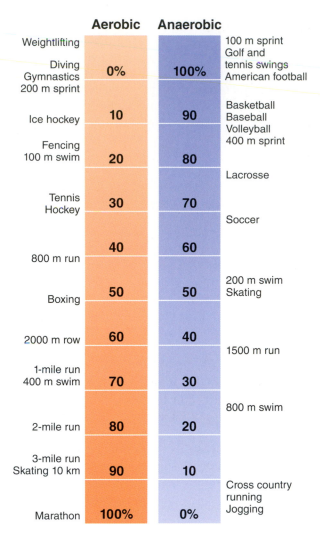

Figure 5.2   Identifying the dominant energy systems used in a given sport allows training programs to be designed to target the specific needs of the athletes.

Cardiovascular System

The heart and blood vessels of the cardiovascular system allow for the movement of blood (and oxygen), hormones, and some waste products throughout the body. To increase the supply of oxygen to skeletal muscles, the cardiovascular system responds immediately to physical activity by increasing the volume of blood pumped, increasing the distribution of blood flow to the active muscles, and directing blood flow away from less active areas. In addition, recurring physical activity results in several physiological adaptations in cardiac function that improve exercise endurance.

Cardiac Output

Cardiac output, defined as the amount of blood pumped out of the heart each minute, is a function of heart rate (number of beats per minute) and **stroke volume** (amount of blood pumped per beat). Resting cardiac output in adults is approximately 4.5 liters for a 50-kilogram female (4.8 qt for a 110 lb female) to 5 liters for a 70-kilogram male (5.3 qt for a 154 lb male) per minute, regardless of training status (Powers et al., 2021). During physical activity, cardiac output rises to increase the delivery of oxygen to active tissues. Oxygen uptake (written as $\dot{V}O_2$), or the amount of oxygen used by muscle tissues, increases in direct proportion to the intensity of exercise until $\dot{V}O_2$max is reached.

At lower exercise intensities, increases in cardiac output result from increases in both heart rate and stroke volume. As your exercise intensity increases above 40% of $\dot{V}O_2$max, your stroke volume plateaus as you reach the limit of how much blood your heart can pump per beat. Your heart rate, however, continues to increase along with oxygen uptake until you reach your maximal heart rate (see figure 5.3). Thus, the capacity of the heart to pump blood—the maximal cardiac output—appears to be one of the primary factors limiting maximal oxygen uptake and maximal exercise intensity. The linear relationship between heart rate and oxygen consumption allows exercise physiologists to use submaximal exercise protocols to estimate $\dot{V}O_2$max. Submaximal assessments are useful in exercise testing because they do not require the client to exercise to maximal capacity, which is appealing to many, and not as much equipment is needed.

Blood Flow Distribution

At rest, the majority of the cardiac output is directed toward your brain and internal organs (e.g., liver and kidneys). Because the oxygen demand of the skeletal muscles is low at rest, they receive only 15% of the cardiac output. When you increase your activity level, your muscles need increased blood flow because they have an increased need for oxygen, nutrients, and to remove waste products. At the onset of activity, your blood vessels constrict (vasoconstrict) in regions of your body that need less blood flow and dilate (vasodilate) in your active skeletal muscles and heart. This shift in distribution moves blood away from your kidneys and digestive organs to allow up to two-thirds of your blood to flow to your skeletal muscles during high-intensity exercise.

Cardiovascular Adaptations to Training

Your ability to exercise at moderate to high intensities for prolonged periods is referred to as your *aerobic* or *cardiovascular* endurance. You have learned that the gold standard for assessing your aerobic endurance is your $\dot{V}O_2$max. Through endurance training, you can increase your $\dot{V}O_2$max by 2% to 50% or 15% on average. Much of the improvement in your $\dot{V}O_2$max results from an increase in stroke volume as the ventricles of the heart increase in size and contract more forcefully. The increase in stroke volume is observed at rest, during submaximal exercise, and during maximal exercise. Resting and submaximal heart rates decrease following endurance training, as you can assess using a heart rate monitor or the heart rate function on your watch. In other words, because of endurance training, your heart can pump more blood with fewer beats.

To determine the effectiveness of an endurance training program or training programs targeting any other component of physical fitness, fitness testing should be conducted before and after the program. This approach establishes a baseline fitness level and allows you to assess changes due to the training program. One way to check the efficacy of your training program is to conduct a graded exercise test (figure 5.4), in which the intensity increases progressively. At each stage of the test, you should notice that your heart rate is lower than it was at that same stage (or level of intensity) before you started the training program.

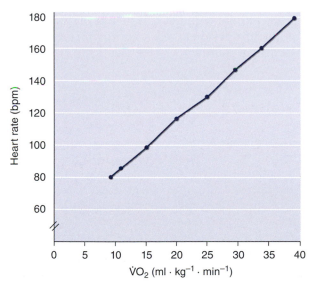

Figure 5.3 Relationship between heart rate and $\dot{V}O_2$ during a graded exercise test.

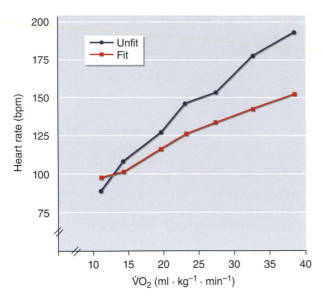

Figure 5.4   Comparison of the heart rate responses of trained and untrained people during a graded exercise test.

Your endurance training regimen will also increase the number of capillaries in your cardiac and skeletal muscles. This results in increased blood flow and oxygen delivery to the active muscles. The increased density of capillaries also slows the blood moving past the muscles, allowing greater time for oxygen to be extracted. Blood pressure at rest and during exercise will also decrease, lowering the stress on the heart.

Despite your best efforts at fitness training, you might find that others remain ahead of you in gains, including some who seem to do so with less effort and training. This point brings us to a discussion of other factors that affect maximal oxygen uptake. One is genetics. Studies of the $\dot{V}O_2$max of identical and

fraternal twins have shown greater variation between fraternal twin pairs than between identical twin pairs (Bouchard et al., 1986). In other words, individuals with identical genes (e.g., identical twins) are more alike in $\dot{V}O_2$max, thus indicating a strong genetic contribution to maximal aerobic capacity.

Another factor that influences $\dot{V}O_2$max is age. Maximal oxygen uptake begins to decrease after age 30 years because of a decrease in maximal heart rate and therefore in maximal cardiac output. The decline in $\dot{V}O_2$max with aging also can result, in part, from a decrease in physical activity. Promisingly, researchers have documented that maintaining high levels of endurance exercise slows the age-related decline in $\dot{V}O_2$max (Valenzuela et al., 2020).

Respiratory System

While the cardiovascular system transports oxygen to skeletal muscles, the respiratory system regulates the exchange of gases, including oxygen and carbon dioxide, between the external environment (air) and the internal environment (inside the body). Inspired air moves from the nasal cavity or mouth through the respiratory passages to the alveoli in the lungs. In the alveoli, oxygen diffuses into the blood in the pulmonary capillaries, and carbon dioxide leaves the blood and is expired from the body. The process of moving air in and out of the lungs is known as **ventilation**. The amount of air exhaled per minute is the minute volume ($\dot{V}_E$); it is the product of the amount of air exhaled per breath (**tidal volume**) and the number of breaths per minute (breathing rate). This is the respiratory equivalent of cardiac output (stroke volume × heart rate). The minute volume of a person at rest is approximately 6 liters (6.3 qt) per minute.

Determining Your Maximum Heart Rate

You might be familiar with the equation 220 – age for predicting maximum heart rate. Population-specific equations now exist for estimating maximum heart rate that might provide superior estimates. If you are working with a group of adult males and females with a wide range of ages and fitness levels, Gellish and colleagues (2007) recommended the equation 207 – (0.7 × age). Gulati and associates (2010) recommended the equation 206 – (0.88 × age) for middle-aged women who do not have symptoms who are referred to stress testing (where heart rate and rhythm and blood pressure are monitored during exercise). Continued research will provide more accurate population-specific equations. It is important to remember that all estimates of maximum heart rate have a degree of error. Because heart rate can be over- or underestimated, it is necessary to always observe how your client responds to exercise programs in which intensity is set using an estimated maximum heart rate.

At the beginning of physical activity, your breathing rate increases in response to stimuli from sensory **receptors** in the moving limbs (e.g., muscle spindles and joint receptors) as well as the motor cortex (i.e., the part of the brain that stimulates muscles to contract). As exercise intensity progressively increases, ventilation will rise linearly. At low exercise intensities, the increase in minute volume results primarily from an increase in the tidal volume while the number of breaths each minute remains constant. At higher exercise intensities, increases occur in both tidal volume and respiratory frequency, leading to bigger and more frequent breaths.

At higher intensities of exercise, your breathing rate will continue to increase (figure 5.5). The point at which ventilation deviates from a linear increase and your breathing frequency begins to increase rapidly is the **ventilatory threshold**. This threshold occurs at exercise intensities between 50% and 75% of $\dot{V}O_2$max. You have likely experienced the feeling of this disproportionately higher breathing frequency. With endurance training, the ventilatory threshold will occur later (at a higher exercise intensity or higher percentage of $\dot{V}O_2$max). Training at or a little above the ventilatory threshold intensity improves aerobic capacity (Ghosh, 2004). Distance running performance is highly correlated with the intensity at which the ventilatory threshold occurs (Ghosh, 2004), so this is an area of interest for many researchers and coaches.

> ### KEY POINT
> During high-intensity exercise, an increasing volume of air is moved in and out of the lungs, first by breathing deeper and then by breathing faster.

Variation in Temperature and the Response to Physical Activity

In addition to studying the primary body systems affected by physical activity, exercise physiologists study the influence of environmental factors on our experience of sport and exercise. One such factor that affects the physiological response to exercise is temperature variation. Our bodies have complex and effective ways of dealing with temperature changes in the outside environment. Through thermoregulation, humans can regulate their internal (core) temperature so that it remains relatively constant over a wide range of environmental temperatures. When you begin to exercise, skeletal muscle contractions produce heat and cause body temperature to rise. In comfortable ambient temperatures, your body core temperature reaches a plateau in 20 to 30 minutes (figure 5.6). The higher the intensity of your physical activity, the higher your core temperature will be when it reaches a plateau.

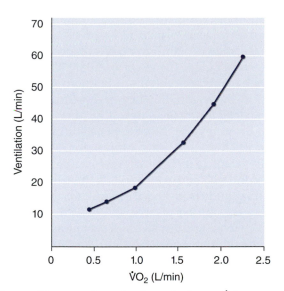

Figure 5.5 Relationship of ventilation to $\dot{V}O_2$ as exercise intensity increases progressively during a graded exercise test.

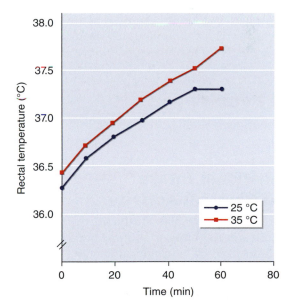

Figure 5.6 Increases in body core (rectal) temperature during exercise in comfortable ambient temperatures (77 °F, or 25 °C) and warm ones (95 °F, or 35 °C).

The main mechanism for losing heat during exercise is the evaporation of sweat from the skin surface. Have you ever wondered why you sweat more profusely during physical activity in warm environments? When air temperature approaches or exceeds skin temperature, the body's cooling mechanism of bringing warm blood to the surface is less effective. In this case, the major avenue of heat loss is the evaporation of sweat; however, this method is limited in hot, humid environments. In this instance, water and salt are lost from the body without cooling the body. This results in increased risk of heat illness (e.g., heat exhaustion and heatstroke) and decreased exercise capacity. The ability to regulate body temperature effectively while exercising in the heat is influenced by several factors including a person's level of cardiovascular fitness, level of hydration, and state of acclimatization to heat. Higher levels of each increase the ability to regulate core body temperature while exercising in the heat.

While we tolerate physical activity in cool environments better by losing heat from blood close to the skin as well as from the evaporation of sweat, cold temperatures (especially in water) also can limit physical performance. In cold water, heat loss from the skin increases dramatically because water is an excellent conductor of heat. Submersion in cold water stimulates shivering and an increase in **metabolic rate** (the rate at which the body uses energy). Swimming in cold water also elevates the metabolic rate, which can lead to an earlier onset of fatigue. Similarly, people whose clothing becomes wet during physical activity in cold environments lose heat more rapidly and are at risk of **hypothermia** (below-normal body temperature).

KEY POINT

Sweating is ineffective at cooling the body during exercise in high heat and humidity.

Nutritional Intake and Physical Activity

Exercise physiologists study nutritional intake because our responses to training and performance are influenced by biochemical energy sources. As a result, interdisciplinary research that blends exercise physiology with dietetics leads to knowledge that improves personal health and supports high-level sport performance.

The three main energy nutrients in foods are carbohydrate, fat, and protein. While most of us have enough stored fat to sustain low-intensity activities for many days, we need carbohydrate to fuel higher-intensity endurance activities. Because the amount of carbohydrate stored in the body is less than 1 pound (0.45 kg), daily carbohydrate intake is crucial to optimal athletic performance. It is also recommended that all athletes consume 1.2 to 2.0 grams of protein per kilogram (0.02-0.03 oz/lb) of body mass (Academy of Nutrition and Dietetics et al., 2016). The recommendations for dietary fat intake do not differ between physically active people and the general population. The following sections highlight the roles of carbohydrate, protein, water, and iron in performing physical activity.

Carbohydrate

Carbohydrate is stored as glycogen primarily in the skeletal muscles and the liver. Normal muscle glycogen stores are depleted in about 90 minutes of continuous exercise at 75% of $\dot{V}O_2$max. The amount of glycogen stored is directly related to the carbohydrate content of one's diet. Coaches and endurance athletes know that muscle glycogen storage is increased by a diet that is high in carbohydrate (70% or more of total calories) consumed for 2 or 3 days. Considering this fact, some endurance athletes use a technique known as *carbohydrate loading* before competition. Carbohydrate loading is most effective for activities that last more than 2 hours (e.g., marathons) and offers no apparent advantage in activities that last less than 1 hour.

Maintaining carbohydrate levels during prolonged (> 60 minutes) vigorous exercise is important because low blood glucose is associated with fatigue. Consumption of carbohydrate drinks containing 6% to 8% carbohydrate by weight during prolonged exercise has been shown to improve performance and delay the onset of fatigue (Sawka et al., 2007). Drinks containing a higher concentration of carbohydrate (more than 10% by weight), however, delay gastric emptying and can lead to gastric distress (Clayton et al., 2014). Glycogen stores also can be depleted by intermittent physical activity and sport participation. Soccer players often deplete their muscle glycogen stores during the second half of a game (Hills & Russell, 2017). In a seminal study using muscle biopsies, Saltin (1973) showed that soccer players with depleted glycogen stores covered less distance and spent less time running after halftime. Generally, if a vigorous activity or exercise session lasts longer than 60 minutes, a sport drink is recommended to supply carbohydrates for energy (and minerals to maintain electrolyte levels) to help sustain performance.

In addition, athletes with low carbohydrate intake (less than 50% of total calories) can progressively deplete their glycogen stores through daily training. To maintain adequate glycogen stores, physically

Energy gels are an easy way to ingest carbohydrate during intense, prolonged (> 60 minutes) exercise.

Peter Mundy/Getty Images

active individuals should consume a diet in which carbohydrate accounts for 55% to 60% of total calories. It is also helpful to eat foods high in carbohydrate immediately following exercise to increase the rate of muscle glycogen storage (Ivy et al., 1989).

Protein

The recommended dietary allowance (RDA) for protein is 0.8 gram per kilogram (0.01 oz/lb) per day. Athletes who perform high-intensity training and competition have elevated protein needs to support repairing and replacing damaged proteins; maintaining the function of metabolic pathways; supporting increases in lean tissue; and aiding the adaptations of bone, tendons, and ligaments to physical activity. As noted, the recommended range of protein consumption during training regimens ranges from 1.2 to 2.0 grams of protein per kilogram (0.02-0.03 oz/lb) of body mass. The higher end of the range is recommended during periods of increased training frequency or duration and when starting new training programs (Academy of Nutrition and Dietetics et al., 2016).

Even though protein requirements are elevated during high-intensity endurance training and strength training, athletes do not automatically need to consume additional protein. In fact, the majority of the population in the United States already meets or exceeds the recommended intake for protein (Berryman et al., 2018). Regardless of the type of diet an athlete consumes, planning is required to ensure that the nutritional intake is meeting the person's health and physical activity needs. No one method of eating will work for every person or every athlete. Therefore, an athlete should meet with a dietitian before revising their diet.

Fluid Intake

Water makes up 55% to 60% of the human body. During physical activity, you lose water through sweating. When you exercise on a warm day, even just walking to class, you will notice that you sweat more than on cooler days. Sweat loss decreases body fluids both within and between cells; it also decreases **plasma volume** (the fluid portion of blood). If fluid is not consumed to replace the water lost through sweat, dehydration will result in a rise in body temperature, leading to greater risk of heat exhaustion and heatstroke. When plasma volume decreases, less blood returns to the heart, which reduces stroke volume. To compensate, heart rate increases to maintain cardiac output. Also, as plasma volume is lost, the blood becomes more viscous (thicker, increasing resistance to flow), which requires the heart to contract with greater force to move the blood through the blood vessels. This change leads to an increase in blood pressure and increases strain on the circulatory system.

KEY POINT

Inadequate fluid intake during and after exercise results in elevated body temperature, greater risk of heat illness, and decrements in performance.

Dehydration is less likely to affect short, high-intensity activities such as sprinting, but failure to replace fluids adequately can decrease performance in endurance activities. It is generally accepted by researchers that endurance performance is impaired beginning at a 2% decrease in body weight due to fluid loss (James et al., 2017). To ensure adequate fluid replacement during physical activity, drink fluids (e.g., water or carbohydrate-electrolyte drink) at regular intervals (every 15 to 20 minutes). During activities lasting less than 1 hour, consuming carbohydrate-electrolyte drinks offers no advantage over plain water. In more prolonged activities, fluids containing a small amount of carbohydrate (6-8%) are more likely than water to be beneficial in enhancing fluid absorption and maintaining blood glucose. Fluid replacement can be difficult to accomplish during play in certain sports, such as soccer and field hockey, where the rules do not allow time-outs during competition except for injury. Thus, players in these sports should drink additional fluid before the contest and during halftime to reduce the fluid deficit. Following activity, it is important to continue to drink fluids even without a feeling of thirst, because thirst is not an accurate indicator of the need for fluid. Organizations such as ACSM, National Athletic Trainer's Association (NATA), and NSCA provide guidelines to athletes and coaches for fluid consumption before and fluid replacement during and following exercise.

Iron Intake

Iron is used to produce **hemoglobin** and **myoglobin**. Hemoglobin is a protein in red blood cells that allows oxygen to be transported from the lungs to areas around the body. Each hemoglobin molecule contains four iron atoms and binds with four oxygen molecules. Myoglobin is a protein involved in the transport of oxygen within muscles that contains a single heme complex and carries one molecule of oxygen.

A person with low iron has a decreased ability to transport oxygen that negatively affects exercise performance. **Anemia** is defined as a hemoglobin concentration below 12 grams per deciliter of blood in women and below 13 grams per deciliter in men. People with anemia experience reductions in $\dot{V}O_2$max and in cardiovascular endurance (Pasricha et al., 2014). Iron deficiency is one of the most common nutritional deficiencies in the United States, especially among adolescent girls and women. The most likely causes of iron depletion in physically active women are inadequate iron intake and excessive blood loss through menses. Heme iron found in meat, fish, and poultry is more highly absorbed from the gastrointestinal tract than is the non-heme iron found in other food sources. Consuming iron-rich foods daily can help avoid iron deficiency.

KEY POINT

Anemia decreases the ability to deliver oxygen to the skeletal muscles during exercise and reduces $\dot{V}O_2$max and cardiovascular endurance.

Physical Activity Guidelines

High-intensity exercise is not necessary to gain health benefits from physical activity. A large body of scientific evidence (e.g., Haskell et al., 2007) indicates conclusively that health benefits are obtained by engaging in moderate-intensity physical activity on a regular basis. Moderate-intensity physical activities include brisk walking (3 to 4 mph, or 4.8 to 6.4 km/h), playing doubles tennis, cycling less than 10 miles per hour (16 km/h), swimming at moderate speed, golfing (pulling a cart), climbing stairs, and mowing the lawn.

The Physical Activity Guidelines for Americans (USDHHS, 2018) are based on a systematic review of the literature and provide physical activity recommendations. The guidelines emphasize any amount of physical activity is valuable in improving health, encouraging greater movement and less sitting throughout the day. To maximize the many health benefits associated with movement, the recommended amount of physical activity for adults is at least 150 to 300 minutes a week of moderate-intensity activity or 75 to 150 minutes a week of vigorous-intensity aerobic physical activity. Combinations of moderate and vigorous activity also can be used to meet the guideline. In addition to aerobic activity, muscle-strengthening activities for the major muscle groups can provide additional health benefits if performed two or more days a week at a moderate or greater intensity. Greater volumes and intensities of physical activity and especially exercise can confer additional health benefits.

KEY POINT

Even small increases in movement time and decreases in sitting time can improve health.

The United States government is not alone in addressing physical activity at the population level. Many countries and organizations (such as the World Health Organization [WHO]) generate science-based recommendations for physical activity and exercise. Physical inactivity is a global concern. A pooled analysis of 96% of the global population indicated more than 25% of adults do not perform enough physical activity (at least 150 min moderate intensity, 75 min vigorous intensity, or an equivalent combination each week), putting 1.4 million adults at risk of hypokinetic diseases (Guthold et al., 2018). Additionally, worldwide, three in four adolescents (aged 11 to 17 years) do not meet this recommendation (WHO, 2018). This is despite 70% of the 168 countries in the Guthold and colleagues (2018) study having a national policy on physical activity.

The global economic burden of physical inactivity is substantial. Direct and indirect (due to lost productivity) health care costs were estimated at US$67.5 billion in 2013 based on information from 142 countries, representing 93.2% of the world's population (Ding et al., 2016). The WHO (2018) has a global action plan targeting a 15% global reduction in physical inactivity for adults and adolescents by 2030.

Exercise Guidelines

Performance-related fitness benefits also can be achieved by performing structured bouts of exercise. To improve the cardiovascular endurance component of fitness, the ACSM (2022) recommends that adults take part in exercise 30 to 60 minutes per day on 5 days per week at moderate intensity or 20 to 30 minutes per day (continuously) on 3 days per week at vigorous intensity. Exercise intensities of 64% to 76% of maximal heart rate are labeled as moderate exercise, and exercise intensities of 77% to 93% of maximal heart rate are considered hard or vigorous exercise. Activities that are aerobic and can be maintained for prolonged periods are best for improving cardiovascular endurance. These activities include running, hiking, walking, swimming, cross-country skiing, bicycling, stair climbing, and rowing. Cardiovascular fitness also can be aided by participation in sports—such as soccer, lacrosse, and tennis—that involve high-intensity, intermittent activities carried out over prolonged periods.

To minimize risk of injury, exercise programs designed to improve fitness and health should begin

Hot Topic

Exercise Guidelines for People Who Have Survived Cancer

With improved detection and treatment methods, the number of people who survive cancer is steadily increasing. More than 15.5 million people in the United States have survived cancer, and, over the next 20 years, this number is expected to double (Siegel et al., 2019). The ACSM International Multidisciplinary Roundtable on Physical Activity and Cancer Prevention and Control compiled evidence-based guidelines for exercise testing, prescription, and delivery for people who have survived cancer (Campbell et al., 2019). Current evidence supports specific exercise prescriptions for decreasing anxiety, depressive symptoms, and fatigue, and improving health-related quality of life and physical function. These prescriptions adjust the four training variables used in exercise prescription: frequency, intensity, time, and type (FITT). Aerobic training at moderate intensity is encouraged 30 minutes, 3 times per week for 8 to 12 weeks. Resistance training is further advised two times per week at 60% of 1RM, including two sets of 8 to 15 repetitions (Campbell et al., 2019). In addition, the risk of cancer-specific and all-cause mortality is specifically reduced in those with early-stage breast, colorectal, and prostate cancers who perform physical activity (Patel et al., 2019). Along with specific characteristics of their clients including cancer type and stage, treatment, and side effects, this information serves as a guideline for fitness professionals to assist people who have survived cancer.

with low- to moderate-intensity activities and gradually increase in intensity and duration of activities as fitness improves. This approach is especially important when working with individuals who are new to fitness programs. For example, an exercise program for sedentary adults might begin with walking at a speed of 3 to 4 miles per hour (4.8-6.4 km/h), which is close to the moderate intensity of 64% of maximal heart rate, and gradually increase in duration from 30 minutes to 60 minutes. If 30 minutes is too long, smaller bouts can be used to build up exercise tolerance. People interested in jogging can begin by walking, progress to alternating between walking and jogging, and then advance to jogging.

Effects of Age on Fitness

With the rapid increase in the older segments of the population, it is important to understand how age affects cardiovascular fitness. This is especially important because a decline in $\dot{V}O_2$max in older individuals can negatively affect the ability to perform everyday tasks and to live independently (Paterson et al., 2004). Although research has shown that remaining physically active slows the decline in $\dot{V}O_2$max that occurs with age (Valenzuela et al., 2020), athletic performance still declines with increasing age (Ganse et al., 2018). Figure 5.7 illustrates the decrease in world record marathon time across age groups. Age-related reductions in the ability to deliver blood and oxygen to active skeletal muscle and the capacity of the active skeletal muscle to use the oxygen it receives have been associated with the decline in $\dot{V}O_2$max (Betik & Hepple, 2008).

Aging (growing) at the younger end of the age spectrum also influences $\dot{V}O_2$max. As children grow, more oxygen is required to supply active tissues during physical activity. When expressed in absolute terms (liters per minute), $\dot{V}O_2$max increases with growth in children. However, when $\dot{V}O_2$max is expressed relative to kilograms of body mass, it remains relatively constant or decreases during growth as children increase in size and gain body fat. With children, the term $\dot{V}O_2$ *peak* is used to describe aerobic fitness, representing the highest value attained in an exercise test conducted to exhaustion. As boys age from 10 to 18 years old, the highest $\dot{V}O_2$ achieved in liters per minute increases linearly. For girls, $\dot{V}O_2$ peak rises steadily to age 13 to 14 years (Armstrong & Welsman, 2019).

Benefits of Physical Activity and Fitness

In general, higher levels of fitness are associated with improved health status. This generally held hypothesis is based on research conducted by exercise physiologists. Nystoriak and Bhatnagar (2018) emphasized in their review that more frequent exercise is strongly associated with a lower risk of developing and dying from cardiovascular disease. Elagizi and associates (2020) characterized cardiovascular fitness as the "single most important" predictor of overall health.

Distinct from one's level of physical fitness, a direct relationship also exists between increasing levels of physical activity and decreased occurrence and death from cardiovascular disease (Kesaniemi et al., 2001). For instance, Kubota and colleagues (2017) found

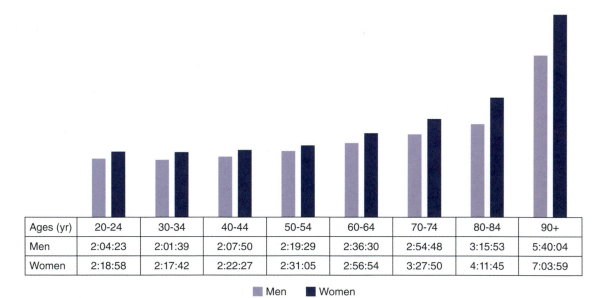

Ages (yr)	20-24	30-34	40-44	50-54	60-64	70-74	80-84	90+
Men	2:04:23	2:01:39	2:07:50	2:19:29	2:36:30	2:54:48	3:15:53	5:40:04
Women	2:18:58	2:17:42	2:22:27	2:31:05	2:56:54	3:27:50	4:11:45	7:03:59

■ Men   ■ Women

Figure 5.7   World record marathon times decrease with increasing age.

more active individuals had a lower lifetime risk of developing cardiovascular disease than those less active. In fact, in a review of research, Reimers and colleagues (2012) found an increase in life expectancy for physically active participants ranging from 0.43 years to 6.9 years.

KEY POINT
Higher levels of physical fitness and physical activity lower the risk of developing and dying from cardiovascular disease.

Physical activity and fitness also play a role in reducing the risk of hypertension. Interestingly, the evidence supports lower-intensity exercise programs (closer to 50% of $\dot{V}O_2$max) as being more effective in reducing blood pressure than those of higher intensity (Fagard, 2001; Kesaniemi, 2001). Endurance training will lead to reductions in resting blood pressure in those diagnosed with hypertension and in those who have normal blood pressure readings. Expected reductions in systolic and diastolic blood pressure range from 5 to 7 mmHg (millimeters of mercury; Pescatello et al., 2004). While this might not initially appear to be a large change, it is important to combine this knowledge with an understanding that decreases in blood pressure as small as 2% are associated with reductions in stroke and coronary artery disease (Pescatello et al., 2004).

The health benefits of physical activity are wide ranging and not only encompass improvements related to physiological function and disease risk, but also benefit quality of life. Moderate-intensity physical activity improves sleep quality such as sleeping longer while in bed, falling asleep faster, and waking up less in all age groups in healthy populations (Wang & Boros, 2021). Symptoms of depression and anxiety are also improved with participation in physical activity. Benefits specific to those who are 65 years of age and older include an improved sense of well-being and the promotion of independent living (Kesaniemi et al., 2001).

KEY POINT
Regular physical activity improves physical and emotional health as well as healthy longevity.

Brain health, related to optimal functioning, is a new area of emphasis within exercise physiology. The Physical Activity Guidelines for Americans (USDHHS, 2018) highlighted evidence of the ben-eficial relationship of exercise with brain health for conditions such as Alzheimer's, cerebral palsy, autism spectrum disorder, and Down syndrome. Based on the available evidence, the ACSM (2022) published guidelines and recommendations for exercise professionals who assess fitness and prepare exercise programs for people who have a condition related to brain health. In addition to overall health benefits, additional benefits are related to the unique characteristics of each condition. For people who have Alzheimer's disease, exercise increased functional ability in the early stages of the disease (Morris et al., 2017). Self-efficacy was increased in people who have Down syndrome (Jo et al., 2018), and time on task (Wong et al., 2015) and motor skill development (Healy et al., 2018) improved in people with autism spectrum disorder. The progression of the motor symptoms of Parkinson's disease was slowed by exercise participation (Schenkman et al., 2018). This is an exciting area of emphasis that has widespread implications for the application of exercise to improve human performance and quality of life.

Keep on Stepping!
Higher daily step counts are associated with lower mortality risk from all causes, regardless of the intensity at which the steps are taken. If your goal is to improve your health, focus on taking more steps per day, regardless of how hard you are working. Many tools such as applications on your phone or fitness watch can help you track your daily steps. In addition to an exercise program, you can make small changes in your daily routine to affect your step count, such as not taking the closest parking spot, not taking the shortest route to class, or planning a short walk after each meal.

Exercise and Weight Control
Participation in daily physical activity helps control body weight. Weight gain and weight loss are determined by the interplay between the consumption and the expenditure of calories. Exercise physiologists have added to our understanding of caloric expenditure by investigating the process by which the body expends calories and uses energy. The metabolic rate—the rate at which the body uses energy—increases in direct proportion to the intensity of activity. Lower-intensity activities such as walking at 3 miles per hour (4.8 km/h) increase the metabolic rate threefold, to a caloric expenditure of about 4 calories per minute. Higher-intensity activities such running at 7 miles per hour (11 km/h) increase the metabolic rate tenfold, expending 12.5 calories per

minute. Total energy expenditure also depends on the duration of physical activity.

Moreover, energy expenditure does not return to resting levels immediately after an exercise bout. Rather, researchers have shown that the metabolic rate remains elevated during recovery from physical activity and exercise. Thus, the total energy expended due to a single bout of exercise is somewhat greater than the energy cost of the activity itself, especially following high-intensity, longer-duration activity.

KEY POINT

Exercise can lower body weight and reduce risk of the comorbidities associated with excess body fat, including coronary heart disease, non-insulin-dependent diabetes mellitus, and hypertension.

Whereas some people seek to maintain body weight through exercise, others want to lose body weight and body fat. When controlling caloric intake, a direct relationship exists between how much exercise is performed and weight loss, with more activity leading to greater reductions in weight (Kesaniemi, 2001). The recommended exercise dose for weight loss is approximately 150 to 250 minutes of weekly physical activity (Donnelly et al., 2009). When caloric intake exceeds caloric expenditure, the excess calories are converted to fat and stored in adipose tissue. Sedentary people are more likely to be overweight or obese, a condition that puts them at increased risk of developing coronary heart disease (Willett et al., 1995); they are also at higher risk of developing non-insulin-dependent diabetes mellitus [NIDDM] (Park, 2020). Reducing body weight is effective in lowering serum triglycerides and reducing blood pressure in those with hypertension. Furthermore, participation in daily physical activity is beneficial not only in reducing the risk of NIDDM, but also in helping those with this condition regulate their blood glucose levels.

Understanding the relationship between exercise and body weight is critical to helping exercise physiologists address the increasing rates of obesity in society. Large-scale population studies have shown that obesity has increased dramatically in the United States since the late 1990s. These population-based investigations often use the field measure presented earlier in the chapter, BMI, as a surrogate measure of obesity. This ratio is calculated using a person's body mass and height:

BMI = body mass in kilograms
÷ the square of height in meters

While BMI is not an assessment of body composition, it is an easy and noninvasive surrogate measure, and elevated values are indicative of obesity-related disease risk. A BMI of more than 30 kilograms per square meter is classified as obesity, and the percentage of adults in the United States classified as obese has increased steadily. In 2017 to 2018, for instance, the age-adjusted rate had reached 42.4% of adults in the United States, with no difference between males and females (Hales et al., 2020). The problem of obesity is also apparent in younger segments of the U.S. population (Hales et al., 2017). The prevalence of obesity in U.S. children aged 2 to 5 years in 2015 to 2016 was 13.9% (Hales et al., 2017). The rates are even higher for children aged 6 to 11 years (18.4%) and for adolescents aged 12 to 19 years (20.6%).

Targeting obesity with physical activity and exercise interventions also has potential to address other chronic diseases that occur with greater frequency in persons who are obese. Obese adults have a higher risk of developing heart disease, hypertension, and NIDDM, diseases that also are associated with physical inactivity. Impaired glucose tolerance, a preliminary stage in the development of NIDDM, was found in 25% of obese children and 21% of obese adolescents (Sinha et al., 2002). Overweight and obesity also are related to an elevated risk of some forms of cancer in adults, including endometrial (inner lining of the uterus) and breast cancer in women and colon cancer in both women and men.

While physical activity and exercise interventions can offer uniquely widespread health benefits leading to improvements in a variety of disease states, a challenge remains in how to get more people to participate regularly in physical activity. Only half of adults in the United States regularly participate in 150 minutes of moderate-intensity or 75 minutes of vigorous-intensity physical activity per week, and about one in four adults participates in no leisure-time physical activity (U.S. Department of Health and Human Services et al., 2015). Women are less likely than men to participate regularly in physical activity, and older adults are less likely to do so than younger adults. Several strategies have been proposed for increasing daily physical activity, including increasing the amount of physical activity in physical education classes; providing more opportunities for children and adolescents to participate in physical activity before, during, and after school; building walking and bicycle paths that are separated from roadways; and increasing opportunities for adult participation in physical activity at worksites, in shopping malls, and in community facilities. Exercise scientists use the body of scientific literature to help address these challenges.

Informing Practice and Research in Exercise Physiology

Due to the dynamic nature of science and knowledge, it is important to be an avid consumer of research to find the best evidence available to answer questions in exercise science. By combining practice knowledge (professional experience) with knowledge obtained from peer-reviewed publications, exercise physiologists can optimize program and intervention designs.

Throughout this chapter, you have learned the value of physical activity and exercise to the health and fitness of different segments of the population. Based on your interests, you can apply your exercise physiology knowledge in many settings. The following are examples of how available research evidence can be used to make practice decisions within exercise physiology.

Fall Prevention in the Elderly

When working with the elderly populations, it is important to know that an estimated one-third of community-dwelling adults (≥ 65 years) in the United States experience falls each year. Approximately 10% of these falls result in serious injury (Sherrington et al., 2019). Exercise scientists have studied exercise interventions to address this public health concern. In 2019, 19,684 people participated in 81 randomized controlled studies (Sherrington et al., 2019). Evidence from these studies teaches us that exercise programs can reduce the rate of falls by 23%. The existing pool of research on this topic can also help you understand how to best design a fall-intervention program for older adults. For reducing fall risk, the most effective programs include a combination of balance and functional exercises (Sherrington et al., 2019). Tai chi also aids in fall reduction. Intervening early (primary prevention), even before individuals are at increased risk of a fall, is also important. Programs can be effective when offered in group settings as well as to individuals and produce the most benefits when offered for 12 weeks or more. You can combine this information with your available resources (space and equipment), your expertise, and the specific characteristics of the group you are working with (age, health status, fall history) to create a program to best serve your population.

Addressing Childhood Obesity

Childhood obesity is another issue targeted by exercise science professionals. In one review of what researchers have published on programs for children, 153 randomized controlled trials were conducted between June 2015 and January 2018 (Brown et al., 2019). As you read the information in this area, you will learn that when working with children, overweight is determined by calculating BMI. This score is generated by a child's height and weight when compared to other children in the same country of the same sex and age. Also, the younger population is generally evaluated in three age groups: 0 to 5 years, 6 to 12 years, and 13 to 18 years. Therefore, knowing the age of the children you will be working with is important when making programming decisions.

Physical activity or exercise interventions can have a modest effect in reducing weight in those aged 6 to 18 years (Brown et al., 2019), so this would be an ideal age range to intervene. In addition to leading to positive changes in body weight, we also know that physical activity habits developed early in life track into adulthood (Telama, 2009). Therefore, the benefits of physical activity interventions you help design and lead for children can have long-term individual and public health implications.

Exercise Following a Stroke

If you are working with older clinical populations, you are likely to interact with a person who has had a stroke. Annually, more than 795,000 people in the United States have a stroke (Virani et al., 2020). Your knowledge of exercise science can benefit this population, because more than 75% of individuals who survive a stroke are not performing the recommended amount of physical activity (American Heart Association, 2021). Existing evidence indicates using a variety of exercise modes offers the best outcomes to individuals who have had a stroke. Benefits such as improved walking speed and distance, improved balance, and reduced fall risk are maximized from the combination of cardiovascular and resistance training (Saunders et al., 2020). Cardiac rehabilitation currently is not reimbursed by insurance companies for recovery from stroke. However, researchers are beginning to accumulate evidence on the benefits of an exercise-based cardiac rehabilitation program for survivors of stroke (Regan et al., 2021). You could be one of the clinical exercise physiologists working to show how cardiac rehabilitation improves the quality of life for those who have had a stroke and helping build the case for insurance reimbursement.

Each of these examples highlights how research evidence from exercise physiology can be used to make evidence-based exercise prescriptions. In addition to evidence gleaned from the available scholarship on a topic, client and patient characteristics and existing resources (such as funding and infrastructure) must also be considered when determining best practices in specific environments and in different segments of society. Once programming or intervention decisions are made, exercise scientists can generate translational knowledge by evaluating these best practices in community, school, and clinical settings to assess efficacy and identify factors that promote or inhibit participation to guide implementation and policy decisions.

Wrap-Up

Exercise physiologists study the acute and chronic responses to physical activity and exercise. Throughout this chapter you acquired an overview of this subdiscipline. You learned how this knowledge can be used to help athletes achieve peak performance,

help people participate safely in physical activity, and conduct research on preventing and treating disease by means of physical activity and exercise.

The benefits that can be gained from regular participation in physical activity and exercise are diverse and can be applied to help people in many segments of the population, from old to young and from healthy to clinical. This provides a wide range of employment opportunities for people with bachelor's to graduate-level degrees within the public, private, and corporate sectors. To learn more about educational, research, and employment opportunities within exercise physiology, you are encouraged to visit the websites of the organizations listed here and become familiar with the journals in which exercise physiology advances are published.

More Information on Exercise Physiology

Organizations

American College of Sports Medicine

American Council on Exercise

American Physiological Society

American Society of Exercise Physiologists

Canadian Society for Exercise Physiology

Exercise & Sports Science Australia

National Strength and Conditioning Association

Journals

ACSM's Health & Fitness Journal

Applied Physiology, Nutrition, and Metabolism

Biology of Sport

British Journal of Sports Medicine

European Journal of Applied Physiology

Exercise and Sport Sciences Reviews

International Journal of Exercise Science

International Journal of Sports Physiology and Performance

Journal of Applied Physiology

Journal of Exercise Physiology Online

Journal of Physical Activity and Health

Journal of Physiology

Journal of Science and Medicine in Sport

Journal of Sports Medicine and Physical Fitness

Journal of Sports Sciences

Journal of Strength and Conditioning Research

Measurement in Physical Education and Exercise Science

Medicine & Science in Sports & Exercise

Pediatric Exercise Science

Research Quarterly for Exercise and Sport

Sports Medicine

Translational Journal of the American College of Sports Medicine

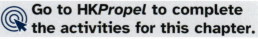 **Go to HK*Propel* to complete the activities for this chapter.**

Review Questions

1. Identify and describe the locations where exercise physiologists might be employed. Explain which opportunity is most appealing to you.

2. Explain how knowledge of exercise physiology can be used to help each of the following: a college athlete, a person trying to lose weight, and a person in the military.

3. Why are exercise physiologists focused on helping people increase physical activity and decrease sedentary time?

4. What are the positive and negative attributes to conducting exercise physiology research in the laboratory and in the field? Explain how one physiological variable that changes during exercise can be assessed in each setting.

5. What physical activities do you perform on a regular basis? What changes in your health or fitness would you expect to see based on the type of activity or exercise you are performing?

6. Explain how physical activity or exercise can improve the health or quality of life of an individual with a clinical condition of your choice.

Suggested Readings

American College of Sports Medicine (ACSM). (2022). *ACSM's guidelines for exercise testing and prescription*. Lippincott Williams & Wilkins.

Borg, G.A. (1982). Psychophysical bases of perceived exertion. *Medicine & Science in Sports & Exercise, 14*(5), 377-381.

Campbell, K.L., Winters-Stone, K.M., Wiskemann, J., May, A.M., Schwartz, A.L., Courneya, K.S., Zucker, D.S., Matthews, C.E., Ligibel, J.A., Gerber, L.H., Morris, G.S., Patel, A.V., Hue, T.F., Perna, F.M., & Schmitz, K. (2019). Exercise guidelines for cancer survivors: Consensus statement from international multidisciplinary roundtable. *Medicine & Science in Sports & Exercise, 51*(11), 2375-2390.

Ekelund, U., Steene-Johannessen, J., Brown, W.J., Fagerland, M.W., Owen, N., Powell, K.E., Bauman, A., Lee, I.M., Lancet Physical Activity Series 2 Executive Committee, & Lancet Sedentary Behaviour Working Group. (2016). Does physical activity attenuate, or even eliminate, the detrimental association of sitting time with mortality? *Lancet, 388*(10051), 1302-1310.

Kraus, W.E., Janz, K.F., Powell, K.E., Campbell, W.W., Jakicic, J.M., Troiano, R.P., Sprow, K., Torres, A., Piercy, L., for the 2018 Physical Activity Guidelines Advisory Committee. (2019). Daily step counts for measuring physical activity exposure and its relation to health. *Medicine & Science in Sports & Exercise, 51*(6), 1206-1212.

Moberg, M., Lindholm, M.E., Reitzner, S.M., Ekblom, B., Sundberg, C.J., Psilander, N. (2020). Exercise induces different molecular responses in trained and untrained human muscle. *Medicine & Science in Sports & Exercise, 52*(8), 1679-1690.

U.S. Department of Health and Human Services (USDHHS). (2018). *Physical activity guidelines for Americans* (2nd ed.). U.S. Department of Health and Human Services.

References

Academy of Nutrition and Dietetics, Dietitians of Canada, & American College of Sports Medicine. (2016). Nutrition and athletic performance. *Medicine & Science in Sports & Exercise, 48*(3), 543-568

American College of Sports Medicine. (2022). *ACSM's guidelines for exercise testing and prescription*. Lippincott Williams & Wilkins.

American Heart Association. (2021, January 27). *Exercise-based cardiac rehab added to stroke recovery improved strength, cardiac endurance*. American Heart Association. https://newsroom.heart.org/ news/ exercise-based-cardiac-rehab-added-to-stroke-recovery-improved-strength-cardiacendurance#:~:text=Despite%20many%20similar%20cardiovascular%20risk,than%2075%25%20of%20all%20U.S

Armstrong, N., & Welsman, J. (2019). Sex-specific longitudinal modeling of youth peak oxygen uptake. *Pediatric Exercise Science, 31*, 204-212.

Åstrand, P.-O. (1991). Influence of Scandinavian scientists in exercise physiology. *Scandinavian Journal of Medicine and Science in Sports, 1*(1), 3-9.

Atherton, P.J., & Smith, K. (2012). Muscle protein synthesis in response to nutrition and exercise. *The Journal of Physiology, 590*(5), 1049-1057.

Bennell, K.L., Marshall, C.J., Dobson, F., Kasza, J., Lonsdale, C., & Hinman, R.S. (2019). Does a web-based exercise programming system improve home exercise adherence for people with musculoskeletal conditions? *American Journal of Physical Medicine & Rehabilitation, 98*(10), 850-858.

Bergstrom, J. (1962). Muscle electrolytes in man. *Scandinavian Journal of Clinical & Laboratory Investigation, 14*, 511-513.

Berryman, J.W. (1992). Exercise and the medical tradition from Hippocrates through antebellum America: A review essay. In J.W. Berryman & R.J. Park (Eds.), *Sport and exercise sciences: Essays in the history of sport medicine* (pp. 1-57). University of Illinois.

Berryman, J.W. (1995). *Out of many, one: A history of the American College of Sports Medicine.* Human Kinetics.

Berryman, C.E., Lieberman, H.R., Fulgoni, III, V.L., & Pasiakos, S.M. (2018). Protein intake trends and conformity with the Dietary Reference Intakes in the United States: Analysis of the National Health and Nutrition Examination Survey, 2001-2014. *The American Journal of Clinical Nutrition, 108*(2), 405-413.

Betik, A.C., & Hepple, R.T. (2008). Determinants of $\dot{V}O_2$max decline with aging: An integrated perspective. *Applied Physiology, Nutrition, and Metabolism, 33*(1), 130-140.

Borg, G.A. (1982). Psychophysical bases of perceived exertion. *Medicine & Science in Sports & Exercise, 14*(5), 377-381.

Bouchard, C., Lesage, R., Lortie, G., Simoneau, J.A., Hamel, P., Boulay, M.R., Pérusse, L., Thériault, G., & Leblanc, C. (1986). Aerobic performance in brothers, dizygotic and monozygotic twins. *Medicine & Science in Sports & Exercise, 18*(6), 639-646.

Bray, M.S., Hagberg, J.M., Pérusse, L., Rankinen, T., Roth, S.M., Wolfarth, B., & Bouchard, C. (2009). The human gene map for performance and health-related fitness phenotypes: The 2006-2007 update. *Medicine & Science in Sports & Exercise, 41*(1), 35-73.

Brooks, G.A. (2012). Bioenergetics of exercising humans. *Comprehensive Physiology, 2*(1), 537-562.

Brown, T., Moore, T.H., Hooper, L., Gao, Y., Zayegh, A., Ijaz, S., Elwenspoek, M., Foxen, S.C., Magee, L., O'Malley, C., Waters, E., & Summerbell, C.D. (2019). Interventions for preventing obesity in children. *The Cochrane Database of Systematic Reviews, 7*(7), CD001871. https://doi.org/10.1002/14651858.CD001871.pub4

Campbell, K.L., Winters-Stone, K.M., Wiskemann, J., May, A.M., Schwartz, A.L., Courneya, K.S., Zucker, D.S., Matthews, C.E., Ligibel, J.A., Gerber, L.H., Morris, G.S., Patel, A.V., Hue, T.F., Perna, F.M., & Schmitz, K. (2019). Exercise guidelines for cancer survivors: Consensus statement from international multidisciplinary roundtable. *Medicine & Science in Sports & Exercise, 51*(11), 2375-2390.

Chau, J.Y., van der Ploeg, H.P., Merom, D., Chey, T., & Bauman, A.E. (2012). Cross-sectional associations between occupational and leisure-time sitting, physical activity and obesity in working adults. *Preventive Medicine, 54*(3-4), 195-200.

Clayton, D.J., Evans, G.H., & James, L.J. (2014). Effect of drink carbohydrate content on post-exercise gastric emptying, rehydration and the calculation of net fluid balance. *International Journal of Sport Nutrition and Exercise Metabolism, 24*(1), 79-89.

Conners, R.T., Morgan, D.W., Fuller, D.K., & Caputo, J.L. (2014). Underwater treadmill training, glycemic control, and health-related fitness in adults with type 2 diabetes. *International Journal of Aquatic Research and Rehabilitation, 8*(4), 382-396.

Dill, D.B. (1967). The Harvard Fatigue Laboratory: Its development, contributions, and demise. In C.B. Chapman (Ed.), *Physiology of muscular exercise* (pp. 161-170). American Heart Association.

Dill, D.B., Talbott, J.H., & Edwards, H.T. (1930). Studies in muscular activity. VI. Responses of several individuals to a fixed task. *Journal of Physiology, 69*(3), 267-305.

Ding, D., Lawson, K.D., Kolbe-Alexander, T.L., Finkelstein, E.A., Katzmarzyk, P.T., van Mechelen, W., Pratt, M., & Lancet Physical Activity Series 2 Executive Committee. (2016). The economic burden of physical inactivity: A global analysis of major non-communicable diseases. *Lancet, 388*(10051), 1311-1324.

Di Pietro, L., Dziura, J., & Blair, S.N. (2004). Estimated change in physical activity level (PAL) and prediction of 5-year weight change in men: The Aerobics Center Longitudinal Study. *International Journal of Obesity and Related Metabolic Disorders, 28*(12), 1541-1547.

Division of Nutrition, Physical Activity, and Obesity, National Center for Chronic Disease Prevention and Health Promotion. (2020, May 13). *Why it matters.* Centers for Disease Control and Prevention. www.cdc.gov/physicalactivity/about-physical-activity/why-it-matters.html

Donnelly, J.E., Blair, S.N., Jakicic, J.M., Manore, M.M., Rankin, J.W., & Smith, B.K. (2009). American College of Sports Medicine position stand. Appropriate physical activity intervention strategies for weight loss and prevention of weight regain for adults. *Medicine & Science in Sports & Exercise, 41*(2), 459-471.

Ekelund, U., Steene-Johannessen, J., Brown, W.J., Fagerland, M.W., Owen, N., Powell, K.E., Bauman, A., Lee, I.M., Lancet Physical Activity Series 2 Executive Committee, & Lancet Sedentary Behaviour Working Group. (2016). Does physical activity attenuate, or even eliminate, the detrimental association of sitting time with mortality? *Lancet, 388*(10051), 1302-1310.

Elagizi, A., Kachur, S., Carbone, S., Lavie, C.J., & Blair, S.N. (2020). A review of obesity, physical activity, and cardiovascular disease. *Current Obesity Reports, 9*(4), 571-581.

Fagard, D. (2001). Exercise characteristics and the blood pressure response to dynamic physical training. *Medicine & Science in Sports & Exercise, 33*(6 Suppl.), S484-S492.

Ganse, B., Ganse, U., Dahl, J., & Degens, H. (2018). Linear decrease in athletic performance during the human life span. *Frontiers in Physiology, 9,* Article 1100. https://doi.org/ 10.3389/fphys.2018.01100

Gellish, R.L., Goslin, B.R., Olson, R.E., McDonald, A., Russi, G.D., & Moudgil, V.K. (2007). Longitudinal modeling of the relationship between age and maximal heart rate. *Medicine & Science in Sports & Exercise, 39*(5), 822-829.

Ghosh, A.K. (2004). Anaerobic threshold: Its concept and role in endurance sport. *The Malaysian Journal of Medical Sciences, 11*(1), 24-36.

Grazioli, E., Dimauro, I., Mercatelli, N., Wang, G., Pitsiladis, Y., Di Luigi, L., & Caporossi, D. (2017). Physical activity in the prevention of human diseases: role of epigenetic modifications. *BMC Genomics, 18*(Suppl 8), 802.

Gulati, M., Shaw, L.J., Thisted, R.A., Black, H.R., Bairey Merz, C.N., & Arnsdorf, M.F. (2010). Heart rate response to exercise stress testing in asymptomatic women: The St. James Women Take Heart Project. *Circulation, 122*(2), 130-137.

Guthold, R., Stevens, G.A., Riley, L.M., & Bull, F.C. (2018). Worldwide trends in insufficient physical activity from 2001 to 2016: A pooled analysis of 358 population-based surveys with 1.9 million participants. *The Lancet. Global Health, 6*(10), e1077-e1086.

Hales, C.M., Carroll, M.D., Fryar, C.D., & Ogden, C.L. (2017). *Prevalence of obesity among adults and youth: United States, 2015-2016.* NCHS data brief, no. 288. National Center for Health Statistics. www.cdc.gov/nchs/data/databriefs/db288.pdf

Hales, C.M., Carroll, M.D., Fryar, C.D., & Ogden, C.L. (2020). *Prevalence of obesity and severe obesity among adults: United States, 2017-2018.* NCHS data brief, no. 360. National Center for Health Statistics. www.cdc.gov/nchs/data/databriefs/db360-h.pdf

Hamilton, M.T., Healy, G.N., Dunstan, D.W., Zderic, T.W., & Owen, N. (2008). Too little exercise and too much sitting: Inactivity physiology and the need for new recommendations on sedentary behavior. *Current Cardiovascular Risk Reports, 2*(4), 292-298.

Hansen, H., Bieler, T., Beyer, N., Kallemose, T., Wilcke, J.T., Østergaard, L.M., Andeassen, H.F., Martinez, G., Lavesen, M., Frølich, A., & Godtfredsen, N.S. (2020). Supervised pulmonary tele-rehabilitation versus pulmonary rehabilitation in severe COPD: A randomized multicenter trial. *Thorax, 75*(5), 413-421.

Haskell, W.L., Lee, I.M., Pate, R.R., Powell, K.E., Blair, S.N., Franklin, B.A., Macera, C.A., Heath, G.W., Thompson, P.D., & Bauman, A. (2007). Physical activity and public health: Updated recommendation for adults from the American College of Sports Medicine and the American Heart Association. *Medicine & Science in Sports & Exercise, 39*(8), 1423-1434.

Healy, S., Nacario, A., Braithwaite, R.E., & Hopper, C. (2018). The effect of physical activity interventions on youth with autism spectrum disorder: A meta-analysis. *Autism Research, 11*(6), 818-833.

Hill, A.V. (1926). *Muscular activity.* Herter Lectures. Williams & Wilkins Company.

Hill, A.V., & Lupton, H. (1923). Muscular exercise, lactic acid, and the supply and utilization of oxygen. *QJM, os-16*(62), 135-171.

Hills, S.P., & Russell, M. (2017). Carbohydrates for soccer: A focus on skilled actions and half-time practices. *Nutrients, 10*(1), 22.

Hong, J., Kong, H.J., & Yoon, H.J. (2018). Web-based telepresence exercise program for community-dwelling elderly women with a high risk of falling: Randomized control trial. *Journal of Medical Internet Research mHealth & uHealth, 6*(5), e132.

Hultman, E. (1967). Physiological role of muscle glycogen in man, with special reference to exercise. *Circulation Research, 21*(Suppl. 1), 99-114.

Huxley, H., & Hanson, J. (1954). Changes in the cross-striations of muscle during contraction and stretch and their structural interpretation. *Nature, 173*(4412), 973-976.

International Health, Racquet & Sportsclub Association. (2021, January). *IHRSA media report: Health and fitness consumer data and industry trends before and during the COVID-19 pandemic.* International Health, Racquet & Sportsclub Association. www.ihrsa.org/ publications/2021-ihrsa-media-report/

Ivy, J.L., Katz, A.L., & Cutler, C.L. (1989). Muscle glycogen resynthesis after exercise: Effect of time on carbohydrate ingestion. *Journal of Applied Physiology, 64*(4), 1480-1485.

Ivy, J.W. (2007). Exercise physiology: A brief history and recommendations regarding content requirements for the kinesiology major. *Quest, 59*(1), 34-41.

James, L.J., Moss, J., Henry, J., Papadopoulou, C., & Mears, S.A. (2017). Hypohydration impairs endurance performance: A blinded study. *Physiological Reports, 5*(12), e13315.

Jo, G., Rossow-Kimball, B., & Lee, Y. (2018). Effects of 12-week combined exercise program on self-efficacy, physical activity level, and health related physical fitness of adults with intellectual disability. *Journal of Exercise Rehabilitation, 14*(2), 175-182.

Katzmarzyk, P.T., Powell, K.E., Jakicic, J.M., Troiano, R.P., Piercy, K., Tennant, B., & 2018 Physical Activity Guidelines Advisory Committee. (2019). Sedentary behavior and health: Update from the 2018 Physical Activity Guidelines Advisory Committee. *Medicine & Science in Sports & Exercise, 51*(6), 1227-1241.

Kenney, W.L., Wilmore, J.H., & Costill, D.L. (2020). *Physiology of sport and exercise* (7th ed.). Human Kinetics.

Kesaniemi, Y.K., Danforth, E., Jr., Jensen, M.D., Kopelman, P.G., Lefèbvre, P., & Reeder, B.A. (2001). Dose-response issues concerning physical activity and health: An evidence-based symposium. *Medicine & Science in Sports & Exercise, 33*(6 Suppl), S351-S358.

Khatra, O., Shadgan, A., Taunton, J., Pakravan, A., & Shadgan, B. (2021). A bibliometric analysis of the top cited articles in sports and exercise medicine. *Orthopaedic Journal of Sports Medicine, 9*(1), 2325967120969902.

Kilpatrick, M., Jung, M., & Little, J. (2014). High-intensity interval training: A review of physiological and psychological responses. *ACSM's Health and Fitness Journal, 18*(5), 11-16.

Kjøbsted, R., Hingst, J.R., Fentz, J., Foretz, M., Sanz, M.N., Pehmøller, C., Shum, M., Marette, A., Mounier, R., Treebak, J.T., Wojtaszewski, J., Viollet, B., & Lantier, L. (2018). AMPK in skeletal muscle function and metabolism. *FASEB Journal, 32*(4), 1741-1777.

Krebs, H.A., Salvin, E., & Johnson, W.A. (1938). The formation of citric and alpha-ketoglutaric acids in the mammalian body. *Biochemical Journal, 32*(1), 113-117.

Krogh, A. (1919). The number and distribution of capillaries in muscles with calculations of the oxygen pressure head necessary for supplying the tissue. *Journal of Physiology, 52*(6), 409-415.

Kroll, W.P. (1982). *Graduate study and research in physical education.* Human Kinetics.

Kubota, Y., Evenson, K.R., Maclehose, R.F., Roetker, N.S., Joshu, C.E., & Folsom, A.R. (2017). Physical activity and lifetime risk of cardiovascular disease and cancer. *Medicine & Science in Sports & Exercise, 49*(8), 1599-1605.

Lawler, P.R., Filion, K.B., & Eisenberg, M.J. (2011). Efficacy of exercise-based cardiac rehabilitation post-myocardial infarction: A systematic review and meta-analysis of randomized controlled trials. *American Heart Journal, 162*(4), 571-584.

Lee, I., Shiroma, E.J., Kamada, M., Bassett, D.R., Matthews, C.E., & Buring, J.E. (2019). Association of step volume and intensity with all-cause mortality in older women. *JAMA Internal Medicine, 179*(8), 1105-1112.

Lutz, N., Clarys, P., Koenig, I., Deliens, T., Taeymans, J., & Verhaeghe, N. (2020). Health economic evaluations of interventions to increase physical activity and decrease sedentary behavior at the workplace: A systematic review. *Scandinavian Journal of Work, Environment & Health, 46*(2), 127-142.

Margaria, R., Edwards, H.T., & Dill, D.B. (1933). The possible mechanisms of contracting and paying the oxygen debt and the role of lactic acid in muscular contraction. *American Journal of Physiology, 106*(3), 689-715.

Matthews, C.E., Chen, K.Y., Freedson, P.S., Buchowski, M.S., Beech, B.M., Pate, R.R., & Troiano, R.P. (2008). Amount of time spent in sedentary behaviors in the United States, 2003-2004. *American Journal of Epidemiology, 167*(7), 875-881.

McArdle, W.D., Katch, F.I., Pechar, G.S., Jacobson, L., & Ruck, S. (1972). Reliability and interrelationships between maximal oxygen intake, physical work capacity and step-test scores in college women. *Medicine & Science in Sports & Exercise, 4*(4), 182-186.

McTiernan, A., Friedenreich, C.M., Katzmarzyk, P.T., Powell, K.E., Macko, R., Buchner, D., Pescatello, L.S., Bloodgood, B., Tennant, B., Vaux-Bjerke, A., George, S.M., Troiano, R.P., Piercy, K.L., & 2018 Physical Activity Guidelines Advisory Committee. (2019). Physical activity in cancer prevention and survival: A systematic review. *Medicine & Science in Sports & Exercise, 51*(6), 1252-1261.

Morris, J.K., Vidoni, E.D., Johnson, D.K., Van Sciver, A., Mahnken, J.D., Honea, R.A., Wilkins, H.M., Brooks, W.M., Billinger, S.A., Swerdlow, R.H., & Burns, J.M. (2017). Aerobic exercise for Alzheimer's disease: A randomized controlled pilot trial. *PloS One, 12*(2), e0170547.

Morris, J.N., Heady, J.A., Raffle, P.A.B., Roberts, C.G., & Parks, J.W. (1953). Coronary heart disease and physical activity of work. *Lancet, 2*(5111), 1111-1120.

National Center for Health Statistics. (2021). *National Health Interview Survey: Exercise or physical activity.* Centers for Disease Control and Prevention. www.cdc.gov/nchs/fastats/exercise.htm

NIH Consensus Development Panel on Physical Activity and Cardiovascular Health. (1996). Physical activity and cardiovascular health. *Journal of the American Medical Association, 276*(3), 241-246.

Nystoriak, M.A., & Bhatnagar, A. (2018). Cardiovascular effects and benefits of exercise. *Frontiers in Cardiovascular Medicine, 5*, 135. https://doi.org/10.3389/fcvm.2018.00135

Pasricha, S.-R., Low, M., Thompson, J., Farrell, A., & De-Regil, L.-M. (2014). Iron supplementation benefits physical performance in women of reproductive age: A systematic review and meta-analysis. *Journal of Nutrition, 144*(6), 906-914.

Patel, A.V., Friedenreich, C.M., Moore, S.C., Hayes, S.C., Silver, J.K., Campbell, K.L., Winters-Stone, K., Gerber, L.H., George, S.M., Fulton, J.E., Denlinger, C., Morris, G.S., Hue, T., Schmitz, K.H., & Matthews, C.E. (2019). American College of Sports Medicine Roundtable Report on Physical Activity, Sedentary Behavior, and Cancer Prevention and Control. *Medicine & Science in Sports & Exercise, 51*(11), 2391-2402.

Park, J.H., Moon, J.H., Kim, H.J., Kong, M.H., & Oh, Y.H. (2020). Sedentary lifestyle: Overview of updated evidence of potential health risks. *Korean Journal of Family Medicine, 41*(6), 365-373.

Paterson, D.H., Govindasamy, D., Vidmar, M., Cunningham, D.A., & Koval, J.J. (2004). Longitudinal study of determinants of dependence in an elderly population. *Journal of the American Geriatrics Society, 52*(10), 1632-1638.

Pescatello, L.S., Franklin, B.A., Fagard, R., Farquhar, W.B., Kelley, G.A., Ray, C.A., & American College of Sports Medicine. (2004). American College of Sports Medicine position stand. Exercise and hypertension. *Medicine & Science in Sports & Exercise, 36*(3), 533-553.

Powers, S.K., Howley, E.T., & Quindry, J. (2021). *Exercise physiology: Theory and application to fitness and performance.* McGraw Hill.

Rall, J.A. (2017). Nobel Laureate A.V. Hill and the refugee scholars, 1933-1945. *Advances in Physiology Education, 41*(2), 248-259.

Regan, E.W., Handlery, R., Stewart, J.C., Pearson, J.L., Wilcox, S., & Fritz, S. (2021). Integrating survivors of stroke into exercise-based cardiac rehabilitation improves endurance and functional strength. *Journal of the American Heart Association, 10*(3), e017907.

Reimers, C.D., Knapp, G., & Reimers, A.K. (2012). Does physical activity increase life expectancy? A review of the literature. *Journal of Aging Research*, Article 243958.

Saltin, B. (1973). Metabolic fundamentals in exercise. *Medicine and Science in Sports, 5*(3), 137-146.

Saunders, D.H., Sanderson, M., Hayes, S., Johnson, L., Kramer, S., Carter, D.D., Jarvis, H., Brazzelli, M., & Mead, G.E. (2020). Physical fitness training for stroke patients. *The Cochrane Database of Systematic Reviews, 3*(3), Article CD003316.

Sawka, M.N., Burke, L.M., Eichner, E.R., Maughan, R.J., Montain, S.J., Stachenfeld, N.S. (2007). American College of Sports Medicine position stand: Exercise and fluid replacement. *Medicine & Science in Sports & Exercise, 39*, 377-90.

Schenkman, M., Moore, C.G., Kohrt, W.M., Hall, D.A., Delitto, A., Comella, C.L., Josbeno, D.A., Christiansen, C.L., Berman, B.D., Kluger, B.M., Melanson, E.L., Jain, S., Robichaud, J.A., Poon, C., & Corcos, D.M. (2018). Effect of high-intensity treadmill exercise on motor symptoms in patients with de novo Parkinson disease: A phase 2 randomized clinical trial. *Journal of the American Medical Association, 75*(2), 219-226.

Sherrington, C., Fairhall, N.J., Wallbank, G.K., Tiedemann, A., Michaleff, Z.A., Howard, K., Clemson, L., Hopewell, S., & Lamb, S.E. (2019). Exercise for preventing falls in older people living in the community. *The Cochrane Database of Systematic Reviews, 1*(1), Article CD012424.

Siegel, R.L., Miller, K.D., & Jemal, A. (2019). Cancer statistics, 2019. *CA: A Cancer Journal for Clinicians, 69*(1), 7-34.

Sinha, R., Fisch, G., Teague, B., Tamborlane, W.V., Banyas, B., Allen, K., Savoye, M., Rieger, V., Taksali, S., Barbetta, G., Sherwin, R.S., & Caprio, S. (2002). Prevalence of impaired glucose tolerance among children and adolescents with marked obesity. *New England Journal of Medicine, 346*(11), 802-810.

Telama, R. (2009). Tracking of physical activity from childhood to adulthood: A review. *Obesity Facts, 3*, 187-195.

Thompson, P.D., Buchner, D., Piña, I.L., Balady, G.J., Williams, M.A., Marcus, B.H., Berra, K., Blair, S.N., Costa, F., Franklin, B., Fletcher, G.F., Gordon, N.F., Pate, R.R., Rodriguez, B.L., Yancey, A.K., Wenger, N.K., American Heart Association Council on Clinical Cardiology Subcommittee on Exercise, Rehabilitation, and Prevention; American Heart Association Council on Nutrition, Physical Activity, and Metabolism Subcommittee on Physical Activity. (2003). Exercise and physical activity in the prevention and treatment of atherosclerotic cardiovascular disease. *Arteriosclerosis, Thrombosis, and Vascular Biology, 23*(8), 42-49.

Thompson, W.R. (2019). Worldwide survey of fitness trends for 2020. *ACSM's Health & Fitness Journal, 23*(6), 10-18.

Thompson, W.R. (2021). Worldwide survey of fitness trends for 2021. *ACSM's Health & Fitness Journal, 25*(1), 10-19.

Tremblay, M.S., Aubert, S., Barnes, J.D., Saunders, T.J., Carson, V., Latimer-Cheung, A.E., Chastin, S., Altenburg, T.M., Chinapaw, M., & SBRN Terminology Consensus Project Participants. (2017). Sedentary Behavior Research Network (SBRN): Terminology Consensus Project process and outcome. *International Journal of Behavioral Nutrition and Physical Activity, 14*(1), 75.

U.S. Department of Health and Human Services (USDHHS). (1996). *Physical activity and health: A report of the surgeon general.* U.S. Department of Health and Human Services, Centers for Disease Control and Prevention, National Center for Chronic Disease Prevention and Health Promotion. https://health.gov/sites/default/files/2019-09/paguide.pdf

U.S. Department of Health and Human Services (USDHHS). (2018). *Physical activity guidelines for Americans* (2nd ed.). U.S. Department of Health and Human Services.

U.S. Department of Health and Human Services, Centers for Disease Control and Prevention, National Center for Chronic Disease Prevention and Health Promotion, & Division of Nutrition, Physical Activity and Obesity. (2015). *Data, trends, and maps.* Centers for Disease Control and Prevention. www.cdc.gov/nccdphp/DNPAO/index.html

Valenzuela, P.L., Maffiuletti, N., Joyner, M.J., Lucia, A., & Lepers, R. (2020). Lifelong endurance exercise as a countermeasure against age-related $\dot{V}O_2$max decline: Physiological overview and insights from masters athletes. *Sports Medicine, 50*(4), 703-716.

Virani, S.S., Alonso, A., Benjamin, E.J., Bittencourt, M.S., Callaway, C.W., Carson, A.P., Chamberlain, A.M., Chang, A.R., Cheng, S., Delling, F.N., Djousse, L., Elkind, M., Ferguson, J.F., Fornage, M., Khan, S.S., Kissela, B.M., Knutson, K.L., Kwan, T.W., Lackland, D.T., Lewis, T.T., . . . American Heart Association Council on Epidemiology and Prevention Statistics Committee and Stroke Statistics Subcommittee. (2020). Heart disease and stroke statistics—2020 update: A report from the American Heart Association. *Circulation, 141*(9), e139–e596.

Wang, F., & Boros, S. (2021). The effect of physical activity on sleep quality: A systematic review. *European Journal of Physiotherapy, 23*(1), 11-18.

Williamson, E., Srikesavan, C., Thompson, J., Tonga, E., Eldridge, L., Adams, J., & Lamb, S.E. (2020). Translating the strengthening and stretching for rheumatoid arthritis of the hand programme from clinical trial to clinical practice: An effectiveness–implementation study. *Hand Therapy, 25*(3), 87-97.

Willett, W.C., Manson, J.E., Stampler, M.J., Colditz, G.A., Rosner, B., Speizer, F.E., & Hennekena, C.H. (1995). Weight, weight change, and coronary disease in women. *Journal of the American Medical Association, 273*(6), 461-465.

Winter, E.M., Abt, G., Brooks, F., Challis, J.H., Fowler, N.E., Knudson, D.V., Knuttgen, H.G., Kraemer, W.J., Lane, A.M., van Mechelen, W., Morton, R.H., Newton, R.U., Williams, C., Yeadon, F.R. (2016). Misuse of "power" and other mechanical terms in sport and exercise science research. *Journal of Strength and Conditioning Research, 30*(1), 292-300.

Wong, C., Odom, S.L., Hume, K.A., Cox, A.W., Fettig, A., Kucharczyk, S., Brock, M.E., Plavnick, J.B., Fleury, V.P., & Schultz, T.R. (2015). Evidence-based practices for children, youth, and young adults with autism spectrum disorder: A comprehensive review. *Journal of Autism and Developmental Disorders, 45*(7), 1951-1966.

World Health Organization. (2018). *Global action plan on physical activity 2018-2030. More active people for a healthier world.* World Health Organization.

CHAPTER 6

Motor Behavior

Katherine T. Thomas and Xiangli Gu

CHAPTER OBJECTIVES

In this chapter, we will

- define motor behavior and contrast motor learning, motor control, and motor development;
- explain how motor skills are learned and the role of feedback, practice, and attention;
- describe how motor skills are controlled in the neuromuscular system;
- trace how the learning and control of motor skills change across the life span; and
- demonstrate how motor learning and control improve during childhood and adolescence and then decline during old age.

At your annual family reunion, you notice several changes in your family members. Some of your older relatives have mobility issues, and one is using a walker. However, other older relatives are moving with ease and are playing cornhole with your young cousins. You are impressed by how quickly young and old have conquered the basics of the game. Your sister arrives, and she has a new baby and a toddler who is just beginning to walk. These observations cause you to wonder about what changes you might see in your family members next year. The baby will be walking and the toddler running, and you worry that the senior cornhole players might not be as active as they are this year. What explains the varied abilities in your family?

Goals of Motor Behavior

- To understand how motor skills are learned, how processes such as feedback and practice improve learning and performance of motor skills, and how response selection and response execution become more efficient and effective
- To understand how motor skills are controlled, how the neuromuscular system functions to coordinate the muscles and limbs to move the body, and how environmental and individual factors affect the mechanisms of response selection and response execution
- To understand how the learning and control of motor skills change (i.e., behavioral and neurological) across the life span, how motor learning and control improve during childhood and adolescence, and how motor learning and control deteriorate with age

This chapter introduces motor behavior, which includes motor learning, motor control, and life span motor development. The chapter will help you begin to understand why motor behavior holds interest for performers who want to improve their movement expertise, for scientists who contribute to scholarly study, and for professional practitioners who seek to help others develop or maintain their motor skills.

Benefits of Motor Behavior Knowledge

Have you ever wondered why you enjoy certain physical activities and not others? Do you simply dislike certain exercises? Often, being skilled at an activity helps us enjoy it. If you pitch a great fastball, run a quick 5K, boast a killer volleyball spike, or hit a terrific backhand in tennis, you are more likely to enjoy these activities. Someone with less skill in these activities might find them frustrating and less enjoyable. Of course, there are notable exceptions; for example, some of us spend many hours and a great deal of money demonstrating that we have little skill at golf when compared with elite golfers—or simply compared to par!

Think about a skill you like and do well—for example, riding a bike. How did you learn to ride a two-wheel bicycle? What types of practice experiences worked best for you? How did you learn to coordinate the movements of your feet in order to pedal and your hands in order to steer, and to do both while balancing? Do you remember early times when you moved slowly and wobbled, could not turn quickly, needed help starting, and often fell rather than stopping in a controlled manner? In what ways did your brain and nervous system develop and adjust so that you could improve your control and coordination?

Practice is an important factor in learning skills across the life span. Most people improve motor skills yet never understand how the nervous system adapts, how it develops or controls movement, or how best to practice in order to improve performance. The study of motor behavior focuses on how skills are learned and controlled, how exercise professionals can help people learn, and how movement changes from birth through the end of life.

People generally believe that practice improves performance; after all, we see evidence of this notion at home, in school, and in the community. For example, parents urge children to practice playing the piano, athletes ask coaches for help in practicing sport skills, and students perform practice exercises to learn keyboard typing. Clearly, however, some forms of practice are superior to others. Thus, important questions about practice include *what*, *how*, and *how much*? Another important question is, "What can be changed with practice?" For example, is an expert better because of practice, or were they born with special talent? Research in motor behavior tells us that parents and teachers are partly right when they say, "Practice makes perfect." More precisely, they might have said, "*Correct* practice makes perfect"; as research tells us, the effectiveness of practice is influenced by many factors.

In the chapter-opening scenario, you observed motor development in your sister's toddler. Certain skills emerge naturally and in a predictable pattern. Most babies walk by their first birthday, and this is one of the motor milestones that is used to determine if a baby is developing normally. You might have noticed that your sister's toddler is taller and heavier than last year. At the same time, you probably noticed that some of your elderly relatives might seem shorter, smaller, and more frail. You also noticed that as some older individuals age, they appear older (physiological age) than others of the same chronological age. Aging is complicated and explained by environmental factors such as diet, disease, and exercise and physical activity in addition to genetics. Now you have observed life span motor development, from birth through aging. Motor control helps us understand how movements are controlled at all ages and is useful in the study of aging. Finally, you observed motor learning as children and adults played cornhole, a game that is relatively new to your family. How do we acquire new skills and maintain skills after periods without practice? Those are key questions in motor learning.

The study of motor behavior provides scientific knowledge that helps us understand the process of the development, control, and learning of motor skills (Schmidt et al., 2019; Thomas, 2006; Ulrich & Reeve, 2005). Understanding this process can be used by casual consumers such as parents or volunteer coaches and by professionals to plan more effective exercise and practice sessions, identify movement problems, and set appropriate expectations across the life span.

What Do Motor Behaviorists Do?

Scholars who study motor behavior are most often employed at universities or medical facilities, but some scholars work in research facilities unassociated with universities or medical facilities. Motor behavior research can be conducted in a university laboratory, in a clinical setting (e.g., a hospital), or in an industrial or military setting.

- *Universities:* The duties of motor behaviorists at universities typically include research, teaching, and service. Their research might focus on learning, control, or development. Their teaching duties can include courses in motor behavior or related courses such as biomechanics or sport psychology, research methods, measurement and evaluation, and pedagogy and youth sport. Service for a university faculty member might include evaluating motor disorders, managing a program for people with motor disorders, and conducting workshops or clinics about motor disorders.

- *Clinical setting:* Medical and educational researchers often study motor learning, motor control, and motor development. They might investigate how the nervous system changes with age in terms of controlling movement, what the best methods are for teaching rehabilitation protocols to patients in physical therapy, how the performance of surgeons can be improved, or what causes movement deterioration in people with Parkinson's and other neuromuscular diseases.

- *Industry:* A motor behavior researcher working in industry might look for the optimal method for training workers to assemble a product.

- *Military:* A military researcher might investigate which training methods best improve fighter pilots' reactions to electronic threat signals.

Whatever the setting, most motor behavior researchers also write grant applications and perform professional service such as reviewing manuscripts for scholarly journals.

Motor behavior produces knowledge of how people achieve motor skills across the life span. The three components of motor behavior are motor learning, motor control, and life span motor development (a developmental view of motor learning and motor control).

Motor Control and Learning in Driving an Automobile

Do you remember learning how to drive a car? What was one task that was difficult for you to master (e.g., braking smoothly, turning, signaling, parallel parking)? Now think about the last time you drove somewhere familiar—for example, to school or work today. Sometimes we cannot remember anything about a familiar drive. At this stage driving is automatic and controlled without very much cognition. Consider how your skill has changed as a result of learning to drive and then continuing to practice. When you were learning to drive, you had to think about every movement—moving your foot from the accelerator to the brake, using the turn signal, turning the steering wheel. To coordinate all those movements, the central nervous system initiates and controls actions. Later, you thought only about the traffic and where you were going; if you were driving on an often-traveled route, you just need to monitor the traffic with intermittent attentiveness.

The study of motor learning helps us understand how we learn movement skills such as driving a car so well that the skill becomes automatic. Motor control is the study of how movements are executed and helps us understand how we move automatically. When you drive a car on a busy highway on a stormy day, however, you use visual and proprioceptive feedback to control the steering wheel, and you monitor the traffic with intense attentiveness. In this scenario, because we may have less practice at driving in storms and heavy traffic, that type of driving is not well learned and therefore is not automatic. Motor control and learning are related but distinct, as you will see in the next sections.

The study of motor learning and the study of motor control are not distinctly different areas, but they tend to ask different questions.

Motor Learning

- The goals of studying motor learning include understanding the influence of feedback, practice, and individual differences, especially as they relate to the retention and transfer of motor skill.
- An example of a research question in motor learning is, How do different types of feedback influence retention and transfer of motor skill performance?

Motor Control

- The goals of studying motor control include understanding how to coordinate the muscles and joints during movement, how to control a sequence of movements, and how to adapt environmental information to plan and adjust movements.
- An example of a research question in motor control is, How are arms and legs coordinated in different ways while performing walking and running?

In addition, scholars often are interested in exploring how motor learning and motor control vary across age groups, from children to senior citizens. This area is life span motor development.

KEY POINT

Motor control and motor learning are related. Motor control is essential for every movement—from poorly skilled to well skilled. Motor learning, on the other hand, is responsible for the shift from poorly skilled to highly skilled movement.

When you think of motor behavior, you might think of sport skills. However, consider the many other types of movements that people use in their daily activities:

- Babies learning to use a fork and spoon
- Dentists learning to control a drill while looking in a mirror
- Surgeons controlling a scalpel and microsurgeons using a laser while viewing a magnified video image of the brain
- Children learning to roller-skate
- A junior football player working on the throwing skill after injury
- A stroke patient relearning the walking skill
- Dancers performing carefully choreographed movements
- Pilots learning to control an airplane
- Young children learning to control a crayon when coloring

EXERCISE SCIENCE COLLEAGUES

Rachael Seidler

© Rachael Seidler

Rachael Seidler is a professor at the University of Florida in the Department of Applied Physiology and Kinesiology. Previously, she had an academic appointment at the University of Michigan. After an undergraduate degree in exercise science from the University of Oregon, an MS in biomechanics, and a PhD in motor control at Arizona State University, she was a postdoctoral associate in the Brain Sciences Center at the University of Minnesota. Dr. Seidler uses a range of neuroimaging and neuromodulation techniques to study motor learning, focusing on the neural control of movement in health and disease. Participants in her research include patients with Parkinson's disease, NASA astronauts, and healthy young and older adults, and such research has been funded by the National Institutes of Health, the National Science Foundation, NASA, the National Space Biomedical Research Institute, and a variety of private foundations. She is the principal investigator in the Neuromotor Behavior Laboratory, where the goal is "to understand the interactions between the brain and movement." She has found that neural processing of sensory information is altered with spaceflight, with more reliance on visual and somatosensory brain regions and with decreased vestibular processing immediately after flight. Her work on aging has leveraged structural and functional MRI, transcranial magnetic stimulation, and assessment of genetic polymorphisms that code for components of the dopaminergic transmission pathway. She has found that age differences in these neural measures are associated with declines in manual and mobility function. She also reports evidence for adaptive compensation, with some individuals showing engagement of additional brain pathways and reduced functional declines.

All of these activities, and many others, involve motor behavior and thus hold interest for researchers and exercise professionals. Thus, the goals of motor behavior study are important not only in sport, but also in the overall field of physical activity and exercise. Understanding the learning, control, and development of movements plays an essential role in our culture and society. People have probably been asking questions about motor behavior since the beginning of time; scientists have been seeking the answers to those questions for more than 100 years.

History of Motor Behavior

Research has a long history in all three areas of motor behavior—motor learning, motor control, and life span motor development—and the focus of research in each area has changed dramatically (for a historical review, see Thomas, 2006; Ulrich & Reeve, 2005). For example, in the late 1800s and early 1900s researchers were interested primarily in studying motor skills as a means to understand the mind; in other words, motor skill was viewed as a way to examine cognition (Abernethy & Sparrow, 1992). Great interest was taken in motor behavior research during the World War II era from 1939 to 1945 (Thomas, 1997) because the military needed to select and effectively train pilots (Adams, 1987). Thus, motor skills were viewed as a necessary component in military efforts but not because understanding motor skills was an important area of study in itself.

Beginning in the 1960s, and with increasing momentum in the early 1970s, the study of motor behavior evolved as a scholarly element of exercise science. With this change, the scholars doing motor behavior research were no longer neurophysiologists or psychologists; they were specialists in physical activity. Franklin Henry's paper on **memory drum theory** (Henry & Rogers, 1960) might have been the first major theoretical or landmark study (Ulrich & Reeve, 2005) from the discipline of exercise science (called *physical education* at the time). Henry's theory stated that reaction time was slower for complex movements because those movements took more planning time. In this view, for example, movements with several segments—involving moving from one position to a second position and then to a third position—required a longer reaction time than did single-segment movements, because the brain required more time to specify the needed information. The current work on motor programs (representations in the brain of plans for movement) evolved from Henry's memory drum theory.

From this early work, Adams (1987) identified five themes that have persisted through the years:

1. Knowledge of results
2. Distribution of practice
3. Transfer of training
4. Retention
5. Individual differences

These five themes provide the foundation for the study of motor learning. We will provide examples of experiments for each of the themes in the section on knowledge of motor learning. Although these themes persist, researchers are now focusing on the motor skills themselves rather than simply studying them in order to understand cognition.

To help scientists conceptualize the brain as the master controller in planning, organizing, selecting, and controlling movements, a model called *information processing* was adopted. This model represents the motor behavior system as a computer. In this view, general commands are sent from the brain (the central processing unit, or CPU) through the spinal cord (the wiring), which probably reduces the complexity of the information into relatively simple commands, and on to the muscles or muscle groups (the printer or screen). (If your brain were really a computer, when you accidentally touched a hot burner on a stove, a message would pop up asking "Are you sure?" before you could pull your hand away!) From this perspective, the goal of motor behavior is to explain response selection (how the skill to be used is selected) and response execution (how the selected skill is performed).

Similarly, motor control research can be traced to research in the late 1800s on the springlike qualities of muscle (Blix, 1892-1895). Other early research included Sherrington's (1906) seminal work on neural control, which is still useful in explaining how the nervous system controls muscles during movement. In both cases, the purpose was to understand not motor behavior but biology.

Current research in motor control and motor learning usually focuses on understanding how the neuromuscular system controls and repeats movements. One purpose of this research is to understand and develop treatments for conditions such as Parkinson's disease and spinal cord injuries; another is to improve performance in complex skills such as dentistry and surgery and in physical activity. In some cases, technological advances combine with motor control to allow patients to use complex prostheses and to allow quadriplegics to operate robotic arms with their brains. This cutting-edge technology is based on motor control theory and research.

Hot Topic

Virtual Reality–Based Rehabilitation

Virtual reality (VR) technology has been used to simulate an engaging environment that users experience as being comparable to the real world, which can enhance sensorimotor recovery and rehabilitation. The enriched virtual environments have great potential to optimize motor learning by manipulating practice conditions. Levin and colleagues (2015), for example, provided a preliminary viewpoint regarding the upper-limb rehabilitation practices—that is, how motor control and motor learning principles can be implemented as guidelines for the organization of practice in the VR training environments. Subsequently, Levin and Demers (2021) further suggested that virtual rehabilitation applications have the potential to manipulate practice conditions to increase kinematic redundancy and engage the learning in patients with neurological lesions. In the Movement and Physical Activity Epidemiology Laboratory at the University of Texas at Arlington, the fellow researchers are testing the Neuro Rehab VR System with older adults to enhance motor functioning and improve quality of life. The VR system uses functional therapy games (e.g., grocery shopping) in a safe, controlled virtual environment to improve the brain's neuroplasticity and rebuild connections. The applications are vast, from helping to prevent falls in older adults to improving motor-cognitive recovery brought on by aging. Virtual rehabilitation can encourage the engagement and provision of higher doses of training compared to traditional therapies.

A developmental approach to motor learning and motor control (also known as *motor development*) originated in developmental psychology and child development. The study of motor development grew from "baby biographies" (many written before 1900) describing changes that occur in the reflexes and movements of infants. The early work used twins to establish the role of environment and heredity in shaping behavior (Bayley, 1935; Dennis & Dennis, 1940; Dennis, 1938; Galton, 1876; Gesell, 1928; McGraw, 1935, 1939). As with motor control and motor learning, the initial research in motor development took place not because scholars were interested in motor skills per se, but because they were using the study of motor skills to understand other areas of interest.

In the 1940s and 1950s developmental psychologists lost interest in the developmental aspects of motor skill. Study in this area might have ended if not for three motor development scholars: Ruth Glassow (Slone, 1984), Larry Rarick (Park et al., 1996), and Anna Espenschade (1960). Their emphasis, which differed from that of the developmental psychologists, focused on how children acquire skills—for example, how fundamental movement patterns (e.g., kicking, catching, hopping) are formed and how motor performance is affected by growth. These three scholars maintained the focus on the developmental nature of motor learning and motor control in their research through the 1950s and 1960s.

Just as research in motor learning and motor control increased around 1970, developmental research addressing the questions of motor learning and motor control also became popular (Clark et al., 2019; Thomas & Thomas, 1989; Whitall et al., 2020). Motor development was considered part of motor

behavior because it studied the same topics—in this case, developmentally. However, two research themes in motor development continued from the years before 1970: the influence of growth and maturation on motor performance and the developmental patterns of fundamental movements. Growth influences the performance of motor skills in children over time; however, at the cessation of growth (e.g., maturity), growth is no longer a factor. Because growth cannot explain all improvement in motor performance during childhood, developmental scientists have turned to motor control, motor learning, and biomechanics for more information. Thus, we now see three lines of motor development research: motor learning in children; motor control in children; and the influence of growth on motor learning, motor control, and performance.

Currently the term **life span motor development** best describes this area of motor behavior research. Development, in this context, refers to long-term (over many years) periods of change. Change in motor skill and physical appearance is rapid from birth to approximately 18 years of age and is evident again at the other end of the life span, during the aging process. Studying the changes from birth through adolescence was called *motor development*. However, as life expectancy grew, the study of change during aging was added.

KEY POINT

Starting in the 1970s motor behavior developed as an independent scholarly field. Exercise scientists studied motor learning, motor control, and life span motor development.

As you can see, the study of motor behavior partially began as a branch of psychology that used movement or physical activity to understand cognition. Motor behavior also incorporates information from biology and zoology because physical activity is also affected by factors such as heredity, aging, and growth (e.g., Thomas & Thomas, 2008). Researchers also apply principles and laws from physics to the study of humans in motion. Thus, what you know about physics, biology (or zoology), and psychology will help you understand motor behavior. However, motor behavior is a recognized area of study in exercise science.

Research Methods for Motor Behavior

It is a complex endeavor to develop understanding of how movements are learned, how they are controlled, and how they change across the life span. In fact, in order to answer one question, it is usually necessary to perform many experiments. Motor behavior researchers have concentrated on techniques for measuring movement speed and accuracy. Motor control and motor learning researchers use technology similar to that used by researchers in biomechanics. Since the 1970s, technology (e.g., computing, high-speed imaging, electromyography [EMG]) has been advanced and adapted to control the testing situation and precisely record and analyze movements for motor behavior studies. Moreover, noninvasive electronic measurement technologies have allowed the use of real-world movements instead of the simple movement tasks invented in the laboratory for research purposes. Motor behavior courses often include laboratory experiences so that students like you can repeat the classic experiments and theories that you study. This type of hands-on learning has demonstrated benefits to learning (Hofstein & Kind, 2012).

Types of Studies

Any particular research question can be answered many ways. Motor behavior research frequently uses three experimental designs or techniques:

1. Between group
2. Within group
3. Descriptive

The between-group design compares two or more groups exposed to different treatments (interventions) but tests them using the same task. For example, a researcher could use this design to answer the question, "Does practicing a simple movement increase its speed?" This research design compares two groups (randomly formed) on the same task (movement speed). The treatment in this example is practice: If two groups practice differing amounts, will their performance on the task be different? In the second design, the within-group design, all participants are exposed to two or more different treatments and are tested on the same task.

Let's consider an example that could be conducted with either design. Suppose that you want to study whether a participant's **reaction time** (how quickly the movement begins after a signal) varies with the size of the target. You could tell the participant that after you say "go," they are to move a stylus as rapidly as possible to a target 30 centimeters (12 in.) away; the target is a circle either 2 or 4 centimeters (0.8 or 1.6 in.) in diameter (figure 6.1). The question is whether the time between hearing the signal and beginning the movement (i.e., the reaction time) changes depending on the target size. In a between-group design, one group could move to the 2-centimeter target and the second group to the 4-centimeter target. In a within-group design, half the participants could move to the 2-centimeter target first and then to the 4-centimeter target, and the other half could move to the 4-centimeter target first and then to the 2-centimeter target. In deciding which design to use, an investigator should consider whether changing target sizes interferes with performance. If so, a between-group design is preferred; if not, the investigator will use a within-group design because they will not have to test as many participants.

The third type of experimental design used in motor behavior research is descriptive research. Here, the investigator measures or observes participants performing a task. Sometimes the investigator

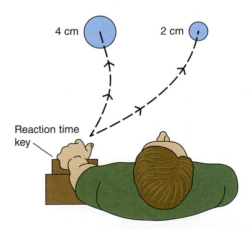

Figure 6.1 A reaction-time study might measure how fast a movement occurs after a signal and whether the movement is affected by the size of the target object.

observes the same participants several times in order to trace changes. For example, an investigator might measure reaction time in the same children when they are 4, 6, and 8 years of age in a longitudinal design. In other cases, the research compares the performance of different groups in a cross-sectional design; for instance, 4-, 6-, and 8-year-old children might be tested for reaction time. Researchers often use this cross-sectional research technique to describe age differences or differences between experts and novices. Descriptive research differs from research using the first two techniques because in this approach the participants receive no treatment.

Studying the Early Stages of Learning

Motor learning research often has used **novel learning tasks** to provide certainty that no participant has tried the task before. This approach eliminates the advantage that some participants might have because they have practiced the task before the experiment. Novel learning tasks tend to be simple—so simple, in fact, that in order to make the task challenging the participant is sometimes asked to wear a blindfold (figure 6.2). Because the task is novel, participants are unlikely to have done it before; because it is simple, participants are able to master it in a short time; and because vision is occluded, learners must rely on what they remember about the movement (e.g., speed, distance, beginning and ending points). Vision is the dominant sense, and when vision is present (and not occluded), the other types of information about movement can be ignored.

You might question whether the results of these experiments can be applied to real-world, complex movement situations. This is a good point. These

simple tasks do allow us to study improvement and have helped us understand a great deal about how movements are learned. However, using simple tasks in research limits what the researcher can learn. One reason is that this approach studies the outcome of the movement (the product) rather than the nature of the movement (the process) itself (Christina, 2017; Rymal, 2018). Novel learning experiments are not helpful for researchers interested in physical activity or sport tasks in which performers have had thousands of trials (e.g., keyboard typing or baseball batting).

> ### KEY POINT
>
> Before the early 1980s motor behavior research often used simple, novel tasks to study early skill learning; such research helped us understand how beginners learn new motor skills.

Studying Expert Performers

To address the limitations of studying novel learning tasks, some scholars have focused their studies on experts. These investigators ask, "What do expert performers do during practice and performance?" One way to answer the question is to compare expert performers with novice performers and evaluate how they differ in terms of perception and knowledge—particularly, decision making, skill, and real-world performance. In these studies, expertise is established according to a criterion such as successful missions for fighter pilots or surgeries for surgeons.

The knowledge and skills of a sport are often unique to that sport (Baker & Young, 2014); for example, knowing about and being skilled in tennis does not help one play soccer. Knowledge does provide an early benefit in sport; for example, the game procedures of tennis and badminton are similar, so a tennis player playing badminton for the first time would have some understanding of the game goal and order of operations. However, the techniques of these two sports might interfere with play in the other sport. Evidence exists for transfer of pattern recognition, a cognitive skill, in sport experts (Chow et al., 2019; Seifert et al., 2018). In another study, video recordings of badminton players were used to determine how age and expertise influence players' ability to predict where a birdie struck by an opponent would land. Participants looked at a video of a badminton player hitting a shot. The video had been altered (e.g., by erasing the head, arm, or racket or by zeroing in on certain motions), and in some cases information was

Figure 6.2   Linear positioning task with blindfolded participant.

erased that was needed for accurately predicting where the birdie would land. As you might guess, expert players could make better, faster predictions with less information than could novice players. Experts also looked at different body parts than novices did and therefore based their decisions on different sources of information. Their advantage seemed to derive from playing experience rather than age.

Experience is an important factor in many settings. Patients are encouraged to ask surgeons how many times they have successfully completed a procedure. Typical microsurgery courses are 1 week or 40 hours; thus, you can see why it would be important to know how much more experience your surgeon has before performing microsurgery on you.

You can see how this finding might help a coach, teacher, or player improve badminton play. It also helps us understand something about skill learning, because researchers have found similar patterns in regard to many sports and in other motor skills. This type of experiment would be nearly impossible to do in a motor learning class because the students might not represent the range of expertise necessary (novice to expert).

Measuring Movements

The tasks used in motor behavior research provide a number of ways to measure movements and their outcomes. For example, suppose that senior citizens perform a movement that involves reaching 12 inches (30 cm) and then grasping and lifting containers of different sizes and weights. We could simply count the number of times that each senior citizen successfully lifts a container (in this task, reaching and grasping are prerequisites to lifting) and use the number of successful lifts as an outcome measurement. By doing so, we might find, for instance, that the number of successful lifts decreased as the objects got smaller or heavier. Although that would be an interesting

finding, our long-term goal probably would be to understand why or how the size or weight of the object influences success. Therefore, we would need to examine the process of the movement.

To examine process, we could use high-speed video of the movement, taken with two cameras, and evaluate how movement differs for 55-, 65-, and 75-year-old participants. From such videos, researchers can measure **kinematics** (location, velocity, acceleration—see chapter 4 on biomechanics). Other measures include muscle activation (EMG) of the arm and hand during the movement, the pinching force between the forefinger and thumb during the grasp, error in the movement, speed of the movement, reaction time before the movement begins, and accuracy of the movement. These process measures might help us understand how the reach, grasp, or lift changed for the various weights and sizes of objects.

Using our hypotheses (educated guesses), we could then begin to answer our original question. For example, we might hypothesize that the reaching and grasping phase would be the same for objects of the same size, even when weight varied, but that the lift would be different when the weight changed. Do you think reach, grasp, or lift would be different as we changed size but kept the weight the same? If we predict that reach or grasp changes, it is critical to measure the process.

Characteristics of Movement Tasks

In addition to deciding what to measure, motor behavior scientists must consider the characteristics of the movements they study (Newell, 2020). For example, some movements are more continuous (e.g., riding a bicycle, balancing on one foot, Zumba), whereas others are more discrete, involving a short period with a distinct beginning and end (e.g., rising to a stand from sitting, lifting an object). Further-

Studying Risk of Falling in Older Adults

A fall is an unintentional event that results in a person ending up on the ground or some other lower level. Many older adults are afraid of falling because they know a fall can mean injuries, disability, major lifestyle changes, loss of independence, or even death. How can we reduce the risk of falling in older adults? A first step is to identify the risk factors for falls in the elderly. Risk factors include dementia or cognitive impairment, loss of muscle mass or strength, reduced success with activities of daily living (ADLs; e.g., bathing, grocery shopping, doing laundry), physical inactivity or low fitness, depression, and impaired balance and gait. How might we study one or more risk factors? Research questions might include the following

- How do participants with high and low scores on ADLs perform on a fitness test?
- How does a balance and walking program affect individuals with dementia or depression?
- Are balance and strength related?

Professional Issues in Exercise Science

Expertise: When to Specialize in a Sport

Research on motor expertise often compares novices with experts to determine how they differ in various sports. The study of motor expertise is an appropriate recent addition to the study of motor development because age and experience are associated with increases in expertise. The study of sport experts has brought forth new thinking about the age at which a young person should begin to specialize in one sport (Côté et al., 2009; Goodway & Robinson, 2015) and what benefits can result from engaging in more varied experiences during childhood and adolescence. Data on elite performers and sport dropouts suggest that participating in more than one sport per year is beneficial up to 15 years of age. At about that age, it is appropriate to begin specializing in a sport; the exception occurs in sports where peak performance comes before physical maturation, such as women's gymnastics. Thus, at about age 15, athletes in most sports can decide to continue just for fun or to specialize with the goal of achieving elite performance.

Early specialization has been influenced by two notions—first, that it is associated with expertise, and second, that 10,000 hours of practice are necessary. However, the work done by Côté and colleagues (2009) presents compelling evidence that for most athletes, sport specialization should not be encouraged until at least age 12 and as late as age 15. Furthermore, a more accurate target for accumulated practice time to reach expert levels of performance would be 3,000 to 4,000 hours as opposed to 10,000 hours. What other exercise science issues should be considered by parents when their children express an early interest in sport specialization (Jayanthi et al., 2013; Waldron et al., 2020)?

Citation style: APA

Côté, J., Lidor, R., & Hackfort, D. (2009). ISSP position stand: To sample or to specialize? Seven postulates about youth sport activities that lead to continued participation and elite performance. *International Journal of Sport and Exercise Psychology, 9*, 7-17.

Goodway, J.D., & Robinson, L.E. (2015). Developmental trajectories in early sport specialization: A case for early sampling from a physical growth and motor development perspective. *Kinesiology Review, 4*(3), 267-278.

Jayanthi, N., Pinkham, C., Dugas, L., Patrick, B., & LaBella, C. (2013). Sports specialization in young athletes: Evidence-based recommendations. *Sports Health, 5*(3), 251-257.

Waldron, S., DeFreese, J.D., Register-Mihalik, J., Pietrosimone, B., & Barczak, N. (2020). The costs and benefits of early sport specialization: A critical review of the literature. *Quest, 72*(1), 1-18.

more, some movements are more open in character, whereas others are more closed. One example of an open movement is grocery shopping and storage; the environmental characteristics change from trial to trial because the items on the list probably change and the contents of the bags is different. In contrast, a closed movement takes place in a more consistent environment, in which the performer tries to do the same thing each time (e.g., rising out of your favorite chair or taking a shower).

Researchers and teachers must identify the characteristics (open or closed, discrete or continuous) of the tasks they study for two reasons:

1. The results of two studies can differ if the task characteristics differ.

2. The characteristics of the task used in the study must match the questions asked.

For example, if you want to know why something happens in bowling, you should not use an open, continuous task to answer the question. Professionals also must know how to draw distinctions between task characteristics because skills of one type might require a different instructional strategy than skills of another type. For example, a physical therapist would approach walking or balance differently than throwing or standing from sitting.

Learning, Retention, and Transfer

The idea of learning—as determined by retention and transfer—is related to the requisite number of hours of practice in the previous discussion about experts. A motor skill can be examined at any point, from the first attempt to well beyond mastery. The goal of most practice is learning, and **learning** involves long-term change in performance; therefore, in motor behavior, learning is usually defined as performance in retention and transfer trials scheduled after practice sessions. **Retention** is measured by performing the task after a time without practice to determine recall. **Transfer** can be tested immediately to determine how practice on one task or in one context affects performance on a different but related task or in a different context. One example that holds meaning for most students relates to the three most threatening words in college: "comprehensive final exam." At the end of the semester, how much do you remember when taking a comprehensive final examination? The purpose of the examination is to determine whether you learned what was taught in the course. For those who have really learned something, a comprehensive final examination is not a problem. As you will see in your motor learning class, retention is also used to measure learning in motor skills.

As previously mentioned, learning can also be measured in terms of transfer to other skills, in which one must do a slightly different version of the task. Experts are better at transferring information or skills than are novices; experts also can perform well after periods without practice (referred to as *retention intervals*). Many coaches feel frustrated when a team learns a play during practice but has trouble executing it in competition. This situation illustrates the lack of transfer; that is, it suggests that the team did not really learn the play since it could not execute when faced with new defensive pressures. The transfer in the practice-to-game example is a change of context by the addition of defense, speed, and other factors that were not present during practice. One example of a task that is well learned based on both retention and transfer is riding a bike. You can ride successfully after a winter off (retention), and you can negotiate new routes and terrain (transfer). Physical therapists face the same challenge as coaches: Can a client use the skills practiced in therapy for additional rehabilitation at home or use the skills practiced at the clinic in the real world?

The amount of transfer depends on several factors such as how similar the context is and the performer's level of skill. You might have learned to drive a car by starting in a simulator. Pilots often train in simulators early in learning and again later to practice dealing with emergency situations that cannot or should not be practiced during an actual flight. Clearly, the closer the simulator experience is to the real-world situation, the greater the transfer. Research shows transfer, especially among more skilled athletes, in identifying and understanding the structures and sequences of another sport (Smeeton et al., 2004). Transfer of pattern recognition was greater between soccer and field hockey than to volleyball. Of course, this transfer is cognitive rather than motor skill transfer. The research on pattern recognition and transfer supports two notions: transfer occurs across sports, and transfer across sports is limited. Experts in one sport do seem to recognize patterns in other sports but not as well as the expert in that sport. The essential question for exercise science professionals is, How much transfer from previous and perhaps well-understood situations is there to new situations? Does your understanding of one

therapy help you with another therapy? How much does spotting one exercise or individual transfer to a different individual or exercise? In those we help, how much transfer should we expect as we move among activities or exercises?

This segues into the last of Adams' (1987) themes, individual differences. Individual differences are the factors that make us unique and therefore help us perform physical activity and exercise in our own way. Some factors contributing to individual differences are body size and shape, past experiences, and genetics. If there were no individual differences, there would be no variation in research results except those due to the treatment. Fortunately, we are unique and tend to celebrate our individual differences; however, those individual differences present a challenge to exercise scientists.

> **KEY POINT**
>
> To study motor skill acquisition, researchers also must study how well skills are retained and how they transfer to other, similar situations.

Motor learning research reveals to us the reason for retention and transfer. Once you learned to ride a bike, you continued to practice until the skill was automatic and overlearned. Getting students to this point is the goal of comprehensive final examinations. By now you also should understand that performance in retention and transfer is the goal of most genuine efforts at learning. Therefore, the variables that affect retention and transfer are important to motor learning. Confusion can arise from the fact that everything we see ourselves doing is performance, but *some* of what we see is also based on learning. Performance can be influenced by variables that do not influence learning. Still, performance variables can hold interest for motor learning students and scholars, as you will see later in this chapter.

Overview of Knowledge in Motor Behavior

The following sections will present selected research from each of the three components of motor behavior: motor learning, motor control, and life span motor development.

Motor Learning

We begin this section with two big ideas or principles of motor learning. Then we will visit the remaining themes identified by Adams (1987): knowledge of results, distribution of practice, and individual differences. The two principles we have selected to represent motor learning hold that

1. correct practice improves performance and supports learning and
2. augmented feedback enhances practice and thereby learning.

These principles were selected for several reasons. First, the variables they emphasize—practice and feedback—are treated at length in many motor learning textbooks (e.g., Larsson, 2020; Schmidt & Lee, 2020). Furthermore, these are likely the two most widely studied variables in motor learning, and scholars agree that they are the two most important variables in motor learning (Schmidt & Lee, 2020). Indeed, two of our favorite historical adages for motor learning are "practice is a necessary but not sufficient condition for learning" and "practice the results of which are known makes perfect." Both of these evidence-based truths indicate that practice and feedback are independently important variables, and the relationship between them is also important.

KEY POINT

Motor learning is an internal state that is relatively permanent; practice is required in order for it to occur, and it is difficult to observe and measure. Thus, transfer and retention tests are used to measure learning.

Consider why these variables are important for you. Assuming that transfer occurs from one learning situation to another—for example, from learning a motor skill to learning the content in anatomy—understanding how to practice for learning could be helpful. As you move toward a career, even if you are not a teacher, you will no doubt be called on to help someone learn something, and understanding how to do so will be to your advantage. For example, you might need to teach an employee how to do their job, or you might want to help your child as they learn a new skill. Physical therapists and occupational therapists also teach clients to perform tasks. Thus, understanding how we learn motor skills is critical in many fields.

The adage "You have to crawl before you can walk" is a good way to think of skill acquisition—an orderly process that sometimes leads to falls.

The skill acquisition process constitutes an orderly progression. The learner begins practice by making many large errors while trying to understand the task. The cognitive demands are great in the early stages of learning; in fact, the task might be more cognitive than motor (Larsson, 2020; Schmidt & Lee, 2020; Schmidt et al., 2019). With practice, the errors become more consistent; that is, rather than making a different error on each trial, performers make the same errors repeatedly. At this point, the demands are less cognitive and more motor oriented, and the errors are smaller and less frequent. Thus, response execution improves. Once the learner can execute the skill with fewer and smaller errors and no longer has to think about the skill while performing it, the skill is considered learned or automatic. Now, instead of thinking about what each body part is doing (response execution), the performer can think about strategy or about the context when deciding which response to select and execute.

Recall that the study of motor learning is an effort to explain and predict conditions that will make skill acquisition easier or faster and make learning more permanent. Such conditions include individual differences between learners, such as speed of movement and coordination. Task differences are also important conditions in skill acquisition because tasks might be either more open (e.g., batting) or more closed (e.g., bowling). Learning also can be affected by environmental conditions such as practice, **feedback (intrinsic and extrinsic)**, and transfer. Another way of thinking about the changes in movement during learning is that the portion that is automatic increases; that is, more of the movement becomes preprogrammed so that the performer is not thinking about the way the body feels during the movement (Albers et al., 2005). At the same time, less of the movement remains under sensory control; that is, sensory information exerts less influence on how the movement is performed. More about this idea is presented in the section on motor control. The idea of automatic or learned skill is especially important in the study of experts because automatic performance allows the expert to focus on the strategy and not the execution of the skill.

At this point, you might wonder how we know when something is learned. Clearly, that is a challenge for motor learning research, theory, and application. In the section on research methods, we defined retention and transfer as ways to demonstrate learning. Adams (1987) identified retention as one of the five themes in motor learning research.

Whereas learning involves the relatively permanent acquisition of a skill, **motor performance** involves the degree to which someone can demonstrate that skill at any given time. In other words, performance is the current observable behavior—a snapshot of what the learner is doing right now. Sometimes performance reflects learning, as when a player can demonstrate a newly acquired skill. At other times, it does not. For instance, most of us have had the experience of turning in an examination and then remembering an answer that we knew but were unable to put on the paper during the exam. We would argue that we had learned the material but just could not produce it for the examination. In such a case, we are saying that performance does not represent learning. Thus, a single measure of performance might not indicate much about learning without other performance measures over time.

One way to distinguish between performance and learning variables is to remember that performance variables produce a temporary effect, whereas learning variables produce a relatively permanent effect. Knowing the difference between these effects is a critical part of motor learning. For instance, you might have trouble typing a term paper after working late into the night and getting very tired. Yet the next day, after getting some rest, you might be able to type rapidly while making few errors. In this case, fatigue depressed your performance, but you had learned the keyboarding skills through practice and could demonstrate this learning once you were rested.

KEY POINT

Because motor performance can be influenced by variables such as fatigue, the best measure of motor learning involves tests of retention and transfer.

Research and Evidence-Based Practice in Exercise Science

Visual Cues in Skill Rehabilitation

Gómez-Jordana and colleagues (2018) demonstrated that virtual footprints presented in an immersive, interactive VR environment can improve walking performance significantly (i.e., step length, cadence, and step velocity) in participants with Parkinson's disease (PD). Their analysis showed clear improvements in the PD patients for all virtual cueing conditions, in particular with respect to step length, step velocity, and overall gait variability. Step length (smaller steps) alteration is a common negative change associated with PD; this study supports the inference that the neural mechanisms behind the control of self-paced movement (e.g., walking) improve when external stimuli (e.g., virtual footsteps) are present to PD patients.

that better learning results when individuals practice skills in larger numbers of shorter practice sessions (distributed schedule) than when they do so in long and fewer sessions (massed schedule). The positive effects of distributed versus massed practice on both skill acquisition and retention in the context of motor skill training also have been highlighted in more recent literature (Krigolson et al., 2020; Kwon et al., 2015; Spruit et al., 2015). For example, Kwon and colleagues found that participants in the distributed practice group (two 12-hour intertrial intervals) were both faster and more accurate than participants in the massed practice group (two 10-minute intertrial intervals) during retention tests of the motor sequential learning. This line of research helps us understand the notion of motor performance—the temporary impact of practice distribution—versus learning.

Feedback—Knowledge of Results and Performance

As you might have experienced, feedback is an integral part of the practice regimen that leads to learning. In fact, one cannot learn a skill correctly without feedback. It guides the learner toward performing the task correctly and reinforces correct performance. Feedback can be intrinsic or extrinsic.

- Intrinsic feedback consists of information about performance that you obtain for yourself as a result of a movement.

- Extrinsic feedback (also called *augmented feedback*) consists of information provided by an outside source such as a referee, judge, teacher, coach, or video recording.

Motor behaviorists categorize extrinsic feedback as either knowledge of results (KR) or knowledge of performance (KP) (figure 6.3). While KR and KP are terms used very specifically in experiments, the distinction is likely arbitrary outside laboratory experiments. The role of extrinsic feedback in learning is to provide accurate information that performers cannot provide for themselves. Experiments on KR and KP might hide outcome information from the performer in order to understand underlying factors or negate experience; in physical activity and exercise, outcomes are rarely hidden from the performer.

As its name implies, knowledge of results consists of information about the result of a movement—for example, "You did nine repetitions." The performer can count, and therefore KR from a personal trainer is not necessary or helpful. KP focuses on the performance and issues like angle, speed or force, acceleration or deceleration, starting or ending location, and so forth. Early in learning motor skills, KP is critical and is provided by a personal trainer, therapist, or other expert; thus, it is extrinsic.

KEY POINT

Knowledge of results is about the outcome of a movement. Knowledge of performance is about the process of movement and is important for beginners.

Figure 6.3   Four kinds of feedback for strength training.

As skill develops, the performer might detect performance information and begin to use that information to correct errors and to repeat correct movements. For instance, a good typist who is told that they keyed in "Hybe" instead of "June" would know that their right hand was one key away from the correct initial position on the home row. A beginner typist, however, might need a cue (e.g., "Check your home position") to correct the error; this typist is using KP feedback.

The ultimate goal of feedback is to help performers detect and correct their own errors. The study of feedback has focused on KP and KR with regard to frequency (how often), precision (how detailed), modality (auditory, kinesthetic, or other source), and processing time. Feedback also can be either reinforcing (e.g., "That was a good effort") or negative (e.g., "You did not try very hard"), which are equally good.

Frequency of feedback can be addressed in terms of the percentage of practice trials for which feedback is given (e.g., 50% or every other trial versus 100% or all trials). On the surface, more feedback might seem better, but performers can become dependent when feedback is too frequent. Constant feedback is also counterproductive if the goal is to detect and correct one's own errors, because it prevents one from learning the processes of detection and correction. Thus, knowledge of performance (KP) should be given more often at the beginning of learning and then gradually reduced (Schmidt & Lee, 2020); this process is often referred to as *fading* the feedback.

In your study of motor behavior, you will learn about two other variables in motor learning besides KP and KR—precision and modality of feedback. You also are likely to learn about the amount of time that children need in order to use feedback.

KEY POINT

Extrinsic feedback contains information that the learner could not obtain alone, should be corrective and provided for about half of the trials, and must be followed by sufficient time to make corrections before trying again.

Practice and feedback are important elements in learning a motor skill. Complex skills require correct execution of the movement and good decisions—for example, what skill to execute and when. In baseball, the batter watches the pitcher for critical information that might give a clue about the pitch. In doing so, the batter ignores the crowd's yelling and tries to gather only relevant advance information about the pitch and where to hit it. Figure 6.4 provides an example of how long the batter has to make decisions about swinging the bat after the pitcher releases the ball. The figure demonstrates that a batter has more decision time (150 milliseconds versus 130 milliseconds) if they wait longer and speed up the swing (140 milliseconds versus 160 milliseconds). The example also demonstrates how critical advanced information is to successful performance—in this case, whether the pitcher is likely to throw a fastball or a breaking ball, as well as the likely pitch location. Thus, you might hear a baseball commentator say, "Three balls, one strike; they'll be looking for a fastball high and inside" (i.e., a certain pitch in a specific location). The batter is trying to reduce the decision to a go or no-go situation—they'll swing if the pitch is what they expect—rather than having to consider all possible options. The batter can do this because in this situation they can decide not to swing even if the pitch is a strike (because another strike will not make an out). If the count were three balls and two strikes, the batter would be less likely to look for a specific pitch. Practice and feedback help the batter learn not only how to swing the bat and how differences in swings produce different results, but also how to increase the odds of success in a variety of contexts. Understanding how to organize practice and provide feedback helps learners become successful in increasingly complex situations. This means the coach, teacher, or therapist must understand the principles of motor learning.

Motor Control

Motor learning research deals with the acquisition of skilled movements as a result of practice, whereas work in motor control research seeks to understand the neural, physical, and behavioral aspects of movement (Schmidt et al., 2019). We begin this section with two principles of motor control. Then, we will review five areas of motor control: degrees of freedom, motor equivalency, serial order of movements, perceptual integration, and skill acquisition. Examples will demonstrate the importance of motor control to exercise science.

Motor control has two principles:

1. The brain uses the central nervous system to initiate and control the muscles that make the desired movements.
2. One goal of most movements is to rely on the decision-making centers in the brain as little as possible once the movement is initiated.

These two principles were developed on the basis of two theories and many research studies. One

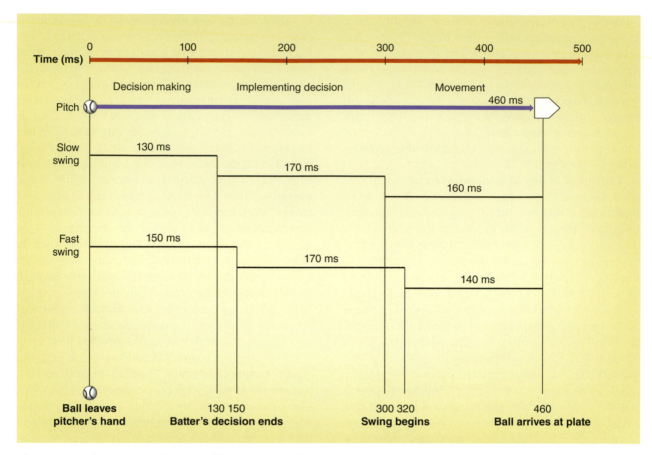

Figure 6.4   Time line showing the critical events in hitting a pitched baseball. The movement time is 160 milliseconds for the slow swing and 140 milliseconds for the fast swing.

Reprinted by permission from R.A. Schmidt and T.D. Lee, *Motor Learning and Performance: From Principles to Application*, 6th ed. (Champaign, IL: Human Kinetics, 2020), 145.

of these theories focuses on what are called *motor programs* and the other on *dynamical systems*. Motor programs (Latash & Zatsiorsky, 2015) provide a theoretical explanation of how we produce and control movements. We can compare a motor program to a computer program that does math problems. First, you select the program to use; this is response selection. The program can add, subtract, multiply, and divide. If you input the relevant numbers and indicate which math operations to perform, the program outputs the answer; this is response execution. A motor program operates in the same way. First, you select the program (response selection) and indicate what it should do (operations). Then the program specifies how to do the skill and sends signals through the spinal cord to the muscles that perform the movement (response execution).

At a minimum, a motor program must do the following things (Schmidt & Lee, 2020):

- Specify the muscles involved in the action
- Select the order of muscle involvement
- Determine the force of muscle contractions
- Specify the relative timing and sequences of contractions
- Determine the duration of contractions

KEY POINT

A motor program is a proposed memory mechanism that allows us to control movements. As motor programs are developed, they become more automatic, thus allowing us to concentrate on using movement in performance situations.

If we had to remember how exactly to create every single movement we execute, our memories would be overloaded. Thus, motor programs explain why we do not have a central storage problem for the great diversity of movement skills we use. Instead of storing in memory each movement that we have ever done, we store groups of movements with similar characteristics. These groups, called *schemata* (Schmidt, 1975), serve as the foundation of motor program theory (Latash & Zatsiorsky, 2015; Schmidt et al., 2019).

Some researchers, however, do not agree with the idea of motor programs drawn from schema theory, and they have proposed another theory. This approach, referred to as *dynamical systems theory* (Fischman, 2007; Haken et al., 1985; Kelso, 1995), suggests a more direct and less cognitive link between motor action and information picked up by the perceptual system. This direct link is called a *coordinated structure*, one characteristic of which is automated movement that relies on very little decision making or central control in the brain.

KEY POINT

Exciting and controversial results are likely to be produced by research that contrasts predictions and key elements of the motor program view of motor behavior with those of the dynamical systems view.

You might be wondering which theory—motor program or dynamical systems—more accurately represents the process of motor control. One decisive test for theories of motor control is to examine how well they explain motor learning and the development of motor expertise (Schmidt et al., 2019). You will have a chance to examine and compare these theories in your motor learning class—and perhaps help advance them if you choose to pursue graduate study in motor control. In motor control research, both theories tend to deal with similar challenges; one is attention, and another is referred to as "the problems of motor control." The next sections will provide some detail on those challenges.

Attention

You might notice that it is easy to carry on a conversation with a passenger while you are driving on a highway that is not busy. But the conversation can get distracted when the traffic becomes heavy because you must switch more attention to the road. Attention seems to be limited in this scenario and might only focus on either external sensory events (other drivers' actions on the highway) or internal mental operations (trying to respond to the passenger). Attention refers to characteristics associated with perceptual, cognitive, and motor efforts related to the performance of motor skills (Magill & Anderson, 2021). Particularly, attention refers to limited information processing capacities. Attention can be divided among two or more tasks simultaneously, or

to the detection of information in the environment. However, because attention is limited, at some point the amount of sensory information or the number of tasks will exceed the amount of attention. At that point, attention must be focused on fewer tasks or less information (Abernethy et al., 2007; Abernethy, 1988).

Until the 1950s researchers provided the theoretical foundation to explain attention as a limited capacity resource in human performance. Theories emphasizing the limits of attentional resources hypothesize that we can perform multiple tasks at the same time if the attention demanded by the tasks does not exceed the available attention capacity. In the 1990s Wolfgang Prinz (1997) proposed the action–effect hypothesis to further show that the learning and performance of skills can be optimized when a learner's attention is directed to the intended outcomes of the action rather than on the movement patterns.

Attention is limited, and attention is necessary to select and control movements. So, how do we manage this challenge? From all the information that is available, we select only what is necessary for success. This is called *selective attention*. You are undoubtedly familiar with the concept: focusing on only important information and ignoring everything else. We do this as we study by ignoring irrelevant noise, when we are carrying a tray loaded with food and drinks, and when we perform a sport skill in front of a crowd. Failing to focus or use selective attention often leads to errors.

KEY POINT

Attention or cognitive capacity is limited. Therefore, we must use it efficiently as we perform motor tasks. Selective attention is used to ignore less important or unimportant sensory information.

Several techniques have been used to examine attention; one example is probe reaction time (PRT). In PRT a stimulus is presented to the performer at various times during a task, and the amount of time required to respond is measured. Reaction time should be slower than when measured as the primary task. Generally, these studies showed PRT was affected more in the early stages of learning and in novices. Another way of looking at selective attention is to compare experts and novices who are asked to do two tasks simultaneously. Expert and novice golfers were asked to putt and do a word search task (Beilock, Wierenga, & Carr, 2002). Both expert and novice putters were able to complete the word search successfully, but only the experts were able to maintain their putting performance. The attentional demands of the secondary task interfered with the novices' performances on the primary task of putting. The expert putters were able to select important information on which to focus and ignore other information. Further, experts likely preprogrammed more of the putt and therefore had additional capacity (attention) to divert to the word search.

Research evidence suggests that the learning or performing of motor skills can be degraded if the learners' attention is directed to their own movement and away from feedback (Wulf, 2007; Wulf & Prinz, 2001). As shown in the distracted driving example, an internal attentional focus on the conversation constrains the motor system by interfering with natural control processes, whereas an external focus allows automatic control processes to regulate the driving movement. It is important to maintain goal-directed attentional control to action-relevant information while diverting attention away from the interference of irrelevant and distracting information such as the conversation.

Remember that a principle of motor control is to preprogram as much of a movement as possible, which means you do not have to focus internally or on the movement. That allows you to focus on the environment. A key finding in motor learning was that extrinsic feedback, particularly KP, was critical for learning and improvement. Now we see that internal feedback, sensory information, is also important for successful performance. Further, attention is limited. This means that we must select the critical sensory information during the movement, program as much of the movement as possible, and after the performance consider the external feedback (KP) to maximize practice.

KEY POINT

A goal of practice is to preprogram more of a movement so executing the skill does not require attention, or requires less attention, during the movement. This allows attention to be directed to other aspects of the task.

In a wide range of perceptual-motor tasks, visual attention to environmental information is essential before carrying out an action. In other words, attention is focused so you can visually select the specific performance-related information in the environment. A physical therapist might use cones as visual cues for a client with Parkinson's disease so as to encourage large steps or to reduce shuffling steps. Rather than saying "big steps" or "pick up your feet," the cones serve to guide step length and step height. The therapist simply says, "Walk through the cones." The positive effects of guiding visual attention in enhancing the acquisition of expert-like perceptual skills in both skilled and novice athletes are supported in the motor control literature (Abernethy, 1988; Abernethy et al., 2012; Ryu et al., 2013).

Researchers have consistently reported evidence that instructions directing learners' attention to the effects of their movements on the environment (e.g., the apparatus; external focus of attention) are more effective than instructions directing their attention to the movements themselves (i.e., internal focus of attention) in sport skills (Wulf et al., 2001; Wulf et al., 2002) and rehabilitation settings (Fasoli et al., 2002). However, a growing body of research also demonstrates that novices and experts can differ in motor skill performance based on the effect of attention focus instruction (Marchant et al., 2009; Perkins-Ceccato et al., 2003). For example, Perkins-Ceccato and colleagues (2003) assessed whether the effects of attentional focusing instructions on golfers' pitching performance depended on their skill level. Their study demonstrated that highly skilled golfers performed better with external attention instruction (e.g., "Concentrate on hitting the ball as close to the target as possible") than with internal focus instruction (e.g., "Concentrate on the form of the golf swing").

An individual has two specific and perhaps less than helpful goals on the first performance of a new task: avoiding being hurt and avoiding looking silly. Neither of these goals assists with the development of a motor program or effective use of attention. During practice learner goals must, and generally do, shift to how the movement feels, what the focus should be, and using feedback. Consider an older adult recovering from a fall. During practice they likely are concerned about how others will perceive them ("Do I look silly?") and avoiding another fall. Because ambient vision is critical to maintaining balance, it would be better for this patient to focus on a visual target than considering what others think. Physical therapists often use a belt or other support tool or prop to reduce the fear of falling. Just as important would be to provide a visual target (a spot on the wall) and to tell the patient that it is normal to worry about another fall. The goal is to focus attention on the important aspects of the task, using ambient vision, other sensory information, and props to relearn balance. The therapist helps the patient selectively attend to ambient vision. This could be combined with feedback, for example, "Good; you are looking at the spot on the wall." Both attentional focus and feedback are important factors in learning.

Human motor performance requires complex context-dependent processing of information that emerges from the selective attentional processes. The theoretical basis for this hypothesis is related to how we code sensory and movement information in memory and is supported by the cognitive neuroscience of attention theory (Deco & Rolls, 2005; Deco et al., 2005).

Five Aspects of Motor Control

In addition to attention, the study of motor control addresses five areas (Rosenbaum, 2010):

1. Degrees of freedom—coordination
2. Motor equivalency
3. Serial order of movements—coarticulation
4. Perceptual integration during movement
5. Skill acquisition

We will describe these in the following sections; however, a motor behavior or motor learning and control class will provide more depth and greater understanding of these concepts.

Degrees of Freedom   One important characteristic of motor control theories is to provide explanations of how the central nervous system functions in movement choice and control when directing movement. **Degrees of freedom** (DoF) refers to the number of things that need to be controlled simultaneously; fewer things mean fewer DoF. DoF is often defined as unique joint rotations that contribute to a movement. Thus, reducing the number of DoF makes controlling movement easier. Studies on the numerous DoF problems in human movements focus on the control strategies of freezing (decreasing) or releasing (increasing) on DoF in motor learning, the possible effects of the class (e.g., discrete skill class of grasping a cup or sitting-to-standing; continuous skill class of walking or handwriting), and the objectives of the skill (e.g., balance, velocity, and accuracy) on DoF control strategies (Guimarães et al., 2020). Beginning drivers often stop completely before making a left turn because this reduces the number of things that must be controlled and increases the safety of the turn. DoF is a notion similar to that of attention, with the essential question for both being how we can handle the numerous complex aspects to execute a movement successfully.

Motor Equivalence   According to Bernstein (1967), **motor equivalence** refers to the capability

of the motor control system to enable a person to perform an action (with a great degree of similarity in characteristics) by adapting to the specific demands of the performance conditions. Research shows that the action of handwriting demonstrates the concept of motor equivalence (Wing, 2000). That is, the specific elements of the handwriting, such as letter forms, writing slant, and relative force for stroke production, show a great degree of similarity whether the person performs the writing skill with the dominant hand, the nondominant hand, or the foot. Try this yourself, and you will see that writing with your dominant and nondominant hand looks more alike than when you compare your writing to someone else's writing. Support for the idea of centrally stored (in the brain) motor programs first came from studies like the handwriting example. The phenomenon of motor equivalence suggests the stability of human movement and has been observed in other skill performances (e.g., pointing movement and multifinger accurate force production; Marteniuk et al., 2000; Mattos et al., 2015) in addition to handwriting. For example, Marteniuk and colleagues (2000) found very similar spatial–temporal movement patterns are produced regardless of the particular muscle groups recruited to achieve a goal while participants are performing the pointing movement. Studies suggest that task complexity might be one parameter leading to the emergence of motor equivalence, which serves as a control strategy to cope with increasing task demands.

Serial Order of Movement   **Serial order** is the sequence and timing of movements so that the movement produced is the one that was intended. Driving a car with a standard transmission requires that the clutch is depressed before the gearshift is moved; when you hear gears grinding you can assume the order was wrong. In speech, "Spoonerisms" demonstrate errors in serial order; "You missed my history lecture" becomes "You hissed my mystery lecture" (Rosenbaum, 2010). Coarticulation refers to the impact of one part of a movement on another part of the movement. The fingers of typists move toward the next key strike before the previous strike is complete. If you could move only one finger at a time your typing speed would be reduced dramatically; thus, coarticulation provides positive benefits as well as explains some errors. When we type we might reverse letters in a word as speed increases, leading to errors in typing.

Recall that Schmidt's schema theory included the order of movements as a key component. Changing the order of the movement parts can create a different skill similar to typing a different word when the order of the finger taps changes during typing. In large movements the amount of variation allowed within a motor program can be defined by the similarity of the movement produced to the planned or expected movement. The overarm (or overhand) throw is important in many sports (e.g., football passing and baseball). The goal for forceful throws is to accelerate with maximum force as close as possible to ball release. This means as a big forward step is taken the hips begin rotating forward, and the last body parts moving forward are the hand and arm with the ball. If the hand precedes the shoulder, the throw looks like a shot put rather than a throw from the outfield to second base. If one took the step after ball release the force would be greatly reduced; for the forward motion of a throw we think "step, turn, throw," not "turn, throw, step." As we practice, we can increase speed and force but always relatively so all parts occur in the same order.

Perceptual-Motor Integration   Perceptual-motor integration refers to the use of sensory information during a movement. While preprogramming produces more rapid movements and demands less attention to sensory or kinesthetic information during the motion, using sensory information can produce more accurate movements. The challenge is to maximize preprogramming while integrating relevant perceptual information. Rosenbaum (2010) separates movements into two phases to demonstrate the importance of perceptual-motor integration: the programmed (ballistic) and sensory phases. However, he also explains that movement itself improves perception, in part, because movement allows us to see, hear, and feel more of our environment.

Skill Acquisition   We will not review skill acquisition (motor learning and performance) as one of the problems of motor control because skill acquisition was presented previously as the focus of motor learning. The brain initiates the planning of movements, and the nervous system then sends signals through the spinal cord to motor units within muscles, which in turn make the movements. What is not known is how the brain represents the information to be sent to the muscles. The study of motor control is an effort to understand what the brain, nervous system, and muscles are doing to direct movements. Just as skill acquisition—including the notions of learning and improvement—is the focus of the study of motor learning, it is critical in the study of motor control as well. Indeed, skill acquisition accentuates the relationship between motor learning and motor control.

Motor Control in Exercise Science

You still might be wondering why understanding motor control is important. The answer is clear for exercise science professionals in neuroscience, physical therapy, or athletic training: anyone who wants to learn, teach, or rehabilitate a skill uses the principles of motor control by trying to use the simplest, most effective movement technique at the early stages of learning.

Changes in the brain affect movement; thus, some studies in motor control examine changes in the brain and the impact of those changes on movement. In patients with Parkinson's disease, for example, walking, gait, and balance are challenging due to changes in the brain (Peterson & Horak, 2016). Research identified structural changes in the brain during space flight; however, those changes did not affect balance because the preflight and postflight balance were not different (Koppelmans et al., 2016). Thus, a basic understanding of the neuromotor mechanisms and processes underlying the control of voluntary human movement is necessary for exercise scientists. If you are planning to become a rehabilitation professional (e.g., athletic trainer, occupational therapist, physical therapist), such knowledge is needed for the assessment of motor dysfunctions and the development of appropriate interventions and treatment. Moreover, some of what you will learn about motor control is not very intuitive and thus is often misunderstood by those who are untrained, leading to less effective instruction, feedback, and practice scheduling.

Motor Control Application in Physical Therapy

Parkinson's disease (PD) is a neurodegenerative disorder that affects specific areas of the brain that produce dopamine. Approximately 60,000 people in the United States are diagnosed with PD each year. Bradykinesia (smaller movements), tremors, and gait and balance problems are characteristics of PD. Physical activity and exercise are recommended for those with PD. A specific therapy provided by a specially certified physical therapist is also recommended: Lee Silverman Voice Treatment BIG (LSVT BIG). LSVT BIG strives to increase the amplitude of movement by exaggerating movements—for example, large arm swings during walking—which retrains the muscles and slows down the progress of the disease. The therapy uses the notion of motor programs and an understanding of motor control to provide better outcomes for PD patients.

Consider the following example. You are sitting at a red traffic light, staring at the light. Cars on your right and left also are waiting for the light to change. One or both of those cars eases forward, and you hit the brake because it feels like you are rolling backward. You did not think about hitting the brake; it was automatic. This process can be explained equally well by either of the two theories of motor control. However, the idea of automatic responses is not well understood by many people. Ambient vision captured the movement of the cars and made you think you were moving. Before you could think "I am moving" your foot hit the brake. Similarly, those who understand eye movements for tracking high-speed objects or how best to use the sensory system (ambient vision) to track fast objects and make rapid and accurate adjustments will perform better than those who do not understand ambient vision. When our ambient vision captures something moving toward us we blink or duck, even when what we see will not hit or hurt us. Our sensory system avoids central processing (thinking about it) and responds quickly as we brake the car and as we track fast-moving objects and duck.

We hear comments alluding to the shift from central (decision-making) control of movements to peripheral (or sensory) control when our fitness clients say things like, "It just felt right." The most rapid and accurate corrections are made when most of the movement is preprogrammed and sensory information is integrated only when necessary at the very end of the movement. In the study of motor control, this issue is a perceptual-motor integration problem: How do we best use sensory information to control movements? If we had to think through each step when running stairs, we would be walking the stairs. Instead, we program the step and adjust at the very end of the step when the foot is landing. Rapid adjustments, like those of the fitness client running stairs, might be explained by positing either that the information is not processed, which eliminates decision-making time (e.g., reaction time), or that no competition exists between responses (Wyble & Rosenbaum, 2016). The idea is that the demands are reduced to the simplest cases possible, thereby increasing the speed of adjustments. More choices or complexity decreases speed. For example, if each step was a different height, running stairs would be more complicated and movement would slow dramatically.

Motor control scientists try to explain what they observe, but they do not always agree on the explanation. We continue working to understand motor control because the range of motor skill performance—from normal to expert—depends on the shift from learning (central control, practice, and feedback) to more automatic movements programmed ahead of time and controlled peripherally.

Life Span Motor Development

As you will recall from the section on the history of motor behavior, life span motor development is a relatively new concept. Historically, motor development was the study of change in movement in infants, children, and adolescents. As the population aged, and probably as scholars aged, change was noted at the other end of the life span and became an area of interest and scholarship. Thus, what was motor development became life span motor development. The opening scenario for this chapter provided examples of observable changes in movement at both ends of the age spectrum.

One hundred years of research in motor development led to the development of four principles of motor development that capture much of the research in motor development while providing a framework for thinking about the changes in movement from birth to maturity. Those principles of motor development hold that

1. children are more alike than different,
2. children are not miniature adults,
3. no one is perfect, and
4. good things are earned (Thomas & Thomas, 2008).

How might we make the first two principles apply to the entire life span?

1. People are more alike than different.
2. Human bodies change in an orderly manner.

People Are More Alike Than Different

Across the life span, humans perform a multitude of movements each day. Walking has been studied extensively at both ends of the age spectrum. Infants begin walking around their first birthday; however, the skill looks different from the walking pattern of older children and adults. Adolf and Hoch (2019) remind us that walking and other motor skills develop within a complex system of factors internal to the infant (embodied) and in the environment (embedded). A challenge is that norms—descriptions of a population—tend to ignore the highly individual development of each infant. The point is that we are unique, and we are also similar. The first steps of infants typically share these characteristics: arms held high, flat-footed short steps, and feet shoulder-width apart. Infants and toddlers often walk like Dr. Frankenstein's monster. Walking in older children and adults is characterized by arms swinging in opposition to the stepping leg and proportional to the step length; heel strike; a pattern of one foot, both feet, one foot; and steps on a line directly under the body (Payne & Isaacs, 2020). When we step on the line directly under the body the step width is zero. In 40% of elderly people, the step width increases to 8 centimeters (3 in.) for females and to 10 centimeters (4 in.) for males from 0 (Snijders et al., 2007). As balance decreases and fear of falling increases, we see the arms held high in older populations.

Walking is critical to many aspects of development. Walking allows infants to see more of the environment, travel in the environment, gain independence, and control objects. Walking is important for elderly people to maintain independence. By 60 to 69 years of age, 10% of the population will have a gait disorder, and by 80 years of age, 60% will have a gait disorder (Pirker & Katzenschlager, 2017). Further, speed of walking declines each year after about age 60. Why does walking change in pattern and speed in the elderly? Three general causes are medical (e.g., heart failure and obesity), neurological (e.g., Parkinson's disease and dementia-related disorders), and nonneurological (e.g., osteoarthritis of knee and hip). Twenty years of research on gait and cognition suggest that these might deteriorate together (Snijders et al., 2007).

Motor skills improve from birth to maturity, remain rather constant in adults, and then decline with aging. The order is generally consistent, but the rate at both ends of the age continuum varies. In the elderly both environmental factors and genetics have more time to affect movement and thus have a greater impact than we see in infants and children. Thus, two 70-year-olds might differ more from each other than two 7-year-olds do. However, humans are more alike than different, particularly in the order of change. In addition, males and females are more alike than different at both ends of the life span. Recall how both males and females demonstrated greater step width, but the difference between men and women was small (2 cm [0.8 in.]). This is true for most aspects of development prior to puberty. After puberty men have an advantage in some activities because of greater muscle mass and more testosterone. However, in old age that advantage decreases, as we will see in the next section.

Human Bodies Change in an Orderly Manner

Physical parameters (e.g., stature, mass, muscle) affect performance of motor skills and health. Changes in our bodies occur at both ends of the life span, although the greatest changes occur between birth and maturity. That is why a principle of motor development is that children are not miniature adults. In the course of the life span, changes in strength and motor coordination are most rapid at the extremes of the age continuum. Growth is less important after adolescence because the changes in physical parameters are less dramatic in adults and smaller in elderly people (Fernihough & McGovern, 2015) than in infants and young children. Clearly, infants, toddlers, and children are smaller than adults. Growth is rapid and affects movement; a child can gain 0.8 inches (2 cm) in stature overnight. Thus, growth is almost always a factor in research with children. Females grow until about 13 or 14 years of age, and males grow until 18 to 20 years. However, in motor learning studies with adults, growth is not a factor. Growth is studied in life span motor development but not in motor learning or control.

Children grow in three physical domains:

- Overall size
- Proportions (e.g., leg length, shoulder breadth, chest depth)
- Body composition (e.g., muscle increases)

Growth influences motor performance, partly because children must contend with the changes in their bodies—an issue that adults do not have to deal with to such a large degree. Adult–child differences also can help explain child–child differences; these are called *individual differences*. For example, size and strength have a positive relationship in that larger children are generally stronger; therefore, as children grow, the increase in size produces increases in strength. However, strength also increases due to neuromuscular efficiency. Because physical growth lies beyond our control, we must understand how it influences performance in order to accommodate the challenges that it brings.

First, we need to examine physical growth and development; see figure 6.5 for a summary of growth in terms of stature and weight and figure 6.6 for changes in proportion. Then we will see how these factors affect overhand throwing skills.

Physical growth is rapid during infancy, generally constant during childhood, and rapid during the growth spurt associated with puberty (Gabbard, 2021; Malina et al., 2004). Children's limbs grow more rapidly than the torso and head, and the increase in proportion of total height attributed to leg length explains some of the improvement in running and jumping performance during childhood. In addition, the shoulders and hips grow wider, and males' shoulders become much broader on average than females' shoulders, which gives males an advantage in certain activities, including throwing. The typical growth pattern supports the principle that children are not miniature adults. However, length and weight (mass) at birth are somewhat predictive of later stature (Fisher et al., 2018). Genetics has more influence on birth length and later stature than birth weight.

Growth is orderly and marked by increases in length, then breadth, and then circumference. (Figure 6.6 shows change in proportion from birth through adulthood.) As a result, children in upper elementary school often look awkward because their legs and arms appear to be long and skinny. After most of the

final length of a limb has been attained, the limb gains thickness as bone circumference and muscles (and fat) increase, thus making the body look more proportional; in other words, children fill out. In fact, males' shoulders and chests usually grow (bones and muscle) until about 30 years of age, whereas most women stop growing by their late teens.

The changes produced by physical growth present three kinds of challenges: mechanical, adaptive, and absolute. The mechanics of movement change because of the different body proportions; the person must adapt to a rapidly changing body. These changes are especially problematic in seasonal sports. Consider a wrestler who experiences a rapid growth spurt from the end of one season to the beginning of the next. His center of gravity has changed, which might influence his balance and the location of optimal points for exerting maximal force. Performance also can be influenced by absolute changes in size. Females gain fat at puberty, which adversely influences performance in most physical activities; in contrast, males gain muscle, which exerts a positive influence on performance. In these ways, physical growth during childhood interacts with all of the other factors that are developing (e.g., motor programs) and therefore must be considered in both instruction and research.

Shifting to the other end of the life span, the bodies of older adults experience predictable changes. What might surprise you is that some of the changes begin to occur earlier than age 65 years, though generally, the changes accelerate after 65 to 70 years of age. Stature begins to decrease by about 0.4 inches (1 cm) per year in humans at about age 40 years and decreases rapidly after age 70 (Shah & Villareal, 2017; Walston, 2020). Measuring stature in older adults can be a challenge because some older adults cannot stand or balance, so knee height or fibula and ulna

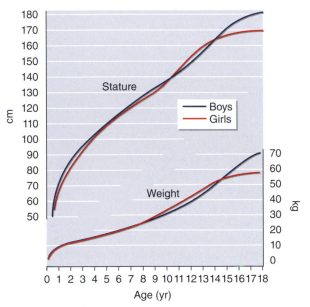

Figure 6.5 Average height and weight curves for American boys and girls.

Adapted by permission from R.M. Malina, "Physical Growth and Maturation," in *Motor Development During Childhood and Adolescence*, edited by J.R. Thomas (Minneapolis, MN: Burgess, 1984), 7. By permission of Jerry R. Thomas.

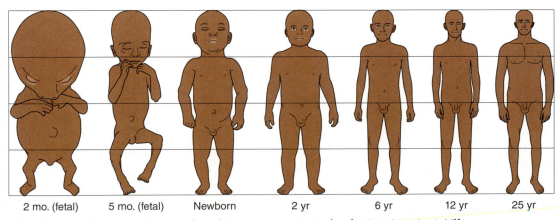

2 mo. (fetal)   5 mo. (fetal)   Newborn   2 yr   6 yr   12 yr   25 yr

Figure 6.6 Changes in form and proportion of the human body during fetal and postnatal life.

Adapted by permission from K.M. Newell, "Physical Constraints to the Development of Motor Skills," in *Motor Development During Childhood and Adolescence*, edited by J.R. Thomas (Minneapolis, MN: Burgess, 1984), 108. By permission of Jerry R. Thomas.

Career Opportunity

Fitness Instructor for Older Adults

Motor behavior provides the information necessary to work effectively with individuals across the life span. Understanding how movements are learned and controlled, how individuals differ in their motor performance, and changes that are age related but not age dependent prepares exercise science professionals to lead programs for children and seniors. As the population ages, career opportunities working with the elderly will continue to expand. Developing, managing, and leading exercise programs for older adults will be critical to improving their health and for successful aging. Health insurance often covers exercise programs for older populations; Silver Sneakers is included on some Medicare plans and is celebrating 30 years of fitness for seniors. The exercises are designed specifically for older populations and are delivered in a variety of settings. For example, Silver Sneakers might be offered in a recreation or fitness center as one component of their programming, at a facility focused on older participants, or online. Positions range from instructor to management, involving the coordination of several classes and instructors. Working with older adults can be rewarding for you and life changing for them.

bone length have been used to estimate stature with some success, generally overestimating stature by approximately 0.2 inches (0.5 cm) (Auyeung et al., 2009). Thus, if we are fortunate enough to live to age 70 years or longer, we are 1 to 3 inches (0.4-1.2 cm) shorter than before (Shah & Villareal, 2017; Walston, 2020). We begin the aging process at around age 30; lean tissue (muscle and organs) decreases, and fat increases. Men tend to gain weight until about age 55, while women gain weight until age 65. Both tend to lose weight later in life; unfortunately, this is probably muscle and bone mass. Lifestyle has a positive impact on these changes because exercise and diet help maintain muscle and bone density and reduce fat. Thus, while change is inevitable and tends to be negative at the old-age end of the spectrum, change is less than during infancy and childhood. Throughout our life span, we can have a positive impact on our bodies through exercise.

No One Is Perfect

The short legs and relatively large heads of infants and toddlers are not ideal for many movements such as running, for which longer legs are an advantage. You can probably think of a sport or movement where a particular body shape or size is likely an advantage, yet the ideal body for an activity does not guarantee success. Further, we can all recall making mistakes as we tried to learn a new skill. Thus, in both respects, no body and nobody (no one) is perfect. Other systems also affect performance at both ends of the life span.

Children differ from adults in terms of neurology, or what we might think of as neural hardware (analogous to a computer with its CPU and monitor). In addition, as adults age, their neuromuscular hardware changes; **sarcopenia** (loss of skeletal muscle), a slowing of the nervous system, and reduced phys-

ical activity can affect both health and performance in older adults (Tarantino et al., 2013). Moreover, attaining, maintaining, or regaining a physically active lifestyle predicts how long an older adult is likely to remain independent (Rikli, 2005; Rikli & Jones, 2013). Thus, in aging, we see a pattern similar to that of childhood—namely, a relationship between physical growth factors (e.g., muscle), practice, and performance.

Task performance by elderly persons can be affected negatively by nervous system deterioration. As people age, they lose neurons in the brain and motor neurons in muscles. These changes result in weakness, slowness, and variability in movement control. On the other hand, older people benefit from experience and can use it to compensate for a loss of speed, strength, and control. Understanding such changes in the central nervous system is important because we often feel we have little opportunity to accelerate or decelerate them; if we live long enough, we are likely to face some decrement in the central nervous system. One might assume that the nervous system is complete at maturity and will only decline; however, the adult hippocampus adds thousands of neurons each day (Shors et al., 2012). Unfortunately, half of those die soon after unless we learn while those neurons are young. The number of new neurons is affected by several factors including physical activity and, of course, age. Fifty percent of those aged 85 and older have Alzheimer's disease (Bishop et al., 2010). This pathological decline is associated with physiological changes such as hypertension and diabetes; the direction of causation is not yet understood. However, the relationship once again suggests the importance of physical activity in brain aging. Children are constantly learning and should be in nearly constant motion; now we can see that this also holds true for those at the other end of the life span (Gabbard, 2021).

One key developmental question regarding motor behavior addresses how the brain and nervous system adjust to increases in cognitive function, body size, and strength during childhood and to decreases in those variables as people age. The brains of children and adolescents are considered plastic, meaning their brains can adapt. While the brains of the elderly adapt too, they do not adapt as well, so older adults must take advantage of experience. Control of movements is a complex phenomenon; while the end results might be similar, children and the elderly achieve success via different paths.

Motor control issues also are important in developmental learning and control. Although the critical issues here are similar to those discussed in the previous section on motor control, one additional question remains: How does growth become integrated into the motor control system? As previously mentioned, children can change rapidly, particularly at puberty (Gabbard, 2021; Malina et al., 2004). How does the motor control system account for changes in size, proportion, and mass? Consider a boy who played baseball from March through July at age 12. By the following March, he is 13 years old, he has grown 4 inches (10 cm) in height, his arms have grown in length, and he weighs more. Moreover, he has not practiced batting or throwing since the preceding season, yet he can still bat and throw successfully. How does his nervous system compensate for his physical changes in order to continue producing coordinated movements?

At the other end of the age continuum, how does a system of motor control that has functioned for many years using one set of parameters account for losses in cognitive function, body mass, flexibility, and strength? No good explanation has been found for these important issues in the developmental aspects of motor learning and control. Some progress, however, is being made (for reviews, see Clark et al., 2019; Payne & Isaacs, 2020). Certainly, the adage "use it or lose it" is applicable and reminds us of the importance of exercise science.

Good Things Are Earned

In situations where movement experience is helpful, those who are elderly might have an advantage, whereas many children have a disadvantage due to lack of experience. The elderly use strategies to make movement safer and more efficient. Sometimes this leads to sedentary lifestyles because, for example, the fear of falling is significantly reduced if one stays seated. As a result, many of the underlying factors that contribute to falls such as weakness and loss of balance are exacerbated by sitting, thus increasing the risk instead of reducing it. Successful aging can be earned over decades of physical activity and other healthy practices. Initiating physical activity as we age is also helpful and recommended for those with sedentary lifestyles.

While the benefits of physical activity are well known, 30.6% of those over 65 years of age reported no physical activity in the past 30 days (Carlson et al., 2018). In 2019, 17% of high school youths did not participate in at least 1 hour of physical activity on 1 day, and 55.9% were not active for 60 minutes on 5 or more days (YRBSS, 2018). Of those same students, 46.1% played video games or used a computer 3 or more hours per day, and 19.8% watched 3 or more hours of television a day. Nearly half (47.8%) had no physical education. How can we improve the participation in physical activity at both ends of the life span?

We can enhance movement and exercise experiences for children, and we can do so in two ways—quality and quantity. Research shows that children with experience can outperform adults with less experience, which means that practice and experience can help children (Chi, 2011). We can improve the quality of experience by using what we know about information processing to help children get more out of practice. The key is that children need practice, and for it to be effective, it must help them retain information, skill, and decision-making ability that they would normally lose. We can enable this type of experience by providing children with adult learning strategies. Children can see their improvement and are thus motivated. Children who do not feel they are improving become frustrated and often quit trying or drop out. Simple strategies such as asking children to identify what they have learned or where they have improved encourages practice. For example, saying, "I can see you working hard, and it is paying off" or "I can see you are getting better," is helpful in the learning process. Unfortunately, what is known about how children learn is not consistently applied to learning experiences.

Children and adolescents might view exercise and fitness as duty (or work) and therefore find it less desirable than activities perceived as play (Kretchmar, 2008). Often sport is considered a primary source for physical activity in youths; however, many athletes drop out of sport. Recall, prior to about 15 years of age, youth should sample a variety of activities (Côté et al., 2009). Some might decide to specialize in one of those sports with a goal of high-level performance. Others might decide to participate at a recreational level. Early specialization leads to higher rates of dropping out, and contrary to popular opinion, sampling does not hinder success at the highest levels of sport. Understanding the thought process of children and youths can help us increase the levels of physical activity moving forward. So, make it fun and encourage children and youths to try lots of different sports over several years with a goal of being physically active.

Can we improve the physical activity patterns in older populations? Sparling and colleagues (2015) suggest a shift from sedentary to light activity rather than sedentary to meeting physical activity guidelines. The idea is incremental activity with goals that sedentary older folks can meet. While intense bouts of activity take less time and produce benefits, some side effects such as injury, fatigue, and soreness might discourage many sedentary people. Sparling and associates (2015) suggest accepting less than the recommended 150 minutes of activity per week and focusing on reducing the time spent sitting. Seven hours of sitting in a chair increases the risk of death. Simple goals like standing and moving during television commercials, standing and pacing while on the phone, taking short walks of 5 or so minutes three times each day work well with goal setting and intrinsic feedback for those who are most sedentary and least active, a group needing change.

Wrap-Up

The study of motor behavior has produced important knowledge for human behavior since the late 1800s. Knowledge developed through motor behavior is important in many areas. Although we often think of it in terms of learning and control of sport skills, our society depends on human movement in many ways—from babies learning to use spoons to surgeons using scalpels, pilots controlling airplanes, children using pencils and pens, dentists using drills, and elderly or injured people working to regain function and skills of independent daily living. Further, we use motor behavior to enhance the quality and length of our lives through physically active lifestyles. The study of motor behavior is aimed at understanding the development, learning, and control of these and other motor skills so that people can use them more effectively.

Knowledge of motor behavior is essential to numerous exercise science professions such as athletic training, physical therapy, and occupational therapy. Your school probably offers at least one undergraduate course in motor learning, motor control, or motor development.

More Information on Motor Behavior

Organizations

American Geriatrics Society

Canadian Society for Psychomotor Learning and Sport Psychology

International Motor Development Research Consortium (I-MDRC)

International Society of Motor Control

North American Society for the Psychology of Sport and Physical Activity

Society for the Neural Control of Movement

Society for Neuroscience

Journals

Human Movement Science

Journal of Motor Behavior

Journal of Motor Learning and Development

Journal of Sport and Exercise Psychology

Kinesiology Review

Motor Control

Pediatric Exercise Science

Research Quarterly for Exercise and Sport

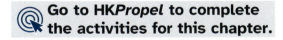

Go to HK*Propel* to complete the activities for this chapter.

Review Questions

1. How does the study of motor behavior differ from sport and exercise psychology?

2. Explain the differences between motor learning and motor control in the field of motor behavior.

3. Why is the change in motor learning and motor control across the life span important?

4. Explain how distributed practice should affect either therapy or fitness training scheduling.

5. In the long term, intrinsic feedback is critical to performance. Explain why and how one goes about shifting from knowledge of performance to intrinsic feedback.

6. Provide an example in which more difficult practice conditions result in better retention and transfer. Why does that happen? How could you plan practices to promote these benefits?

7. Discuss when it might be best to provide either knowledge of performance or knowledge of results to a person learning a motor skill. Does the person's age, sex, or ability or disability matter?

Suggested Readings

Côté, J., Lidor, R., & Hackfort, D. (2009). ISSP position stand: To sample or to specialize? Seven postulates about youth sport activities that lead to continued participation and elite performance. *International Journal of Sport and Exercise Psychology, 9*, 7-17.

Payne, G., & Issacs, L.G. (2020). *Human motor development: A lifespan approach* (10th ed.). Routledge.

Schmidt, R.A., Lee, T.D., Winstein, C. Wulf, G., & Zelaznik, H. (2019). *Motor control and learning: A behavioral emphasis* (6th ed.). Human Kinetics.

Thomas, J.R., & Thomas, K.T. (2008). Principles of motor development for elementary school physical education. *Elementary School Journal, 108*, 181-195.

Wulf, G. (2013). Attentional focus and motor learning: A review of 15 years. *International Review of Sport and Exercise Psychology, 6*, 77-104.

References

Abernethy, B. (1988). Visual search in sport and ergonomics: Its relationship to selective attention and performer expertise. *Human Performance, 1*(4), 205-235.

Abernethy, B., & Sparrow, W.A. (1992). The rise and fall of dominant paradigms in motor behavior research. In J.J. Summers (Ed.), *Approaches to the study of motor control and learning* (pp. 3-45). Elsevier.

Abernethy, B., Schorer, J., Jackson, R.C., & Hagemann, N. (2012). Perceptual training methods compared: The relative efficacy of different approaches to enhancing sport-specific anticipation. *Journal of Experimental Psychology: Applied, 18*(2), 143-153.

Abernethy, B., Maxwell, J.P., Masters, R.S.W., van der Kamp, J., & Jackson, R.C. (2007). Attentional processes in skill learning and expert performance. In G. Tenenbaum & R.C. Eklund (Eds.), *Handbook of sport psychology* (pp. 245-263). John Wiley & Sons.

Adams, J.A. (1987). Historical review and appraisal of research on the learning, retention, and transfer of human motor skills. Psychological Bulletin, 101, 41-74.

Adolf, K.E., & Hoch, J.E. (2019). Motor development: Embodied, embedded, enculturated, and enabling. *Annual Review of Psychology, 70*, 141-164.

Albers, A.S., Thomas J.R., & Thomas, K.T. (2005). Development of rapid aiming movements: Index of difficulty and movement substructure. *Human Movement, 6*(1), 5-11.

Ashford, D., Bennett, S.J., & Davids, K. (2006). Observational modeling effects for movement dynamics and movement outcome measures across differing task constraints: A meta-analysis. *Journal of Motor Behavior, 38*(3), 185-205.

Auyeung, T.W., Lee, J.S.W., Kwok, T., Leung, J.L.P.C., Leung, P.C., & Woo, J. (2009). Estimation of stature by measuring fibula and ulna bone length in 2443 older adults. *The Journal of Nutrition, Health & Aging, 13*(10), 931-936.

Baker, J., & Young, B. (2014). 20 years later: Deliberate practice and the development of expertise in sport. *International Review of Sport and Exercise Psychology, 7*(1) 135-157.

Bayley, N. (1935). The development of motor abilities during the first three years. *Monographs of the Society for Research in Child Development, 1*(1), 1-26.

Beilock, S.L., Wierenga, S.A., & Carr, T.H. (2002). Expertise, attention, and memory in sensorimotor skill execution: Impact of novel task constraints on dual-task performance and episodic memory. *Quarterly Journal of Experimental Psychology, 55*(4), 1211-1240.

Bernstein, N. (1967). *The coordination and regulation of movements.* Pergamon Press.

Bishop, N.A., Lu, T., & Yankner, B.A. (2010). Neural mechanisms of ageing and cognitive decline. *Nature, 464*(7288), 529-535.

Blix, M. (1892–1895). Die lange und spannung des muskels. *Skandinavische Archiv Physiologie, 3,* 295-318.

Brady, F. (2008). The contextual interference effect and sport skills. *Perceptual and motor skills, 106*(2), 461-472.

Carlson, S.A., E., Adams, K., Yang, Z., & Fulton, J.E. Percentage of deaths associated with inadequate physical activity in the United States. *Preventing Chronic Disease, 15,* E38.

Chi, M.T.H. (2011). Theoretical perspectives, methodological approaches, and trends in the study of expertise. In Y. Li & G. Kaiser (Eds.), *Expertise in mathematics instruction* (pp. 17-39). Springer.

Chow, J.Y., Shuttleworth, R., Davids, K., & Araújo, D. (2019). Ecological dynamics and transfer from practice to performance in sport. In N.J. Hodges & A.M. Williams (Eds.), *Skill acquisition in sport* (pp. 330-344). Routledge.

Christina, R.W. (2017). Motor control and learning in the North American Society for the Psychology of Sport and Physical Activity (NASPSPA): The first 40 years. *Kinesiology Review, 6*(3), 221-231.

Clark, J.E., Bardid, F., Getchell, N., Robinson, L.E., Schott, N., & Whitall, J. (2019). Reflections on motor development research across the 20th century: Six empirical studies that changed the field. *Journal of Motor Learning and Development, 8*(2), 438-454.

Côté, J., Lidor, R., & Hackfort, D. (2009). ISSP position stand: To sample or to specialize? Seven postulates about youth sport activities that lead to continued participation and elite performance. *International Journal of Sport and Exercise Psychology, 9,* 7-17.

Dail, T.K., & Christina, R.W. (2004). Distribution of practice and metacognition in learning and long-term retention of a discrete motor task. *Research Quarterly for Exercise and Sport, 75*(2), 148-155.

Deco, G., & Rolls, E.T. (2005). Attention, short-term memory, and action selection: A unifying theory. *Progress in Neurobiology, 76*(4), 236-256.

Deco, G., Rolls, E.T., & Zihl, J. (2005). A neurodynamical model of visual attention. In L. Itti, G. Rees, & J.K. Tsotsos (Eds.), *Neurobiology of attention* (pp. 593-599). Academic Press.

Dennis, W. (1938). Infant development under conditions of restricted practice and a minimum of social stimulation: A preliminary report. *Journal of Genetic Psychology, 53,* 149-158.

Dennis, W., & Dennis, M. (1940). The effect of cradling practices on the age of walking in Hopi children. *Journal of Genetic Psychology, 56,* 77-86.

Espenschade, A. (1960). Motor development. In W.R. Johnson (Ed.), *Science and medicine of exercise and sport* (pp. 419-439). Harper & Row.

Fasoli, S.E., Trombly, C.A., Tickle-Degnen, L., & Verfaellie, M.H. (2002). Effect of instructions on functional reach in persons with and without cerebrovascular accident. *American Journal of Occupational Therapy, 56*(4), 380-390.

Fernihough, A., & McGovern, M.E. (2015). Physical stature decline and the health status of the elderly population in England. *Economics and Human Biology, 16,* 30-44.

Fischman, M.G. (2007). Motor learning and control foundations of kinesiology: Defining the academic core. *Quest, 59,* 67-76.

Fisher, R.P. Corley, B.M. Huibregtse, C.A. Derom, R.F. Vlietinck, M. (2018). Associations between birth size and later height from infancy through adulthood: An individual based pooled analysis of 28 twin cohorts participating in the CODATwins project. *Early Human Development, 120,* 53-60.

Gabbard, C. (2021). *Lifelong motor development.* Lippincott Williams & Wilkins.

Galton, F. (1876). The history of twins as a criterion of the relative power of nature. *Anthropological Institute Journal, 5,* 391-406.

Gesell, A. (1928). *Infancy and human growth.* New York: Macmillan.

Gómez-Jordana, L.I., Stafford, J., Peper, C.L.E., & Craig, C.M. (2018). Virtual footprints can improve walking performance in people with Parkinson's disease. *Frontiers in Neurology, 9,* 681.

Goodway, J.D., & Robinson, L.E. (2015). Developmental trajectories in early sport specialization: A case for early sampling from a physical growth and motor development perspective. *Kinesiology Review, 4*(3), 267-278.

Guimarães, A.N., Ugrinowitsch, H., Dascal, J.B., Porto, A.B., & Okazaki, V.H.A. (2020). Freezing degrees of freedom during motor learning: A systematic review. *Motor Control, 24*(3), 457-471.

Haken, H., Kelso, J.A.S., & Bunz, H. (1985). A theoretical model of phase transitions in human hand movements. *Biological Cybernetics, 51*, 347-356.

Henry, F.M., & Rogers, D.E. (1960). Increased response latency for complicated movements and a "memory drum" theory of neuromotor reaction. *Research Quarterly for Exercise and Sport, 31*, 448-458.

Hofstein, A., & Kind, P.M. (2012). Learning in and from science laboratories. In *Second international handbook of science education* (pp. 189-207). Springer Netherlands.

Jayanthi, N., Pinkham, C., Dugas, L., Patrick, B., & LaBella, C. (2013). Sports specialization in young athletes: Evidence-based recommendations. *Sports Health, 5*(3), 251-257.

Kelso, J.A.S. (1995). *Dynamic patterns: The self-organization of brain and behavior.* MIT Press.

Koppelmans, V., Bloomberg, J.J., Mulavara, A.P., & Seidler, R.D. (2016). Brain structural plasticity with spaceflight. *npj Microgravity, 2*, 2.

Kretchmar, R.S. (2008). The increasing utility of elementary school physical education: A mixed blessing and unique challenge. *The Elementary School Journal, 180*(3), 161-170.

Krigolson, O.E., Ferguson, T.D., Colino, F.L., and Binsted, G. (2020). Distribution of practice combined with observational learning has time dependent effects on motor skill acquisition. *Perceptual and Motor Skills, 128*(2), 885-899.

Kwon, Y.H., Kwon, J.W., & Lee, M.H. (2015). Effectiveness of motor sequential learning according to practice schedules in healthy adults: Distributed practice versus massed practice. *Journal of Physical Therapy Science, 27*(3), 769-772.

Larsson, H. (2020). Introduction to new perspectives of movement learning. In H. Larsson (Ed.), *Learning movements* (pp. 1-16). Routledge.

Latash, M.L., & Zatsiorsky, V. (2015). *Biomechanics and motor control: Defining central concepts.* Academic Press.

Levin, M.F., & Demers, M. (2021). Motor learning in neurological rehabilitation. *Disability and Rehabilitation, 43*(24), 3445-3453.

Levin, M.F., Weiss, P.L., & Keshner, E.A. (2015). Emergence of virtual reality as a tool for upper limb rehabilitation: Incorporation of motor control and motor learning principles. *Physical Therapy, 95*(3), 415-425.

Locke, E.A., & Latham, G.P. (2016). The application of goal setting to sports. *Journal of Sport Psychology, 7*, 205-222.

Magill, R., & Anderson, D. (2021). *Motor learning and control: Concepts and applications* (13th ed.). McGraw Hill.

Malina, R.M. (1984). Physical growth and maturation. In J.R. Thomas (Ed.), *Motor development during childhood and adolescence* (pp. 2-26). Burgess.

Malina, R.M., Bouchard, C., & Bar-Or, O. (2004). *Growth, maturation, and physical activity* (2nd ed.). Human Kinetics.

Marchant, D.C., Clough, P.J., Crawshaw, M., & Levy, A. (2009). Novice motor skill performance and task experience is influenced by attentional focusing instructions and instruction preferences. *International Journal of Sport and Exercise Psychology, 7*(4), 488-502.

Marteniuk, R.G., Ivens, C.J., & Bertram, C.P. (2000). Evidence of motor equivalence in a pointing task involving locomotion. *Motor Control, 4*(2), 165-184.

Mattos, D., Schöner, G., Zatsiorsky, V.M., & Latash, M.L. (2015). Motor equivalence during multi-finger accurate force production. *Experimental Brain Research, 233*(2), 487-502.

McGraw, M.B. (1935). *Growth: A study of Johnny and Jimmy.* Appleton-Century-Crofts.

McGraw, M.B. (1939). Later development of children specially trained during infancy: Johnny and Jimmy at school age. *Child Development, 10*, 1-19.

Newell, K.M. (1984). Physical constraints to development of motor skills. In J.R. Thomas (Ed.), *Motor development during childhood and adolescence* (pp. 105-122). Burgess.

Newell, K.M. (2020). What are fundamental motor skills and what is fundamental about them? *Journal of Motor Learning and Development, 8*(2), 280-314.

Park, R., Seefeldt, V., Malina, R.M., & Broadhead, G.D. (1996). In memoriam: G. Lawrance Rarick. *Journal of Physical Education, Recreation & Dance, 67*(1), 16.

Payne, V.G., and Isaacs, L.D. (2020). *Human motor development* (10th ed.). Routledge.

Perkins-Ceccato, N., Steve, R., Passmore, S.R., & Lee, T.D. (2003). Effects of focus of attention depend on golfers' skill. *Journal of Sports Sciences, 21,* 593-600.

Perreault, M.E., & French, K.E. (2015). External-focus feedback benefits free-throw learning in children. *Research Quarterly for Exercise and Sport, 86*(4), 422-427.

Peterson, D.S., & Horak, F.B. (2016). Neural control of walking in people with parkinsonism. *Physiology, 31*(2), 95-107.

Pirker, W., & Katzenschlager, R. (2017). Gait disorders in adults and the elderly: A clinical guide. *Wiener Klinische Wochenschrift, 129*(3-4), 81-95.

Prinz, W. (1997). Perception and action planning. *European Journal of Cognitive Psychology, 9*(2), 129-154.

Rikli, R. (2005). Movement and mobility influence on successful aging: Addressing the issue of low physical activity. *Quest, 57,* 46-66.

Rikli, R.E., & Jones, C.J. (2013). Development and validation of criterion-referenced clinically relevant fitness standards for maintaining physical independence in later years. *Gerontologist, 53*(2), 255-267.

Rosenbaum, D.A. (2010). *Human motor control* (2nd ed.). Elsevier.

Rymal, A.M. (2018). Let's make it real: A commentary on observation research. *Journal of Motor Learning and Development, 6*(1), 73-80.

Ryu, D., Kim, S., Abernethy, B., & Mann, D.L. (2013). Guiding attention aids the acquisition of anticipatory skill in novice soccer goalkeepers. *Research Quarterly for Exercise and Sport, 84*(2), 252-262

Schmidt, R.A. (1975). A schema theory of discrete motor skill learning. *Psychological Review, 82,* 225-260

Schmidt, R.A., & Lee, T.D. (2020). *Motor learning and performance: From principles to application* (6th ed.). Human Kinetics.

Schmidt, R.A., Lee, T.D., Winstein, C.J., Wulf, G., & Zelaznik, H.N. (2019). *Motor control and learning: A behavioral emphasis* (6th ed.). Human Kinetics.

Schmidt, R.A., & Wrisberg, C.A. (2008). *Motor learning and performance: A situation-based learning approach* (4th ed.). Human Kinetics.

Seifert, L., Papet, V., Strafford, B.W., Coughlan, E.K., & Davids, K. (2018). Skill transfer, expertise and talent development: An ecological dynamics perspective. *Movement & Sport Sciences, 102,* 39-49.

Shah, K., & Villareal, D.T. (2017). Obesity. In H.M. Fillit, K. Rockwood, & J. Young (Eds.), *Brocklehurst's textbook of geriatric medicine and gerontology* (8th ed.). Elsevier.

Shea, C.H., Lai, Q., Black, C., & Park J. (2000). Spacing practice sessions across days benefits the learning of motor skills. *Human Movement Studies, 19,* 737-760.

Sherrington, C.S. (1906). *The integrative action of the nervous system.* Yale University Press.

Shors, T.J., Anderson, M.L., Curlik Ii, D.M., & Nokia, M.S. (2012). Use it or lose it: How neurogenesis keeps the brain fit for learning. *Behavioural Brain Research, 227*(2), 450-458.

Slone, M.R. (1984). *Ruth B. Glassow: The cutting edge.* The Academy Papers. www.nationalacademyofkinesiology.org/.../TAP_20_CuttingEdgeinPEandExerciseSci

Smeeton, N.J., Ward, P., & Williams, A.M. (2004). Do pattern recognition skills transfer across sports? A preliminary analysis. *Journal of Sports Sciences, 22,* 205-213.

Snijders, A.H., van de Warrenburg, B.P., Giladi, N., Bloem, B.R. (2007). Neurological gait disorders in elderly people: Clinical approach and classification. *Lancet Neurology, 6*(1), 63-74.

Sparling, P.B., Howard, B.J., Dunstan, D.W., & Owen, N. (2015). Recommendations for physical activity in older adults. *BMJ, 350.* https://doi.org/10.1136/bmj.h100

Spruit, E.N., Band, G.P., & Hamming, J.F. (2015). Increasing efficiency of surgical training: Effects of spacing practice on skill acquisition and retention in laparoscopy training. *Surgical Endoscopy, 29*(8), 2235-2243.

Tarantino, U., Baldi, J., Celi, M., Rao, C., Liuni, F.M., Iundusi, R., & Gasbarra, E. (2013). Osteoporosis and sarcopenia: The connections. *Aging Clinical and Experimental Research, 25*(1), 93-95.

Thomas, J.R. (1997). History of motor behavior. In J.D. Massengale & R.A. Swanson (Eds.), *History of exercise and sport sciences* (pp. 203-292). Human Kinetics.

Thomas, J.R. (2006). Motor behavior: From telegraph keys and twins to linear slides and stepping. *Quest, 58,* 112-127.

Thomas, J.R., & Thomas, K.T. (1989). What is motor development: Where does it belong? *Quest, 41,* 203-212.

Thomas, J.R., & Thomas, K.T. (2008). Principles of motor development for elementary school physical education. *Elementary School Journal, 108,* 181-195.

Ulrich, B., & Reeve, T.G. (2005). Studies in motor behavior: 75 years of research in motor development, learning, and control. *Research Quarterly for Exercise and Sport, 76,* S62-S70.

Waldron, S., DeFreese, J.D., Register-Mihalik, J., Pietrosimone, B., & Barczak, N. (2020). The costs and benefits of early sport specialization: A critical review of the literature. *Quest, 72*(1), 1-18.

Walston, J.D. (2020). Common clinical sequelae of aging. In L. Goldman, & A.I. Schafer (Eds.), *Goldman-Cecil Medicine* (26th ed.). Elsevier.

Whitall, J., Schott, N., Robinson, L.E., Bardid, F., & Clark, J.E. (2020). Motor development research: I. The lessons of history revisited (the 18th to the 20th century). *Journal of Motor Learning and Development, 2*(2), 345-362.

Wing, A.M. (2000). Motor control: Mechanisms of motor equivalence in handwriting. *Current Biology, 10*(6), R245-R248.

Wulf, G. (2007). *Attention and motor skill learning.* Human Kinetics.

Wulf, G., & Prinz, W. (2001). Directing attention to movement effects enhances learning: A review. *Psychonomic Bulletin & Review, 8*(4), 648-660.

Wulf, G., McNevin, N., & Shea, C.H. (2001). The automaticity of complex motor skill learning as a function of attentional focus. *Quarterly Journal of Experimental Psychology, 54*(4), 1143-1154.

Wulf, G., McConnel, N., Gärtner, M., & Schwarz, A. (2002). Enhancing the learning of sport skills through external-focus feedback. *Journal of Motor Behavior, 34*(2), 171-182.

Wyble, B.P., & Rosenbaum, D.A. (2016). Are motor adjustments quick because they don't require detection or because they escape competition? *Motor Control, 20,* 182-186.

YRBSS. (2018). Division of Adolescent and School Health, National Center for HIV/AIDS, Viral Hepatitis, STD, and TB Prevention. www.cdc.gov/healthyyouth/data/yrbs/index.htm

CHAPTER 7

Sport and Exercise Psychology

Lindsay E. Kipp

CHAPTER OBJECTIVES

In this chapter, we will

- discuss what professionals do in careers related to sport and exercise psychology;
- describe how sport and exercise psychology evolved within exercise science;
- explain how self-perceptions, including self-efficacy, perceived competence, and self-esteem, are related to individuals' experiences in exercise and sport;
- describe how exercise science professionals can influence individuals' motivation for exercise and sport;
- explain characteristics of groups and how to develop group cohesion in exercise and sport; and
- describe how exercise can improve health-related quality of life.

Laura is a recent college graduate who is working a full-time job. She played soccer in high school and enjoyed going to the gym in college. But now she cannot find the motivation to exercise. She does not want to get up early enough to go to the gym before work, and she feels too tired to exercise after work. Laura recently heard about her company's employee wellness program and wants to learn more. As an exercise science professional, you can play a role in helping people like Laura find their motivation to exercise. For example, a corporate fitness director might plan, advertise, and help implement a walking challenge or a lunchtime yoga group. A working knowledge of sport and exercise psychology principles will be important for the success of these programs. For example, how will you help your clients develop confidence and motivation so that they adhere to the program? What strategies will you use to develop social connections and enjoyment throughout the program?

The author acknowledges the contributions of Robin Vealey to portions of this chapter.

CHAPTER 8

Physical Activity Epidemiology

Duck-chul Lee

CHAPTER OBJECTIVES

In this chapter, we will

- discuss the national physical activity guidelines and plan;
- provide the definition of, career opportunities in, and the history of physical activity epidemiology;
- outline common epidemiological research methods and how to measure physical activity in epidemiological studies; and
- discuss the importance of physical activity epidemiology in disease prevention and health promotion.

John is from an active family and was a high school athlete. He decided to major in kinesiology and took all of the core courses such as exercise physiology, exercise psychology, motor control, biomechanics, and so on. In his last year before graduation, he realized that all the knowledge and information about physical activity and health that he learned primarily focused on the individual level, but it lacked how to apply it to a large number of people at the population level. Then, he took a physical activity epidemiology course and found that US national physical activity guidelines have been developed mostly based on large epidemiological evidence on the effects of physical activity on chronic disease prevention and premature mortality risk reduction. John also found several exciting research evidence and data suggesting that physical activity, especially objectively measured physical activity, might be more important than medications for people with chronic diseases such as diabetes and hypertension in terms of longevity. After graduation, he got a job in a state department of health to promote physical activity in various populations. He believes that physical activity is one of the most important lifestyle factors to improve health of all ages.

Goals of Physical Activity Epidemiology

- To identify the distribution and determinants of physical activity
- To determine the associations of physical activity with disease prevention, treatment, and longevity
- To develop physical activity guidelines and public health policy
- To apply effective strategies to promote physical activity in various populations

Physical activity and exercise play key roles in physical, mental, and social health; help prevent, delay, and treat chronic diseases (e.g., cardiovascular disease, diabetes, obesity); and contribute to healthy aging and longevity throughout the life span. How do we know this? Physical activity epidemiologists have been trying to find the answers to the following common questions about physical activity and health: "Who is active?" "Do active people get sick less often and live longer?" "What is the minimum and optimum amount and kind of physical activity for maximum health benefits?" and "How can we promote physical activity at the population level?" To investigate these important topics and produce evidence-based scientific data and knowledge, physical activity epidemiologists conduct large human research studies using various epidemiological research methods and statistical analyses. They also systematically collect data and contribute to developing physical activity recommendations in various populations for disease prevention and health promotion. Most exercise science professionals help their clients and the family and friends of that client, but physical activity epidemiologists in exercise science positively influence the health and longevity of whole populations of people throughout the world.

Epidemiology is "the study of the distribution and determinants of health-related states or events in specified populations, and the application of this study to the control of health problems" (Last, 1988, p. 141). A recent branch of epidemiology aligned with kinesiology and exercise science is called *physical activity epidemiology*. In a foundational study Caspersen described physical activity epidemiology as a two-part field:

- First, as a science, "it studies the association of physical activity, as a health-related behavior, with disease and other health outcomes; the distribution and the determinants of physical activity behavior(s); and the interrelationship of physical activity with other behaviors."

- Second, as a practice, "it applies that knowledge to the prevention and control of disease and the promotion of health." (Caspersen, 1989, p. 425)

Based on these descriptions and concepts, Lee and Brellenthin concisely defined **physical activity epidemiology** as "the study of the distribution and determinants of physical activity, its associations with health-related outcomes, and the application of this study to disease prevention and health promotion" (Lee & Brellenthin, 2023, p. 324).

Physical activity and exercise can be either an exposure (independent variable) or outcome (dependent variable) depending on the study design and purpose of a study. When studying the associations of physical activity with disease and other health outcomes, physical activity is considered an exposure. When studying the determinants of physical activity, physical activity is an outcome. However, the associations of physical activity with various health-related outcomes (e.g., heart attack, cancer, mortality) are the central part of physical activity epidemiology. Findings from these health-related studies provide important data to develop effective public health strategies and policies. From both clinical and public health perspectives, physical activity epidemiology is an important growing area of science and practice for the prevention of disease and health promotion.

Benefits of Physical Activity Epidemiology Knowledge

Physical activity epidemiology has two important applied science missions:

- To develop and continue to enhance physical activity guidelines
- To develop effective, evidence-based strategies and plans to promote physical activity from both scientific and public health perspectives

Physical activity epidemiology generates data and useful knowledge to complete these two missions, which benefits our health and quality of life. Note this chapter will use the term *physical activity*, which includes the more focused term *exercise*, since much of physical activity epidemiology focuses on the health and longevity benefits of most forms of human movements, not just programmed exercise.

Key Terms in Epidemiology

Application: The application of the established understanding of the causal factors related to disease is a major goal of public health. Other names for the application of research are *translation* and *dissemination*. Public health disseminators translate knowledge from epidemiological studies to help increase physical activity among individuals and within social groups and community organizations.

Determinants: Determinants of disease are often called *risk factors* because they increase a person's risk for disease. In physical activity epidemiology, the goals are usually to test the hypothesis that activity is or is not a determinant for a particular disease outcome, or to identify the determinant of physical activity behaviors such as sex, age, race, income, occupation, and environment.

Distribution: Frequency (how often the disease occurs) and patterns of disease occurrence (who developed the disease, where, and when) in a population. In physical activity epidemiology, distribution can refer to the prevalence of meeting the physical activity guidelines (as opposed to the occurrence of disease) in a certain population, area, and time.

Exercise: A form of physical activity that is prescribed, structured, repetitive, and performed with the goal of improving health or fitness. All exercise is physical activity, but not all physical activity is exercise.

Exposure: The independent variable (e.g., a risk factor) that is tested for its association with the outcome of interest.

Outcome: The dependent variable (also commonly called *event* or *endpoint*) that is monitored during a study and that occurs as a result of exposure or intervention.

Physical activity: Any bodily movement produced by the contraction of skeletal muscle that increases energy expenditure above a basal level. In the Physical Activity Guidelines, physical activity generally refers to the subset of physical activity that enhances health.

KEY POINT

Physical activity epidemiology is the study of the distribution and determinants of physical activity, its associations with health-related outcomes, and the application of this study to disease prevention and health promotion.

National Physical Activity Guidelines

One of the most important practical outcomes of the accumulated knowledge of physical activity epidemiology are national physical activity guidelines. Many countries—mostly high-income countries such as the United Kingdom, Australia, and Canada—have set national guidelines and plans to promote physical activity to improve health and healthy longevity. The U.S. National Physical Activity Guidelines provide specific recommendations for the types and amounts of physical activity for individuals of all ages, from preschool-aged children to older adults, as well as specific guidelines for women during pregnancy and adults with chronic health conditions or disabilities, as briefly summarized in table 8.1 (USDHHS, 2018).

The health benefits of exercise and physical activity are studied by researchers in various fields of exercise science, such as exercise physiologists, biomechanists, exercise psychologists, and physical activity epidemiologists. However, physical activity epidemiology focuses more on hard clinical outcomes (e.g., risk or rates of heart attack, diabetes, different types of cancer, and premature mortality) using large (population rather than sample level) and long-term prospective study designs such as cohort study. (See chapter 9 for more on study designs.) Findings from these large studies specifically contribute to the development of the physical activity guidelines for minimum and best amounts of different types of exercise and physical activity in various populations. Thus, the knowledge and information developed by physical activity epidemiologists have direct implications for human health and longevity.

KEY POINT

The U.S. National Physical Activity Guidelines recommend at least 150 minutes of moderate-intensity, 75 minutes of vigorous-intensity, or an equivalent combination of moderate- and vigorous-intensity aerobic activity a week for substantial health benefits for adults. For additional health benefits, adults also should do muscle-strengthening activities on 2 or more days a week.

National Physical Activity Plan

Another well-documented outcome from the knowledge of physical activity epidemiology is the U.S. National Physical Activity Plan, which recommends detailed evidence-based policies, programs, and initiatives designed specifically to promote physical activity in nine different societal sectors (National Physical Activity Plan Alliance, 2016):

Table 8.1 **Key Physical Activity Guidelines for Americans**

Population	Key guidelines
Preschool-aged children (ages 3 through 5 years)	Be active throughout the day to enhance growth and development through active play including a variety of activity types.
Children and adolescents (ages 6 through 17 years)	Do 60 minutes or more of moderate to vigorous activity daily including aerobic, muscle-strengthening, and bone-strengthening activities.
Adults (ages 18 through 64 years)	Do at least 150 minutes a week of moderate-intensity aerobic activity, 75 minutes a week of vigorous-intensity aerobic activity, or an equivalent combination plus at least 2 or more days a week of muscle-strengthening activity.
Older adults (ages 65 and older)	Follow the same guidelines for adults for aerobic and muscle-strengthening activities. In addition, include balance training, consider these people's fitness levels, and understand the effects of chronic conditions and their ability to do activities safely.
Women during pregnancy	Follow the same guidelines for adults for aerobic and muscle-strengthening activities during pregnancy and the postpartum period under the care of their health care provider.
Adults with chronic health conditions or disabilities	Follow the same guidelines for adults for aerobic and muscle-strengthening activities according to their abilities; avoid inactivity.

Key Terms in Physical Activity Epidemiology

The U.S. National Physical Activity Guidelines are written by multidisciplinary panels of scholars and clinicians (e.g., epidemiology, kinesiology or exercise science, physiology, and various specializations of medicine). Given the diversity of scientific fields, the guidelines define terms in a glossary. Note the meanings of selected terms from that glossary and how they relate to meanings in exercise science.

Intensity: Intensity refers to how much work (not scientific meaning or mechanical work) is being performed or the magnitude of the effort required to perform an activity or exercise. Intensity can be expressed either in absolute or relative terms. Moderate-intensity physical activity is done at 3.0 to 5.9 METs on an absolute scale and 5 or 6 on a scale of 1 to 10 on a relative scale to an individual's personal capacity. Vigorous-intensity physical activity is done at 6.0 or more METs on an absolute scale and 7 or 8 on a scale of 1 to 10 on a relative scale to an individual's personal capacity.

Metabolic equivalent of task (MET): MET refers to the energy expenditure required to carry out a specific activity, and 1 MET is the rate of energy expenditure while sitting at rest. This generally corresponds to an oxygen uptake of 3.5 milliliters per kilogram of body weight per minute. Physical activities frequently are classified by their intensity using the MET value as a reference.

Muscle-strengthening activities: Physical activity or exercise that increases skeletal muscular strength, power, endurance, and mass.

Physical fitness: The ability to carry out daily tasks with vigor and alertness, without undue fatigue, and with ample energy to enjoy leisure time pursuits and respond to emergencies. Physical fitness includes several components: cardiorespiratory fitness (endurance or aerobic power), musculoskeletal fitness, flexibility, balance, and speed of movement.

Adapted from U.S. Department of Health and Human Services. *Physical Activity Guidelines for Americans*, 2nd edition. (Washington, DC: U.S. Department of Health and Human Services, 2018).

1. Business and industry
2. Community recreation, fitness, and parks
3. Education
4. Faith-based settings
5. Health care
6. Mass media
7. Public health
8. Sport
9. Transportation, land use, and community design

The plan was originally released in 2010 and updated in 2016 by the National Physical Activity Plan Alliance, a nonprofit organization with representatives from numerous scientific and professional societies (including several from exercise science), in partnership with the U.S. Department of Health and Human Services (USDHHS). The plan is based on a socioecological model of health behavior that physical activity is determined by various factors at the personal, family, institutional, community, and policy levels. Each sector includes broad strategies to promote physical activity in the U.S. population, and each strategy outlines specific tactics for communities, organizations, and agencies. Here are some examples:

Business and Industry Sector

- "Create or enhance access to places for employees to engage in physical activity before, during, and after work hours."
- "Conduct periodic worksite-based health screenings that measure physical activity and fitness levels of workers."

Education Sector

- "Support adoption of school design strategies to support active transport."
- "Support adoption of policies requiring that students at all levels be given physical activity breaks during the school day."

The plan provides recommendations for policy and practices, specifically in the nine societal sectors in the population, to further support and facilitate the Physical Activity Guidelines for Americans.

KEY POINT

The U.S. National Physical Activity Plan recommends detailed evidence-based policies, programs, and initiatives designed specifically to promote physical activity in the U.S. population, and it supports and facilitates the Physical Activity Guidelines for Americans.

Hot Topic

Health Club Membership and Meeting the Physical Activity Guidelines

Health clubs are a popular place to increase physical activity and improve health in the United States and around the world. Many companies cover or subsidize the cost of health club memberships for their employees. Insurance coverage for a health club membership has become more common. Some provinces of Canada provide reimbursement for licensed kinesiologists to provide evidence-based exercise programs. While the scientific evidence of the large benefits of physical activity are clear, less evidence exists (surprisingly) as to whether having a health club membership actually helps people increase physical activity or improves health. Dr. Schroeder and her colleagues (2017) provide evidence-based data examining the relationship between a health club membership and the prevalence of meeting the physical activity guidelines and cardiovascular health indicators in 405 adults aged 30 to 64 years. The authors found that health club members

- have 17, 10, and 14 times increased odds of meeting the recommended aerobic, muscle-strengthening, and both aerobic plus muscle-strengthening activity guidelines, respectively;
- have a 4.8 beats per minute lower resting heart rate and 2.1 milliliters per kilogram of body weight per minute higher cardiorespiratory fitness; and
- spend 1.4 hours per day being less sedentary compared to nonmembers.

In addition, health club members with more than 1 year of membership have more favorable results in physical activity and cardiovascular health (e.g., 1.5 inches [3.8 cm] lower waist circumference in men). This study shows the potential benefits of supporting health club membership as an effective public health strategy although prospective studies are required to confirm the causality between health club membership and meeting the physical activity guidelines and health.

What Do Physical Activity Epidemiologists Do?

Epidemiology is a cornerstone of public health, which promotes and protects the health of people and provides the scientific backbone to quantify health problems; identifies and monitors disease risk factors; designs and conducts human research projects; measures the strength of the associations between risk factors and health outcomes; and produces evidence-based data.

Although epidemiology is a longstanding central component of public health research, physical activity epidemiology was not born from public health; rather, it established its own research area. Researchers in medicine and kinesiology (e.g., Drs. Morris, Paffenbarger, and Blair) who were interested in physical activity, fitness, and health adopted epidemiological research methods (e.g., prospective cohort study) and established the new field of physical activity epidemiology over the last half century (see History of Physical Activity Epidemiology section below). Physical activity epidemiology recently contributed to expanding the scope of public health by adding physical activity as a relatively new and powerful lifestyle risk factor in addition to the traditional risk factors such as diet, smoking, and alcohol abuse. Physical activity epidemiology is also a growing subdiscipline of exercise science because of increasing emphasis on improving health of the general population and not just performance in sport.

Physical activity epidemiologists actively collaborate with investigators from various related disciplines including exercise physiologists, sports medicine doctors, cardiac rehabilitation specialists, other epidemiologists (e.g., in diet, cancer, cardiovascular disease), public health scientists and practitioners, psychologists, physicians, transportation engineers and city planners, policy makers, and community advocates.

Most physical activity epidemiologists have a doctoral degree in kinesiology, epidemiology, or medicine. In addition, most physical activity epidemiologists have completed a postdoctoral research fellowship after their doctoral studies. It is common that physical activity epidemiologists have different backgrounds for their doctoral program (e.g., exercise physiology, exercise psychology, physical activity measurement, or epidemiology from public health). They then develop skills, experience, and knowledge of physical activity epidemiology (e.g., how to conduct epidemiological research, analyze large cohort data, and design and execute clinical trials) during their postdoctoral training on a specific research topic such as cardiovascular disease, cancer, diabetes, or obesity.

Although primary interests are different between physical activity epidemiologists, they all have physical activity at the center of their research, which is one of the key behavior components of chronic disease prevention from a public health perspective, in addition to diet, cigarette smoking, drug and alcohol use,

Career Opportunity

Safe Routes to School Program

Physical activity epidemiologists work at public health departments and agencies for national, state, and local governments (e.g., Centers for Disease Control and Prevention), nonprofit organizations (e.g., bicycle organization to promote physical activity–friendly environment or policy change), and in hospitals (e.g., preventive medicine). They also work at colleges and universities as researchers to conduct and produce meaningful public health data as well as teachers to train future physical activity epidemiologists. Common job descriptions for physical activity epidemiology positions at state departments of health or public health include the ability to analyze and interpret surveillance data to make data-driven recommendations; plan, manage, implement, and evaluate health promotion (e.g., physical activity) programs; give presentations to various groups with excellent written and oral communication skills; and collaborate effectively with a team of various partners (e.g., local health departments, community coalitions, and partner agencies). The Safe Routes to School program is a good example of physical activity epidemiologists making new contributions to encourage children and their families to walk and bicycle to school. In this case, physical activity epidemiologists in the state department of public health work together with experts from city planning, the department of transportation, the school district, and children and their parents. The physical activity epidemiologist's goal is to lower the causes of illness and premature death in their state through prevention, early detection, and management of chronic diseases, and through promotion of a healthy lifestyle including physical activity.

Janet Fulton

Courtesy of Janet Fulton.

Dr. Fulton is a physical activity epidemiologist and chief of the Physical Activity and Health Branch at the Centers for Disease Control and Prevention (CDC) in Atlanta, Georgia. Her research interests include the epidemiology of physical activity and chronic diseases, the measurement and quantification of physical activity, and population-based promotion of physical activity. She earned her PhD in epidemiology at the University of Texas Health Science Center Houston, School of Public Health. She has published over 150 scientific articles on physical activity and health. Dr. Fulton was the science coordinator and a member of the writing group for the first 2008 U.S. Physical Activity Guidelines for Americans and a member of Executive Secretaries for the second 2018 U.S. Physical Activity Guidelines for Americans at the CDC. She also served as a consultant to the World Health Organization for the Global Recommendations on Physical Activity for Health. She is a fellow of the American College of Sports Medicine and the American Heart Association. She practices what she preaches by staying active and enjoying numerous outdoor activities, trains future physical activity epidemiologists at the CDC, and is a wonderful colleague working happily with others in the field, which is another important feature to be a successful physical activity epidemiologist.

and sleep. One example of the relationship between physical activity epidemiology and public health is Healthy People 2030, which was first developed by the USDHHS in 1979 and is updated every 10 years (USDHHS, 2020). Healthy People 2030 sets data-driven national objectives to improve health and includes physical activity objectives for 2030 focusing on reducing inactivity and increasing physical activity to levels recommended in the U.S. Physical Activity Guidelines.

As scientists, physical activity epidemiologists investigate the distribution and determinants of physical activity and its associations with health-related outcomes. For example, they look around at the various environmental factors from natural environment to individual determinants that influence physical activity such as weather, air, parks, transportation, culture, income, gender, age, and beliefs. They also study how much risk of developing chronic diseases (e.g., cardiovascular disease) could be prevented by being physically active and fit in various populations (e.g., individuals with overweight or obesity). As practitioners, they work to prevent and control the disease by promoting physical activity in populations. Thus, physical activity epidemiology is considered a habilitative or preventive science rather than laboratory or rehabilitative science. As public health workers, physical activity epidemiologists not only conduct physical activity and health research, but also produce important scientific data to develop and enhance national physical activity guidelines

and plans for public health policy makers, health professionals, and individuals.

> **KEY POINT**
>
> Physical activity is one of the key behavior components of chronic disease prevention from a public health perspective, in addition to diet, cigarette smoking, drug and alcohol use, and sleep.

History of Physical Activity Epidemiology

In the mid-20th century, heart disease was a rising public health problem in developed countries; thus, early pioneers of physical activity epidemiology were interested in and conducted large observational cohort studies to investigate physical inactivity as one of the important determinants of heart disease. Since the 1970s, physical activity epidemiologists have documented meaningful associations of physical activity as well as physical fitness with many health outcomes including type 2 diabetes, several forms of cancer, obesity, mental health conditions (e.g., depression), bone health (e.g., osteoporosis), cognition, quality of life, and cause-specific mortality in addition to heart disease.

Two well-known pioneers of physical activity epidemiology are Jeremiah "Jerry" Morris (1910-2009) and Ralph Paffenbarger Jr. (1922-2007). Dr. Morris,

Professional Issues in Exercise Science

Association of Daily Step Volume and Intensity With Mortality

Wearable devices such as smart watches track daily steps and are becoming more popular. Although 10,000 steps per day is a commonly known health goal, it can be difficult to achieve and scientific data supporting it is limited. Thus, it is still unclear how many steps per day are required for good health and lower mortality. Lee and colleagues[1] examined the association of number of steps per day with all-cause mortality in over 16,000 older women in the United States. They observed that as few as approximately 4,400 steps per day was significantly associated with lower mortality risk compared to the least active women who had approximately 2,700 steps per day. They also found that the risk of premature mortality decreased with higher daily steps until approximately 7,500 steps per day, after which the risk leveled off. Since walking is easy to do for most individuals at all ages and all levels of fitness, and the number of steps is also easy to understand, the findings from this study have the potential to enhance physical activity guidelines and improve public health programs and policies. Exercise science professionals should stay current with the latest research consensus when prescribing activity levels based on step counts for specific clients.

Citation style: AMA

1. Lee IM, Shiroma EJ, Kamada M, Bassett DR, Matthews CE, Buring JE. Association of step volume and intensity with all-cause mortality in older women. *JAMA Intern Med.* 2019;179(8):1105-1112.

who was a British epidemiologist at the London School of Hygiene and Tropical Medicine, is often considered the first physical activity epidemiologist who studied the association between physical activity and cardiovascular disease. In his famous London bus driver study, Dr. Morris found a significantly higher risk of developing heart disease in the inactive bus drivers who sat all day, compared to the active conductors who climbed the stairs up and down to collect fares from passengers on double-decker buses (Morris et al., 1953). It is noteworthy that Dr. Morris and his colleagues had a novel hypothesis that physical inactivity could cause heart disease, given that people in London in the 1950s were mostly active and a sedentary lifestyle was not yet common.

Dr. Paffenbarger, who was a professor at both Stanford University School of Medicine and Harvard University School of Public Health, conducted several large epidemiological studies. One of his landmark studies is the Harvard Alumni Health Study that started in the 1960s. Over several decades, Dr. Paffenbarger investigated personal characteristics, lifestyle factors including physical activity, chronic diseases, and death in over 20,000 Harvard University alumni. He developed and mailed well-structured physical activity questionnaires to Harvard alumni periodically to collect data. Based on the responses from the survey, he found that low levels of leisure time physical activity were associated with increased risk of developing a number of chronic diseases as well as cardiovascular disease morbidity and mortality (Paffenbarger et al., 1986).

Another pioneer of physical activity epidemiology is Dr. Steven Blair (b. 1939), who established the Aerobics Center Longitudinal Study in 1970. Earlier epidemiological studies used self-reported physical activity, which creates a limitation in the study design because people generally overreport their physical activity, leading to measurement errors. Dr. Blair, however, used objectively measured cardiorespiratory fitness from a maximal treadmill exercise test as an indicator of recent aerobic physical activity. He found that individuals with low levels of cardiorespiratory fitness had increased risk of developing chronic diseases and premature mortality (Blair et al., 1989). The associations between cardiorespiratory fitness and health outcomes were mostly stronger than the associations between self-reported physical activity and health outcomes. Advances in technology in objective monitoring of aerobic physical activity (e.g., accelerometers, GPS, pedometers) have further driven the field to generate strong and new evidence on the health benefits of physical activity.

Research Methods in Physical Activity Epidemiology

Epidemiological research methods often are divided into two common broad categories: observational and experimental. Initial hypotheses are developed and tested using observational study designs such as cross-sectional, case-control, and cohort studies. Mortality or disease **prevalence** or **incidence** is often the dependent (outcome) variable for these investigations. Observational study designs have been used more heavily in the past, and a solid body of evidence has been amassed to support their basic methods.

Research and Evidence-Based Practice in Exercise Science

Effect of the COVID-19 Pandemic on Physical Activity and Sedentary Behavior in Older Adults

Older adults are more vulnerable to COVID-19, with higher hospitalization and death after infection. Physical activity is one of the important lifestyle factors that affect health and longevity in older adults. However, whether COVID-19 has an effect on physical activity in older adults is uncertain. Dr. Lefferts and her colleagues (2022) at Iowa State University conducted an epidemiological research study and provided evidence-based data on physical activity in 387 older adults (average age of 75 years) before the COVID-19 pandemic, right after the COVID-19 pandemic, and after COVID-19 vaccines were available. The authors found that physical activity was greatly reduced and sitting time was significantly increased during the initial 3 months of the COVID-19 pandemic. However, physical activity and sitting time returned to prepandemic levels 1 year later after vaccines were available, and 97% of participants got vaccinated. This finding suggests that the pandemic did not have long-lasting effects on habitual physical activity and sedentary behavior in older adults. The findings further support the importance of vaccination in maintaining a physically active lifestyle during the pandemic to obtain all the associated health benefits in older adults.

After observational studies have consistently demonstrated an exposure–disease link (e.g., physical activity–health outcome), in conjunction with supportive laboratory evidence (e.g., potential mechanisms), experimental studies can then be initiated to test, through a rigorous experimental design, the validity of the observational findings. Outcomes might be mortality, disease incidence, or an intermediate end point such as blood cholesterol levels or blood pressure. In the case of physical activity intervention research (e.g., to promote physical activity), the dependent outcome variable could be the physical activity level of either a person or a community. See chapter 9 for more details on specific designs of research within the observational and experimental categories.

Criteria for Causality in Epidemiology

Science usually does not prove **causality**, which is the primary limitation of observational studies in epidemiology. However, establishing inferences of causal links between potential risk factors (e.g., physical inactivity or low fitness) and health outcomes (e.g., heart attack or cancer) is critically important to make effective public health policies. Randomized controlled trial is considered a gold standard to establish evidence for inference of a causal relationship between risk factor and health outcome. However, sometimes it is neither practical nor ethical to conduct a randomized controlled trial to test the validity of observational findings. For example, given the strong associations between smoking and lung cancer from observational studies, it would be unethical and impractical to randomize participants to either smoking or nonsmoking control groups

for several years to investigate the long-term effects of smoking on cancer mortality. In cases like these, certain criteria or guidelines, if met, strengthen the inference that an observed association between an exposure and outcome is causal. These guidelines, called Hill's criteria, were first outlined by epidemiologist and statistician Austin Bradford Hill in 1965 as part of the process that led to the Surgeon General's Report on Smoking and Health (Hill, 1965). While Hill outlined nine criteria in his seminal report, the five most established and agreed-upon criteria are as follows:

1. *Strength of the association:* The rate of the outcome (e.g., lung cancer) is greater in the exposed group (e.g., smokers) compared with the nonexposed group (e.g., nonsmokers). The larger the association between the exposure and outcome, the more likely it is to be causal.

2. *Consistency:* The association between the exposure and outcome is observed regardless of other factors such as sex, age, race, investigator, or methods.

3. *Temporality:* The exposure precedes the outcome with appropriate delay to allow the outcome to occur (e.g., disease progression).

4. *Biological gradient (dose–response):* Greater amounts or degrees of the exposure are associated with a higher rate of the outcome.

5. *Plausibility:* The observed association has a plausible underlying mechanism that is consistent with existing biological knowledge.

The associations of physical activity with health outcomes are commonly expressed as ratios: **odds ratios** and **relative risk**. These ratios clearly show how much benefit results from physical activity in observational studies on the risk of disease, disability, or death. These epidemiological measures and study designs are described in detail with examples in the following sections.

Observational Study Designs

Observational research examines existing differences in factors (e.g., physical activity, dietary habits, smoking) that can cause disease within a population. Some people in a population choose to be physically active, whereas others do not. Epidemiologists use these naturally occurring differences in a population to observe, and therefore understand, the effect of these differences on specific disease outcomes. There are three common types of observational studies: cross-sectional studies, case-control studies, and cohort studies.

Cross-Sectional Study

The cross-sectional study design is perhaps the most frequently conducted type of study examining the relationship between physical activity and health outcomes at a single time point. Cross-sectional studies examine the associations of physical activity with both intermediate end points (e.g., cardiovascular risk factors such as blood pressure) and incidence of disease (e.g., heart attack). For example, researchers simply assemble two groups of people—highly active people (e.g., athletes) and sedentary control people—and then measure blood pressure. Researchers then compare blood pressures between the two groups to see if blood pressure is lower in active people compared to sedentary people.

Cross-sectional studies offer both advantages and disadvantages.

Advantages

- Can be conducted in large samples in a cost-effective manner in a relatively short time since no follow-up of study participants is required
- Can control other factors that might affect the relationship of interest. For example, to control for the potential effect of body fat on the physical activity–blood pressure relationship, one could recruit active and sedentary people with similar body fat levels. Comparisons of the two groups would then be independent of body fat.
- Can assess prevalence of physical activity (exposure) and hypertension (outcome)
- Can generate hypotheses for possible longitudinal associations (e.g., between physical activity and heart attack)

Disadvantages

- Impossible to know if physical inactivity was actually responsible for high blood pressure since both are measured at the same time (no causality)
- Not appropriate for hypothesis testing
- Not suitable for rare diseases

Scientifically sound results can come only from more advanced observational study designs such as case-control or cohort studies, or from experimental study designs such as the randomized controlled trial.

> **KEY POINT**
>
> Cross-sectional studies are quick, easy, cost effective, and good to generate a hypothesis, yet it is impossible to establish a cause-and-effect association between exposure and outcome because the temporal sequence is unknown when both exposure and outcome are measured at the same time.

Case-Control Study

The case-control study aims to identify factors that are causally related to disease outcomes. In this design, people with disease (i.e., cases) and without disease (i.e., controls) are recruited into the study over the same period. Frequently, the disease-free control participants are selected to match cases with respect to potentially important other factors such as age, sex, or ethnicity. Both cases and controls are then asked about their exposures (e.g., physical inactivity) using in-person interviews or self-administered questionnaires. Thus, the objective of this retrospective exposure assessment in case-control studies is to identify factors that influenced the natural history of the disease during its induction period (i.e., before disease onset).

The case-control epidemiological design has been particularly important in studying the relationship between physical activity and relatively rare diseases such as cancer. An interesting case-control study examining the relationship between physical activity and breast cancer in pre- and postmenopausal women was published by Kobayashi and colleagues (2013). Investigators enrolled 1,110 cases with breast cancer and 1,172 controls without breast cancer. Women in the control group were matched to cases according to age. Upon recruitment, each woman was interviewed about the type, duration, frequency, and intensity of her physical activity participation. Using these data, women were categorized into four levels of physical activity exposure. The least active women reported no regular leisure time physical activity (0 hours per week). Active women were categorized into three categories based on the tertile cutoff points for controls reporting some physical activity (>0 hours per week). The most active postmenopausal women (highest tertile) were at a significantly lower risk of having breast cancer compared to the least active women with no leisure time physical activity. The beneficial association of physical activity was only evident for the postmenopausal but not the premenopausal women, suggesting physical activity in mid- to later life might be more strongly associated with reduced risk of breast cancer.

Case-control studies offer both advantages and disadvantages.

Advantages

- Efficient for investigating rare diseases because they avoid the need for extremely large populations
- Can provide valid estimates of exposure–disease relationships in a shorter time and with less monetary expense
- Can study multiple exposures (e.g., physical activity, smoking, diet) for a single disease outcome (e.g., breast cancer)
- Useful for initial development and testing of a hypothesis

Disadvantages

- Recall bias—that is, exposure information is obtained after the disease has been diagnosed, so people who have just been diagnosed with a major disease might recall their exposures differently than control participants do
- Not possible to determine absolute risk (e.g., incident rates are not available without knowing the exact follow-up time between exposure and outcome)
- Since the control group needs to be similar to the cases (e.g., similar age), the control group might not be representative of the general population (selection bias)
- Can only study one disease at a time

KEY POINT

Case-control studies are specifically useful for investigating rare diseases; recall and selection biases are major disadvantages.

Cohort Study

The terms *follow-up studies*, *prospective studies*, and *longitudinal studies* all have been used to describe the cohort study. In this design, a large disease-free population is defined, and assessment of relevant exposures is obtained. Baseline data are used to categorize the cohort into levels of exposure (e.g., low, medium, and high levels of physical activity). After this baseline assessment has been made, the follow-up period begins. Because chronic diseases such as colon and breast cancer are relatively rare, the follow-up period can last from as little as 1 or 2 years to more than 20 years. At the end of the follow-up period, the number of people within the cohort who died or who were diagnosed with the disease outcome of interest during follow-up is counted to compare which group developed more disease and deaths. The analysis of cohort studies is relatively simple. Because we are interested in how various levels of baseline exposure (e.g., different levels of physical activity) predict disease occurrence, the basic analysis consists simply of calculating disease rates for the levels of exposure. For example, the mortality rates among people reporting regular exercise would be compared with the mortality rates of nonexercisers.

Disease rates in cohort studies often are expressed relative to person-years of follow-up. One person-year represents 1 year of observation for one person during the follow-up period. Then, the incidence rate is calculated as the number of events divided by the total person-years of observation in the study population. For example, if 10 new events occur during the study follow-up with a total of 100 person-years, the incidence rate is 10 per 100 person-years (or 1 per 1,000 person-years). More frequently, estimates of the exposure–disease relationship are expressed as relative risk (hazard ratio).

As an example, Lee and colleagues (2014) reported a cohort study examining the association between leisure time running and mortality in more than 55,000 adults. At baseline, participants completed physical activity questionnaires including leisure time running. Participants then were categorized into runners and nonrunners. During 15 years of follow-up, the authors observed 3,413 all-cause and 1,217 cardiovascular deaths. Compared with nonrunners, the authors found 30% (with a relative risk of 0.70) and 45% (with a relative risk of 0.55) lower risks of all-cause and cardiovascular disease mortality, respectively, with a 3-year expanded life expectancy in runners. Those associations were consistent regardless of gender, age group, body mass index, health conditions, smoking status, and alcohol intake. This study clearly shows the significant mortality benefits of running in a large cohort study with a long-term follow-up (figure 8.1).

Cohort studies offer both advantages and disadvantages.

Advantages

- Allows for estimation of true absolute risk of developing a disease since the exposure is measured before disease onset (temporal sequence is clearly defined)

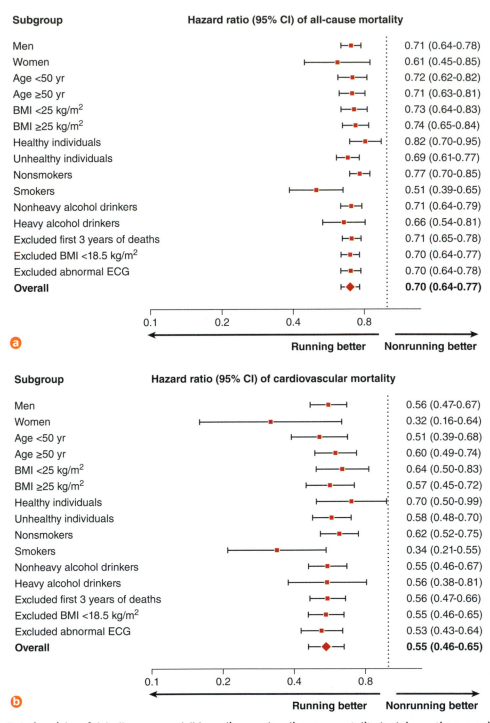

Figure 8.1   Relative risks of *(a)* all-cause and *(b)* cardiovascular disease mortality by leisure time running for 55,137 adults aged 18 to 100 years. The reference group for all analyses includes nonrunners. BMI = body mass index, CI = confidence interval, ECG = electrocardiogram.

Reprinted from D.C. Lee, X. Sui, E.G. Artero, et al., "Long-term Effects of Changes in Cardiorespiratory Fitness and Body Mass Index on All-Cause and Cardiovascular Disease Mortality in Men: The Aerobics Center Longitudinal Study," *Circulation* 124, no. 23 (2011): 2483-2490.

- Good for relatively rare exposure to specific conditions (e.g., if you wanted to study the effects of high-mileage cycling, you could specifically recruit a cohort of cyclists)
- Can study multiple disease outcomes (e.g., the effects of physical activity on various health outcomes such as cancers, heart attack, stroke, hypertension, and mortality)
- Can measure change in exposure (e.g., the effects of staying inactive, becoming active, becoming inactive, and staying active over time on the risk of developing heart attack)

Disadvantages

- Difficult and costly to conduct because of the challenges of following large numbers of people over long periods of time
- Cannot study diseases that occur infrequently, so even a large cohort might not produce enough cases for meaningful analyses
- Differences in other factors (e.g., age) between groups could cause biases (e.g., younger people are more active in general, thus they live longer because they are younger but not because they are active)

KEY POINT

Cohort studies are useful for rare exposures with multiple health outcomes and can correctly classify disease end points with true absolute risk of developing a disease due to its prospective study design. However, it is costly, time consuming, and difficult to study diseases that occur infrequently.

Experimental Study Designs

Experimental designs allow researchers to identify the effects of a specific intervention on a health outcome in a group of people (experimental group) while simultaneously monitoring changes in the same health outcome among people not receiving the intervention (control or comparison group). Randomized controlled trials that are focused on changing health at the individual level are considered the gold standard to investigate possible cause-and-effect associations (causality). Because denying treatment that is known to cause health benefits is illegal, randomized trials for physical activity often provide a form of physical activity to the comparison group (e.g., stretching or yoga) that is not expected to have the same effects as the intervention treatment provided to the intervention group or that compares different types of treatment.

An example of a randomized controlled trial involving a physical activity intervention is Project Active, which compared the effects of a personal, lifestyle physical activity intervention with a traditional, structured exercise intervention (Dunn et al., 1999). Participants were 235 initially sedentary and apparently healthy adults who were randomized into either a lifestyle or a structured intervention group. The hypothesis was that no difference in the physical activity and cardiorespiratory fitness would exist between the two groups at the end of the treatment. Both groups were instructed in programmatically different but physiologically equivalent methods to increase their levels of daily physical activity. Results showed increases in physical activity and cardiorespiratory fitness at the end of the 6-month intensive intervention, but these changes were similar between the two intervention styles.

Randomized controlled trials that are focused on changing behaviors in communities are referred to as *community intervention*. The rationale for community-level interventions is as follows:

- Targeting everyone might prevent more cases of disease than targeting just high-risk people.
- Environmental modifications can be easier to accomplish than large-scale voluntary behavior change.
- Risk-related behaviors are socially influenced.
- Community interventions reach people in their native habitat.
- Community interventions can be logistically simpler and less costly on a per-person basis.

In community intervention, whole communities are randomized to receive multiple treatments in the form of mass media campaigns, school-based programs, point-of-decision prompts (e.g., posting signs to encourage individuals to use the stairs instead of the elevator), and related activities. The comparison (i.e., control) communities do not receive the intervention but often share several demographic characteristics with the treatment community, which serves to show the effect of naturally occurring changes in community behaviors over time. Data collection and analysis include monitoring the extent to which an intervention is implemented as intended and the effects of the intervention on community behaviors (process evaluation), individual-level changes in behavior and health outcomes (individual evaluation), and changes in the environment influenced by the intervention (community-level indicator evaluation). Data analysis must account for community-level and individual-level variations in behaviors. An advantage of community trials is that they allow researchers to see population changes in behaviors as a result of interventions. However, this benefit is tempered by the difficulty of changing individual behaviors and the amount of time that can be required (sometimes many years) to see changes in community behaviors.

A famous example of a community trial is the Stanford Five-City Project (Winkleby et al., 1996), wherein intervention activities designed to reduce cardiovascular disease risk factors were delivered to two communities from 1979 to 1985. The physical activity interventions were targeted toward increasing moderate- and vigorous-intensity physical activity in intervention communities. The interventions used electronic and print media, individual and community activities, and school-based functions. Results showed modest but statistically significant improvements in physical activity in the intervention communities. Men were more likely to participate in vigorous-intensity activities, and women reported spending more time in moderate-intensity activities. Similar changes were not observed in the three control communities.

Experimental study designs offer both advantages and disadvantages.

Advantages

- Randomization ensures that participants are comparable with respect to all important factors (e.g., age, sex, body mass index) between the experimental and control groups, except for the factor being manipulated (e.g., treatment such as exercise).
- Investigators have control over the research process to prevent biases.
- Participants are randomly assigned to experimental or control condition.

Disadvantages

- These studies are expensive and time consuming to conduct, especially when the trial is large and long.
- Recruitment is difficult and compliance is challenging (e.g., participants do not do what they are asked to do).
- Due to smaller sample size compared to the large observational studies, the ability to apply the findings to the general population is limited (limited generalizability).

KEY POINT

A randomized controlled trial is the gold standard to investigate causality between a risk factor and health outcome (e.g., exercise on blood pressure) without major biases and confounding factors. However, randomized controlled trials are costly and often are difficult with regard to recruitment, compliance, retention, and generalizability.

Physical Activity Assessment

Physical activity can be measured using a variety of methods ranging from direct measurement of physiological variables (e.g., body heat, heart rate, oxygen consumption) as indicators of physical activity to asking people to rate how active they recall being during the past week or year. From the 1950s to the 1980s, job titles were used to classify occupational physical activity. However, with the changing profile of the labor market, occupational titles no longer reflect the physical requirements of a job, eliminating the use of job titles to classify occupational energy expenditure.

Because of the exceptionally large number of people in many epidemiological studies, self-administered questionnaires or brief interviews often are used to capture activity done in different domains (e.g., at work, home, transportation, leisure settings). Questionnaires can be classified as global, short recall, and quantitative histories depending on the length and complexity of the items.

- *Global questionnaires* are instruments of one to four items that present a general classification of a person's habitual activity patterns. Because these surveys can take little time to administer (less than 2 minutes), they are preferred for use in large epidemiological studies.
- *Short recall questionnaires* generally have 5 to 15 items and reflect recent physical activity patterns (during the past week or month). Short recall questionnaires are recommended for surveillance activities and observational studies designed to assess the prevalence of adults and children who obtain national recommendations for physical activity.
- *Quantitative history questionnaires* are detailed instruments that have from 15 to 60 items and reflect the intensity, frequency, and duration of activity patterns in various categories, such as occupation, household, sports, transportation,

and leisure activities. Because of their length and complexity, they are usually administered by an interviewer. Quantitative history questionnaires are appropriate for studies designed to examine issues of dose–response in various populations with a wide variety of physical activity patterns.

One of the most commonly used questionnaires is the International Physical Activity Questionnaire (IPAQ), which can be short or long forms (versions) using either telephone or self-administered methods. The seven-item short form is an example of a short recall questionnaire including four sections on vigorous activity, moderate activity, walking, and sitting. The 27-item long form is an example of a quantitative history questionnaire including five activity domains of occupation, transportation, housework, leisure time, and sitting. The IPAQ has been developed in different languages across 12 countries and provides data processing and a scoring protocol to calculate total MET minutes per week and identify individuals who meet the U.S. physical activity guidelines.

Therefore, the IPAQ can be used to compare physical activity data internationally and between studies.

Since their introduction as an objective measure of free-living physical activity in the early 1980s, waist-mounted activity monitors (accelerometers and pedometers) have become a staple of the physical activity assessment repertoire. They have been used extensively in the validation of self-reported physical activity surveys, as outcome measures of physical activity in intervention studies, and in research designed to identify the psychosocial and environmental correlates of physical activity behaviors. The great advantage of objective physical activity measures is that they overcome some of the limitations inherent in self-report methods that rely on information reported by participants. Activity monitors result in far fewer reporting errors or errors introduced by interviewers, and real-time data collection and automated data reduction can provide a rich description of the activity profiles of people and populations. Activity monitors offer a relatively simple and efficient method of measure-

Activity monitors provide accurate data, which helps reduce reporting errors.

Phynart Studio/E+/Getty Images

ment that is suitable for small intervention studies and observational studies of intermediate size (e.g., fewer than 5,000).

The ActiGraph is a small battery-operated accelerometer-based motion sensor commonly worn on the waist. It provides a computerized record of the intensity and duration of ambulatory movement, presented as movement counts. The ActiGraph accelerometer is used in the U.S. National Health and Nutrition Examination Surveys (NHANES) to augment physical activity questionnaires in a subsample of participants. Activity count levels above 1,950 counts per minute often reflect purposeful moderate-intensity walking in the range of 3 to 4 miles per hour (4.8 to 6.4 km/h) (Freedson et al., 1998). Objective information can be used to track changes in activity patterns in response to an intervention or to describe differences in the activity profiles of different populations using the same objective standard. A number of other accelerometer-based activity monitors that are comparable to the ActiGraph devices are commercially available.

Pedometers also have been used in epidemiological studies to measure the amount of accumulated steps in free-living settings. Pedometers are used extensively in health promotion programs to provide feedback to participants about their activity levels or in community trials to measure the effect of walking behavior interventions. For example, to quantify levels of walking among residents in a small southern community, Tudor-Locke and colleagues (2004) mailed pedometers to 209 adults residing at randomly selected households who agreed to wear the pedometer for 7 consecutive days and record their accumulated steps in a logbook each day.

In general, physical activity questionnaires are more feasible, especially in larger observational studies, and objective physical activity measures are more accurate and valid (when compared to direct calorimetry for energy expenditure in a smaller mechanistic study). Epidemiologists usually value feasibility over accuracy to accommodate extensive sample sizes, while physiologists often value accuracy over feasibility with smaller numbers of participants. A major limitation in using questionnaires is recall bias that people, especially children or older adults with cognitive impairment, do not accurately remember their previous physical activity and tend to overreport their activities. However, questionnaires are inexpensive and generally easy to administer. Researchers also can identify the type of physical activity (e.g., walking, bicycling, tennis, weightlifting) and detailed contextual information about where (e.g., home, work, school) and when (e.g., leisure time or transportation) the physical activity occurred. Major limitations in objective measures of physical activity include the high cost to purchase activity monitors, difficulties in identifying the type of activities, lack of contextual information, considerable time and effort to download and process the data, and compliance issues including correct usage and wear over several days. However, activity monitors (e.g., smart watches and smartphone apps) are becoming more affordable, popular, and accurate for both researchers and the public, although the frequent technological updates and new commercial versions of these monitors make it difficult to compare the results between studies over time.

Other objective measures for physical activity are cardiorespiratory fitness and muscular strength that are often used to reflect recent aerobic and muscle-strengthening activities, respectively. The best method of physical activity assessment depends on research priorities such as study size, population, purpose, and outcome measures.

KEY POINT

Physical activity questionnaires are inexpensive, easy to administer, and feasible in a large study that can also identify types and contextual information of physical activity. However, recall bias is a major limitation in questionnaires. Objective physical activity measures are more accurate and valid but are expensive, time consuming in downloading and processing the data, and difficult to know the type and contextual information of physical activity.

Overview of Knowledge in Physical Activity Epidemiology

Based on a systematic review of research, the World Health Organization has concluded that physical inactivity is the fourth leading global risk factor for mortality after high blood pressure, cigarette smoking, and high blood glucose (World Health Organization, 2009). Compelling evidence from physical activity epidemiology supports that physically active individuals have lower risk of developing chronic diseases, such as cardiovascular disease, and premature mortality. Importantly, physically fit individuals with higher cardiorespiratory fitness or muscular strength generally have even larger health benefits.

Physical Activity and Health

It is well documented that physical activity is important to reduce the risk of chronic diseases and increase longevity. Lee, Shiroma, and colleagues (2012) quantified the effect of physical inactivity on major noncommunicable diseases including cardiovascular disease and mortality at the population level. They selected the most recent research studies on the subject and performed a meta-analysis (combines numerical results of these studies; see chapter 9) of their study outcomes. The authors found that physical inactivity causes 6% of the burden of disease from coronary heart disease, 7% of type 2 diabetes, 10% of breast cancer, 10% of colon cancer, and 9% of premature mortality (causing more than 5.3 million deaths) worldwide. These results suggest, for example, that 6% of the burden of disease worldwide due to coronary heart disease can be eliminated if all inactive people became active. They also estimated that the elimination of physical inactivity would increase the life expectancy of the world's population by 0.7 years. This large meta-analysis study indicated that physical inactivity is a major lifestyle risk factor for chronic diseases and mortality, and removal of this unhealthy behavior could improve health worldwide.

Another large, important study examined whether different amounts of physical activity were associated with cardiovascular disease and mortality in a prospective cohort study of over 130,000 people from 17 countries at different economic levels (Lear et al., 2017). Using the International Physical Activity Questionnaire (IPAQ) to assess physical activity, the researchers invited people between the ages of 35 and 70 years from urban and rural areas in each country to reflect the geographical diversity. They found that compared to low physical activity (<150 minutes per week of moderate-intensity activity), moderate (150-750 minutes per week) and high (>750 minutes per week) physical activity were associated with 20% and 35% reduced risks of all-cause mortality, respectively, and 14% and 25% reduced risks of major cardiovascular diseases (including incident heart attack, stroke, and heart failure) and cardiovascular disease mortality, respectively. These associations were found consistently in high-income, middle-income, and low-income countries. This study provided robust evidence to increase physical activity in countries of different socioeconomic backgrounds for the prevention of cardiovascular disease and premature mortality.

Research on the health benefits of physical activity have focused more on aerobic physical activity, and less data is available on the health benefits of resistance and muscle-strengthening activities, especially on cardiovascular disease and mortality. Liu and associates (2019) investigated the associations of resistance exercise independent of and combined with aerobic exercise on the risk of total cardiovascular events (cardiovascular morbidity and mortality combined) and all-cause mortality in a large prospective cohort study. They used self-reported physical activity to measure both aerobic and resistance exercise in over 12,000 men and women. During a mean follow-up of 5 years, they found that resistance exercise was associated with 40% to 70% lower risk of total cardiovascular events independent of aerobic exercise. They further found that even one time or less than 1 hour per week of resistance exercise was associated with lower risk of total cardiovascular events regardless of meeting the aerobic physical activity guidelines. They reported similar results on all-cause mortality. In separate studies, researchers have found that resistance exercise reduces risk of developing metabolic syndrome (Bakker at al., 2017), hypercholesterolemia (Bakker et al., 2018), and

Financial Benefits of Being Active

Sedentary behavior adversely affects not only health, but also the economic burden of medical care and disability in individuals and communities, which is important to help policy makers quantify the impact of physical activity on health care costs. Ding and colleagues (2016) found that the estimated health care costs attributable to physical inactivity on major noncommunicable diseases (e.g., coronary heart disease, stroke, cancer) was $53.8 billion worldwide using large data from 142 countries, representing 93% of the world's population. Another study by Anderson and associates (2005) reported that the average annual health care costs were $5,783 in inactive (0 day/week of physical activity) versus $4,240 in active individuals (≥4 days per week of physical activity) in over 4,000 U.S. adults aged at least 40 years old. Overall, health care costs were about 20% to 40% lower in active individuals compared with inactive individuals, although it varies depending on the health care system, age group, and definition of physical activity between studies. These data clearly indicate monetary savings are significant by being physically active in addition to the health benefits gained from being active.

obesity (Brellenthin, 2021), independent of aerobic exercise. These data are important and clearly support the current physical activity guidelines recommending both aerobic and muscle-strengthening activities for the greatest health benefits.

KEY POINT

It is well documented that physical activity including both aerobic and muscle-strengthening activities are important to prevent chronic diseases and increase longevity.

Physical Fitness and Health

As mentioned earlier, self-reported physical activity has measurement error because people tend to over-report their activity levels. One option to minimize this issue is to use objective measures such as cardiorespiratory fitness and muscular strength as indicators of recent aerobic and muscle-strengthening activities, respectively.

The American Heart Association (AHA) summarized the extensive epidemiological and clinical research and provided strong scientific evidence that lower levels of cardiorespiratory fitness are associated with a higher incidence of cardiovascular disease, all-cause mortality, and mortality attributable to various cancers (Ross et al., 2016). They also suggested that lower cardiorespiratory fitness is a stronger predictor

of all-cause mortality than established traditional risk factors such as smoking, hypertension, high cholesterol, and type 2 diabetes. Furthermore, the AHA reported that numerous epidemiological studies have demonstrated that more than half of the reduction in all-cause and cardiovascular disease mortality occurs when moving from the least fit group to the next least fit group. This indicates that encouraging sedentary individuals to engage in even modest amounts of physical activity could provide considerable health benefits.

One popular research topic is the impact of fitness versus fatness on health. It has been controversial and continuously debated given both fitness and fatness are important clinical risk factors that affect the development of health recommendations and policies. However, most studies have been limited to a single assessment of fitness and fatness and assumed no changes over time. To address this issue, a team of researchers examined the independent and combined associations of changes in both fitness and body mass index with mortality in a cohort study of over 14,000 men with 11 years of follow-up (Lee et al., 2011). As shown in figure 8.2, they found that individuals who lost fitness had 65% to 106% increased risk of all-cause mortality regardless of changes in body mass index compared with individuals who gained fitness and lowered their body mass index (ideal change). In additional analyses, they also observed that even increases in percent body fat, measured by hydro-

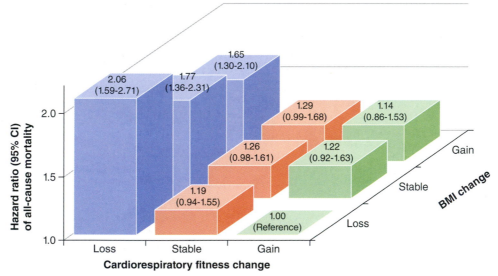

Figure 8.2   Relative risk (95% confidence intervals) of all-cause mortality by changes in cardiorespiratory fitness and body mass index in 14,358 men aged 20 years or older (mean age was 44 years). All data were adjusted for age, examination year, parental cardiovascular disease, body mass index, maximal METs at baseline, changes in lifestyle factors, changes in medical conditions, and the number of clinic visits between the baseline and last examinations in calculations of changes in fitness and body mass index. MET = Metabolic equivalent of task.

Reprinted from D.C. Lee, X. Sui, E.G. Artero, et al., "Long-term Effects of Changes in Cardiorespiratory Fitness and Body Mass Index on All-Cause and Cardiovascular Disease Mortality in Men: The Aerobics Center Longitudinal Study," *Circulation* 124, no. 23 (2011): 2483-2490.

densitometry (underwater weighing) or skinfold method, were not associated with all-cause mortality after adjusting for change in fitness.

They further found that every 1-MET increase in cardiorespiratory fitness was associated with 15% and 19% reduced risk of all-cause and cardiovascular disease mortality, respectively. The reduced risk of cardiovascular disease mortality was further supported by the follow-up study (Lee, Sui, et al., 2012). This study found that every 1-MET improvement in cardiorespiratory fitness was associated with 7%, 22%, and 12% lower risk of developing hypertension, metabolic syndrome, and hypercholesterolemia, respectively, which are all established traditional cardiovascular disease risk factors. These data from both studies on the importance of cardiorespiratory fitness have a direct public health impact since extensive attention has been given to weight loss, which is challenging for most overweight and obese individuals who constitute two-thirds of the U.S. population. These studies highlight that maintaining and improving cardiorespiratory fitness is important for reducing premature mortality regardless of weight and fatness change; thus, increased attention needs to be placed on cardiorespiratory fitness, which can be improved by increasing physical activity.

Similar to aerobic physical activity, the current evidence used for the development of physical activity guidelines on muscle-strengthening activity is based on self-reported resistance exercise in large population studies since there is no reliable objective measure of lifestyle muscle-strengthening activity. One way to resolve this issue is to use objectively measured muscular strength, because muscular strength improves by performing regular resistance exercise. Compared to the knowledge and evidence of the health benefits of cardiorespiratory fitness, we still know less about the health benefits of muscular strength since muscular strength was historically considered mostly in relation to sport performance for athletes. However, it is a growing field of research in exercise science in many populations (e.g., youth and the elderly). Several recent studies have suggested significant benefits of high levels of muscular strength on the prevention of chronic diseases and premature mortality.

One of these studies reported the associations of handgrip strength, which is a simple, quick, and inexpensive measure as a proxy of total body strength, with various chronic diseases and cardiovascular and all-cause mortality in a large cohort study from 17 countries (Leong et al., 2015). They included nearly 140,000 men and women, and over 3,300 participants died during an average follow-up of 4 years.

They found that every 5-kilogram (11 lbs) reduction in grip strength was associated with 16%, 17%, 7%, and 9% increased risks of all-cause mortality, cardiovascular mortality, heart attack, and stroke, respectively. They also found that grip strength was a stronger predictor of all-cause and cardiovascular mortality than systolic blood pressure.

In another large meta-analysis from 38 cohort studies that included approximately 2 million healthy men and women and over 63,000 deaths, the researchers found that higher levels of handgrip strength were associated with a 31% reduced risk of all-cause mortality with a stronger association in women than men (García-Hermoso et al., 2018). They also found that adults with higher levels of muscular strength measured by knee extension strength test had a 14% lower risk of death. This study indicates that higher levels of upper- and lower-body muscular strength are associated with a lower risk of mortality in the healthy adult population. The ease and low cost of measuring static grip strength with a hand dynamometer and positive correlation with other strength measures make it a good surrogate measure of muscular strength for epidemiologic studies.

Many people, including physicians, believe that people with hypertension should not perform resistance exercise due to potential risks associated with holding one's breath while straining (Valsalva maneuver). Hypertension is one of the most common chronic diseases and increases the risk of premature mortality. Dr. Artero and his colleagues (2011) assessed the impact of muscular strength measured by 1 repetition maximum (1RM) of bench and leg press on all-cause mortality in 1,506 men with hypertension aged 40 years and older. During an average follow-up of 18 years in their cohort study, 183 deaths occurred. The authors found that men in the upper third of muscular strength had 44% lower risk of all-cause mortality compared to men in the lower third of strength, after controlling for potential confounders (e.g., age, smoking, body mass index, blood pressure, total cholesterol, diabetes, physical activity), even including cardiorespiratory fitness. In the joint analysis by both muscular strength and cardiorespiratory fitness, they found lowest risk of mortality in men with higher strength and fitness. This study has an important public health message that men with hypertension should follow the current physical activity guidelines and engage in both aerobic and resistance exercise not only to reduce or maintain blood pressure, but also to improve cardiorespiratory fitness and muscular strength for mortality benefits.

KEY POINT

Both cardiorespiratory fitness and muscular strength are strong and independent predictors of chronic diseases and premature mortality in various populations. To improve fitness and strength, meeting the current physical activity guidelines, which recommend both aerobic and resistance exercise, should be emphasized.

Wrap-Up

Knowledge and evidence developed from physical activity epidemiology made significant contributions to the development of Physical Activity Guidelines for Americans and Physical Activity Plan for health care professionals, policy makers, and individuals. Physical activity epidemiology is an important, growing area of exercise science and practice from both clinical and public health perspectives. It investigates the distribution and determinants of physical activity and how much the risk of chronic diseases could be reduced by being physically active and fit, as well as works to prevent and control these diseases by promoting physical activity in populations. Physical activity epidemiology started to grow in the 1950s in an effort to prevent cardiovascular disease by establishing large prospective cohort studies. Earlier studies used self-reported physical activity and focused on health benefits of aerobic activity, then expanded to objectively measured activity monitors (e.g., accelerometers) and started to look at the benefits of physical fitness and muscle-strengthening activity on various health outcomes. To establish causal relationships between risk factors (e.g., physical inactivity) and health outcomes in epidemiology, the following criteria are important to test and meet: strength of the association, consistency, temporality, biological gradient (dose–response), and plausibility. Epidemiological research studies follow observational or experimental study designs. Observational studies are good to establish and develop initial hypothesis, and experimental studies are good to test the hypothesis following more rigorous research methods to control biases and confounders. It is clear and well documented that higher levels of physical activity (including both aerobic and muscle-strengthening activities) and physical fitness (including both cardiorespiratory fitness and muscular strength) are independently associated with lower risks of developing chronic diseases and premature mortality.

More Information on Physical Activity Epidemiology

Organizations

Centers for Disease Control and Prevention (CDC)

Epidemiology and Prevention | Lifestyle and Cardiometabolic Health section of the American Heart Association (AHA)

Physical Activity Alliance (PAA)

Physical Activity Section of the American Public Health Association (APHA)

World Health Organization (WHO)

Journals

American Journal of Epidemiology

Annals of Epidemiology

British Journal of Sports Medicine

Journal of Physical Activity and Health

Medicine & Science in Sports & Exercise

Research Quarterly for Exercise and Sport

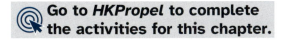 **Go to *HKPropel* to complete the activities for this chapter.**

Review Questions

1. What are the 2018 U.S. Physical Activity Guidelines for adults?

2. What is the definition of physical activity epidemiology, and how is it related to exercise science?

3. What are the five most established and agreed-upon criteria of Hill's criteria for causality in epidemiology?

4. Describe the major differences between observational studies and randomized controlled trials.

5. What are the associations between physical activity and fitness with chronic diseases and premature mortality?

Suggested Readings

Bouchard, C., Blair, S.N., & Haskell, W.L. (Eds.). (2012). *Physical activity and health* (2nd ed.). Human Kinetics.

Dishman, R. K., Heath, G. W., Schmidt, M.D., & Lee, I.M. (Eds.). (2022). *Physical activity epidemiology* (3rd ed.). Human Kinetics.

Keadle, S.K., Bustamante, E.E., & Buman, M.P. (2021). Physical activity and public health: Four decades of progress. *Kinesiology Review, 10*(3), 319-330.

U.S. Department of Health and Human Services (USDHHS). (2018). *Physical activity guidelines for Americans* (2nd ed.). U.S. Department of Health and Human Services.

World Health Organization. (2020). *WHO guidelines on physical activity and sedentary behaviour.* World Health Organization.

References

Anderson, L.H., Martinson, B.C., Crain, A.L., Pronk, N.P., Whitebird, R.R., O'Connor, P.J., & Fine, L.J. (2005). Health care charges associated with physical inactivity, overweight, and obesity. *Preventing Chronic Disease, 2*(4), A09.

Artero, E.G., Lee, D.C., Ruiz, J.R., Sui, X., Ortega, F.B., Church, T.S., Lavie, C.J., Castillo, M.J., Blair, S.N. (2011). A prospective study of muscular strength and all-cause mortality in men with hypertension. *Journal of the American College of Cardiology, 57*(18), 1831-1837.

Bakker, E.A., Lee, D.C., Sui, X., Artero, E.G., Ruiz, J.R., Eijsvogels, T.M.H., Lavie, C.J., & Blair, S.N. (2017). Association of resistance exercise, independent of and combined with aerobic exercise, with the incidence of metabolic syndrome. *Mayo Clinic Proceedings, 92*(8), 1214-1222.

Bakker, E.A., Lee, D.C., Sui, X., Eijsvogels, T.M.H., Ortega, F.B., Lee, I.M., Lavie, C.J., & Blair, S.N. (2018). Association of resistance exercise with the incidence of hypercholesterolemia in men. *Mayo Clinic Proceedings, 93*(4), 419-428.

Blair, S.N., Kohl, H.W., III, Paffenbarger, R.S., Jr., Clark, D.G., Cooper, K.H., & Gibbons, L.W. (1989). Physical fitness and all-cause mortality. A prospective study of healthy men and women. *Journal of the American Medical Association, 262*(17), 2395-2401.

Brellenthin, A.G., Lee, D.C., Bennie, J.A., Sui, X., & Blair, S.N. (2021). Resistance exercise, alone and in combination with aerobic exercise, and obesity in Dallas, Texas, US: A prospective cohort study. *PloS Medicine, 18*(6), e1003687.

Caspersen, C.J. (1989). Physical activity epidemiology: Concepts, methods, and applications to exercise science. *Exercise and Sport Sciences Reviews, 17,* 423-474.

Ding, D., Lawson, K.D., Kolbe-Alexander, T.L., Finkelstein, E.A., Katzmarzyk, P.T., van Mechelen, W., Pratt, M., & Lancet Physical Activity Series 2 Executive Committee. (2016). The economic burden of physical inactivity: A global analysis of major non-communicable diseases. *Lancet, 388*(10051), 1311-1324.

Dunn, A.L., Marcus, B.H., Kampert, J.B., Garcia, M.E., Kohl, H.W., III, & Blair, S.N. (1999). Comparison of lifestyle and structured interventions to increase physical activity and cardiorespiratory fitness: A randomized trial. *Journal of the American Medical Association, 281*(4), 327-334.

Freedson, P.S., Melanson, E., & Sirard, J. (1998). Calibration of the Computer Science and Applications, Inc. accelerometer. *Medicine & Science in Sport & Exercise, 30*(5), 777-781.

García-Hermoso, A., Cavero-Redondo, I., Ramírez-Vélez, R., Ruiz, J.R., Ortega, F.B., Lee, D.C., & Martínez-Vizcaíno, V. (2018). Muscular strength as a predictor of all-cause mortality in an apparently healthy population: A systematic review and meta-analysis of data from approximately 2 million men and women. *Archives of Physical Medicine and Rehabilitation, 99*(10), 2100-2113.

Hill, A.B. (1965). The environment and disease: Association or causation? *Journal of the Royal Society of Medicine, 58,* 295-300.

Kobayashi, L.C., Janssen, I., Richardson, H., Lai, A.S., Spinelli, J.J., & Aronson, K.J. (2013). Moderate-to-vigorous intensity physical activity across the life course and risk of pre- and post-menopausal breast cancer. *Breast Cancer Research and Treatment, 139,* 851-861.

Last, J.M. (1988). *A dictionary of epidemiology.* Oxford University Press.

Lear, S.A., Hu, W., Rangarajan, S., Gasevic, D., Leong, D., Iqbal, R., Casanova, A., Swaminathan, S., Anjana, R.M., Kumar, R., Rosengren, A., Wei, L., Yang, W., Chuangshi, W., Huaxing, L., Nair, S., Diaz, R., Swidon, H., Gupta, R., . . . Yusuf, S. (2017). The effect of physical activity on mortality and cardiovascular disease in

130 000 people from 17 high-income, middle-income, and low-income countries: The PURE study. *Lancet, 390*(10113), 2643-2654.

Lee, D.C., & Brellenthin, A.G. (2023). Physical activity epidemiology research. In J.R. Thomas, P.E. Martin, J.L. Etnier, & S.J. Silverman (Eds.), *Research methods in physical activity* (8th ed., pp. 321-346). Human Kinetics.

Lee, D.C., Pate, R.R., Lavie, C.J., Sui, X., Church, T.S., & Blair, S.N. (2014). Leisure-time running reduces all-cause and cardiovascular mortality risk. *Journal of the American College of Cardiology, 64*(5), 472-481.

Lee, D.C., Sui, X., Church, T.S., Lavie, C.J., Jackson, A.S., & Blair, S.N. (2012). Changes in fitness and fatness on the development of cardiovascular disease risk factors hypertension, metabolic syndrome, and hypercholesterolemia. *Journal of the American College of Cardiology, 59*(7), 665-672.

Lee, I.M., Shiroma, E.J., Lobelo, F., Puska, P., Blair, S.N., Katzmarzyk, P.T., & Lancet Physical Activity Series Working Group. (2012). Effect of physical inactivity on major non-communicable diseases worldwide: An analysis of burden of disease and life expectancy. *Lancet, 380*(9838), 219-229.

Lee, D.C., Sui, X., Artero, E.G., Lee, I.M., Church, T.S., McAuley, P.A., Stanford, F.C., Kohl, H.W., III, & Blair, S.N. (2011). Long-term effects of changes in cardiorespiratory fitness and body mass index on all-cause and cardiovascular disease mortality in men: The Aerobics Center Longitudinal Study. *Circulation, 124*(23), 2483-2490.

Lee IM, Shiroma EJ, Kamada M, Bassett DR, Matthews CE, Buring JE. Association of step volume and intensity with all-cause mortality in older women. *JAMA Intern Med.* 2019;179(8):1105-1112.

Lefferts, E.C., Saavedra, J.M., Song, B.K., & Lee, D.C. (2022). Effect of the COVID-19 pandemic on physical activity and sedentary behavior in older adults. *Journal of Clinical Medicine, 11*(6), 1568.

Leong, D.P., Teo, K.K., Rangarajan, S., Lopez-Jaramillo, P., Avezum, A., Jr, Orlandini, A., Seron, P., Ahmed, S.H., Rosengren, A., Kelishadi, R., Rahman, O., Swaminathan, S., Iqbal, R., Gupta, R., Lear, S.A., Oguz, A., Yusoff, K., Zatonska, K., Chifamba, J., . . . Prospective Urban Rural Epidemiology (PURE) Study investigators. (2015). Prognostic value of grip strength: Findings from the Prospective Urban Rural Epidemiology (PURE) study. *Lancet, 386*(9990), 266-273.

Liu, Y., Lee, D.C., Li, Y., Zhu, W., Zhang, R., Sui, X., Lavie, C.J., & Blair, S.N. (2019). Associations of resistance exercise with cardiovascular disease morbidity and mortality. *Medicine & Science in Sport & Exercise, 51*(3), 499-508.

Morris, J.N., Heady, J.A., Raffle, P.A., Roberts, C.G., & Parks, J.W. (1953). Coronary heart-disease and physical activity of work. *Lancet, 262*(6795), 1053-1057.

National Physical Activity Plan Alliance. (2016). *U.S. National Physical Activity Plan*. https://paamovewithus.org/national-physical-activity-plan/

Paffenbarger, R.S., Jr, Hyde, R.T., Wing, A.L., & Hsieh, C.C. (1986). Physical activity, all-cause mortality, and longevity of college alumni. *New England Journal of Medicine, 314*(10), 605-613.

Ross, R., Blair, S.N., Arena, R., Church, T.S., Després, J.P., Franklin, B.A., Haskell, W.L., Kaminsky, L.A., Levine, B.D., Lavie, C.J., Myers, J., Niebauer, J., Sallis, R., Sawada, S.S., Sui, X., Wisløff, U., American Heart Association Physical Activity Committee of the Council on Lifestyle and Cardiometabolic Health, Council on Clinical Cardiology, Council on Epidemiology and Prevention, . . . Stroke Council. (2016). Importance of assessing cardiorespiratory fitness in clinical practice: A case for fitness as a clinical vital sign: A scientific statement from the American Heart Association. *Circulation, 134*(24), e653-e699.

Schroeder, E.C., Welk, G.J., Franke, W.D., & Lee, D.C. (2017). Associations of health club membership with physical activity and cardiovascular health. *PLoS One, 12*(1), e0170471.

Tudor-Locke, C., Williams, J.E., Reis, J.P., & Pluto, D. (2004). Utility of pedometers for assessing physical activity construct validity. *Sports Medicine, 34*(5), 281-291.

U.S. Department of Health and Human Services (USDHHS). (2018). *Physical activity guidelines for Americans* (2nd ed.). U.S. Department of Health and Human Services.

U.S. Department of Health and Human Services (USDHHS). (2020). *Healthy people 2030*. U.S. Department of Health and Human Services.

Winkleby, M.A., Taylor, C.B., Jatulis, D., & Fortmann, S.P. (1996). The long-term effects of a cardiovascular disease prevention trial: The Stanford Five-City Project. *American Journal of Public Health, 86*, 1773-1779.

World Health Organization. (2009). *Global health risks: Mortality and burden of disease attributable to selected major risks*. World Health Organization.

PART III
Related Professional Subdisciplines

Exercise science knowledge often is integrated with other scientific knowledge to assist clients in promoting health, longevity, and improving function. These applied subdisciplines of exercise science provide the translational evidence that is integrated with basic exercise science evidence to work with clients in real-world settings. Part III introduces you to research methods common in exercise science and how the evidence from that research is integrated with professional experience and client or patient wishes in evidence-based practice (chapter 9). In addition, two general areas of applied, professional research related to exercise science are presented. Chapter 10 highlights the growing knowledge base on sport performance from research in strength and conditioning, sport nutrition, and sport analytics. In chapter 11 the extensive applied research and knowledge in medical and allied health fields related to therapeutic exercise are summarized.

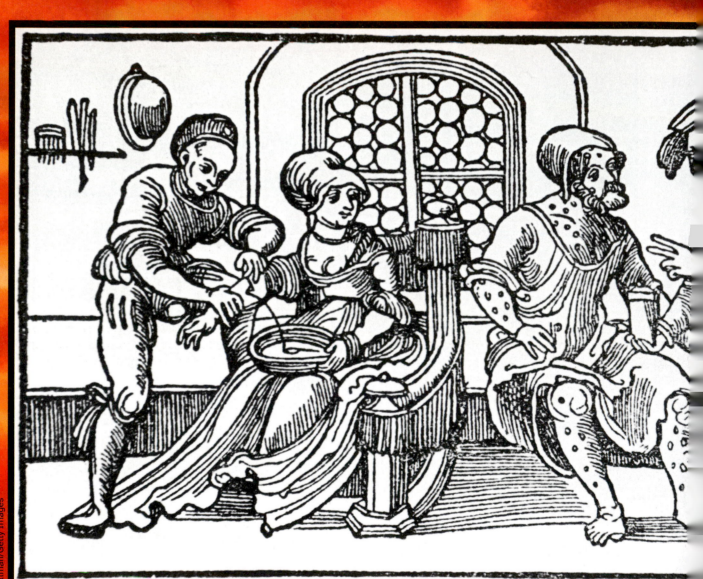

CHAPTER 9

Research and Evidence-Based Practice

James L. Farnsworth II and Natalie L. Myers

CHAPTER OBJECTIVES

In this chapter, we will

- introduce you to the fundamentals of evidence-based practice,
- help you develop an appreciation for research by understanding its role in evidence-based practice,
- review the different types of research designs and how they affect the decision-making process, and
- describe the process for identifying and evaluating the strength of research evidence to support evidence-based practice in exercise science and kinesiology professions.

Prior to the 1850s many believed that the majority of illnesses and conditions were caused by an imbalance in a person's humors. The concept of humors can be traced back to ancient Egyptian medicine but became popularized thanks to Hippocrates and other Greek physicians such as Galen who believed that the body consisted of four basic humors (blood, yellow bile, black bile, and phlegm). A common 18th-century treatment known as *bloodletting* involved draining some of the patient's blood. Although many began to criticize humoralism beginning in the 17th century, it was not until 1858 that the practice was demolished definitively. Many other historical medical practices were once considered the "best" treatment, such as using heroin as a cough suppressant or to relieve asthma symptoms, using arsenic and mercury to treat syphilis, or using cocaine drops to relieve pain from a toothache. Although some of these treatments seem absurd, at the time they were considered the best available or standard of treatment.

In the early 21st century, it is hard for many people familiar with advanced medical imaging, genetic testing, and robotic surgery to appreciate the rapid improvement in science and its application in medicine and exercise. Recall that chapter 1 noted the difference between doctoral (PhD) degrees and professional degrees (e.g., MD, DPT, DAT). PhDs were awarded several hundred years before physicians began borrowing the term *doctor*, a word that means teacher. The rapid advancement of biomedical sciences after the 1900s and low quality of many medical schools highlighted by the Flexner (1910) report spurred numerous reforms. Human anatomy, biology, physiology, and numerous pathological conditions are exceedingly complex, making many diseases and injuries difficult to treat. It was not until the 1990s that medicine initiated systematic methods of evaluating and combining the vast amount of scientific and medical evidence for treatment. This development of evidence-based practice, which has spread to other allied health and professional fields, means that we now understand that better and safer options are available to treat and manage our patients and clients. In the future we might discover that some of our modern medical practices are as ineffective or outdated as those of the 18th century! In this chapter you will learn about evidence-based practice, research principles, how to identify research, and the application of research to assist with making evidence-based decisions in your profession.

Benefits of Research and Evidence-Based Practice in Exercise Science

Many exercise and sport science professionals conduct research because they want to help clients and positively affect practice in their field. Researchers also are inquisitive by nature and like to find answers to questions. In this professional field it is particularly important "to address questions that have the potential to improve performance" (Bishop et al., 2006, p. 167). These questions about performance can be influential to other professionals such as coaches, athletes, and administrators. Research is a fantastic way to show proof of concept and outcomes that can be used to leverage decision making or even change outdated ways of thinking. Research in sports medicine and exercise science also focuses on questions about factors influencing risk of injury and the reduction of and treatment of injury.

Having a sound background in research methods and evaluating evidence allows you to bridge the gap between the controlled research environment and the messier and complex real-world environment of the clinic, field, and gym. Even if the act of conducting the actual research does not interest you, understanding where to search, how to search, and how to interpret and critically appraise exercise science research can improve professional practice and client interaction. The understanding of research also allows you to engage in evidence-based practice.

> ## KEY POINT
> Exercise science research generally focuses on two main benefits: improving performance and reducing risk of injury.

Evidence-Based Practice

Evidence-based medicine (EBM) is a phrase grounded in European philosophical mid-19th-century origins. A consistent definition of EBM did not arise in the scientific literature until the 1990s with the conscientious, explicit, and judicious use of current

Career Opportunity

Exercise Science Researcher

Have you ever thought about why we exercise or train a specific way? Did you ever dream of being a detective when you were younger? If you answered yes to any of the previous questions, you might be interested in a career as an exercise science researcher. Research often recalls a negative stereotype of someone in a white coat working in a lab; however, research is so much more. Many exercise science researchers work directly with athletes to discover the most effective training methods or ways to reduce injury risk. Career opportunities as a research scientist are plentiful, with jobs available across the country in specific research institutes or university settings. Working in a university setting provides you with the opportunity not only to conduct research, but also to teach the upcoming generation of professionals in the field. To get a job working as a university professor, most institutions require you to have completed a doctoral degree in a related field. In exercise science doctoral program education begins after completing your master's degree and typically takes anywhere from 3 to 5 years to finish. If this career is something that interests you, consider asking your classroom instructor about some of their experiences with teaching and conducting research.

best evidence in making decisions about the care of individual patients (Sackett et al., 1996). In the 1990s EBM most often was recognized in the context of medicine, yet professionals outside of medicine were beginning to embrace and implement an evidence-based approach to practice and learning. Consequently, Dawes and colleagues (2005) advocated for a transition from EBM to **evidence-based practice (EBP)** to encompass not only health care professionals, but all professionals adopting an evidence-based approach to professional practice. EBP is a problem-solving approach that integrates the best available research, professional experience, and client values into practice. The purpose of EBP is to promote collaborative and inquisitive thinking from the professional to ensure quality care and a positive and an effective impact on the client.

Researchers promoting EBP typically recommend a five-step process that allows for EBP implementation (Dawes et al., 2005; Johnson, 2008; Melnyk et al., 2011; Sackett et al., 2000). The details of each step of EBP are described in table 9.1.

KEY POINT

Evidence-based practice is a problem-solving approach that integrates the best available research, professional experience, and client values into practice. Integrating EBP into practice allows for well-informed decision making and the integration of the client into that decision-making process.

Table 9.1 The Five Steps of Evidence-Based Practice

Steps of EBP	Explanation	Examples of each step
1. Ask an answerable question	The goal is to improve practice, so the question should be clearly worded. Questions are often categorized into background questions (who, what, where, when, how, and why) or foreground questions (specific questions that affect treatment). Foreground questions often are formed using a format called PICO(T) (Lin et al., 2010).	*Background question:* What wearable technology is valid and reliable for monitoring step count during running? *Foreground question:* Is it possible for a running-related training program to change footstrike patterns in novice runners?
2. Find the best evidence	This is the information-gathering phase. Search for the best available evidence through trustworthy sources. Consider a database that contains peer-reviewed literature. The goal is to find high levels of evidence (described in figure 9.1).	Use sources such as PubMed, EBSCO Host, or Google Scholar to find the evidence. Types of questions can be answered with different levels of research evidence. Levels of evidence are scored on a scale of 1 to 5 according to the Oxford Center for Evidence-Based Medicine 2011 Levels of Evidence Table (OCEBM Levels of Evidence Working Group, 2011). Level 1 represents the highest level of evidence. Using the questions above: The best evidence to answer the background question would come from a meta-analysis or systematic review (level 1 evidence). However, this type of evidence might not be available and can be answered with cohort studies (level 3 evidence). The best evidence to answer the foreground question would be with a systematic review of randomized controlled trials (level 1 evidence). However, this type of evidence might not be available and can be answered with a single randomized controlled trial (level 2 evidence) or a cohort study (level 3 evidence).

> continued

Table 9.1   The Five Steps of Evidence-Based Practice   >*continued*

Steps of EBP	Explanation	Examples of each step
3. Evaluate the evidence	The evidence can be evaluated by analyzing the research design to determine if the study yielded high internal[a] and external[b] validity. The goal is to determine if the results of the study are trustworthy or if errors exist within the study that should be considered as part of the interpretation of the results. Questionnaires are available to professionals that will help them appraise the evidence (described in the Critically Reviewing Research Literature section of this chapter).	Several types of critical appraisal tools exist based on the study design implemented by the researcher. Joanna Briggs Institute (JBI) is an international research organization that has developed critical appraisal checklists for several types of research designs (Joanna Briggs Institute, 2021). Let us imagine both your background and foreground question led you to find level 3 evidence that was designed as cohort studies. You can use the JBI critical appraisal checklist for cohort studies to assist in assessing the internal and external validity of the published paper.
4. Apply information in combination with professional experience and client values	While the research is an important part of EBP, the professional must also incorporate their professional experience while also paying attention to what is important to the client. Consider using personal expertise and the evidence to inform the client and listen to the client to understand their personal values and goals.	From your own experience as a runner, you use RunScribe wearable technology and usually recommend this to your clients as it is reliable and valid. But your client expresses that they do not like any extra weight on the foot while running. Consequently, RunScribe is no longer a viable option for this client. You do some investigating and find researchers have found Sensoria fitness socks (commercially available wearable technology) to be valid and reliable when assessing step count. In this example the professional combined all areas of EBP to make an informed decision.
5. Evaluate outcomes	Be sure to evaluate how the client is responding. This also includes personal reflection from the professional. It is important to determine if the application of the information or intervention was effective.	Ask the client if the technology is meeting expectations. Here are a few example questions: Is the fitness sock comfortable? Is the information produced from the technology easy for you to interpret?

[a]Internal validity refers to the correctness of a conclusion. How confident can we be in the findings of the study based on the methodology of the study?

[b]External validity refers to the generalizability of the results. Can the results be generalized to different populations?

Research

Research is a systematic inquiry that allows scholars, scientists, and professionals to answer specific questions by gathering data with the goal of establishing generalizable inferences and knowledge. Thomas and colleagues (2015) identify five types of research:

- Analytical research is logical analysis that attempts to explain complex phenomena like in history and philosophy.
- Descriptive research strives to define the current state or nature of phenomena.
- Experimental or quasiexperimental research strives to establish evidence for mechanisms or causal effects.
- Qualitative research is in-depth observation, description, and analysis of human thoughts and behaviors.
- Mixed-methods research combines both qualitative and quantitative (experimental, descriptive, analytical) approaches.

Most research falls on a continuum based on the ultimate goal. **Basic science** (fundamental or bench science) focuses on theory and the ultimate mechanisms of reality or function, while **applied science** answers questions about real-world application of knowledge. While basic research is curiosity driven and strives to solve specific theoretical problems, it contributes to the "stock of knowledge from which the applied initiatives are drawn" (Henard & McFadyen, 2005, p. 503). In other words, while basic research helps humans understand the true nature of the world, it often leads to subsequent applied research and eventual real-world use in technology and treatment.

The location of basic and applied research can be done in either a laboratory setting or a field setting. Laboratory-based research is often conducted under "sterile" or lab "bench" conditions, allowing for tightly controlled investigations, while field research is conducted in the real world outside of a laboratory (Aziz, 2017). Some examples of field research locations are in a classroom, clinic, park, or weight room.

We need both kinds of research. Basic research uses great control in order to isolate potential causative effects and mechanisms, while applied research is used to see if basic research theories actually work in the real-world. The COVID-19 pandemic clearly showed the need for this kind of applied research, sometimes called *translational research*. Translational research examines if interventions we know work in well-controlled clinical trial conditions (efficacy) also will be achieved in the real-world (effectiveness for people and costs) conditions, where many factors are involved.

KEY POINT

Basic research focuses on theory or the ultimate mechanisms of reality and function, while applied research focuses on more immediate solutions to real-world problems. These types of research are performed in both the laboratory and the field.

Whatever the kind of research, basic or applied, it must be reported, reviewed, and discussed by peers so that the evidence and knowledge can be integrated with previous studies to be considered research. Publication of original research reports in scientific journals and proceedings are the building blocks of theoretical and applied knowledge. A clinician or professional who collects data on their clients to guide their treatment over time is not doing research until they share their results with their peers, who can evaluate it for integration into the body of knowledge of their field.

An example of basic research is an exercise physiologist who is interested in determining if vitamin D deficiency is related to metabolic factors in people with obesity. A biomechanist doing applied research might want to determine if people with low back pain can change lifting technique following an instructional program. An exercise science professional might do descriptive research collecting performance data before, during, and at the end of specific training programs. All three of these studies become research when they are shared with the researchers' peers. This reporting of research is often verbal at a professional conference and then, ultimately, in a published written report following peer review.

When learning about the fundamentals of research it is important to discuss the two broad systematic inquiry techniques: qualitative and quantitative research. Most research is quantitative and dominates the biological, physical, and medical sciences. **Quantitative** research is the collection of numerical data that is based on precise measurements using instruments or tests with known validity and reliability. Some research in anthropology, behavior, education, and other fields is qualitative. **Qualitative** researchers explore the meaning of a phenomenon or process as understood through the participants, in a natural setting. Qualitative researchers work to investigate smaller numbers of participants, focusing in depth on their feelings, perceptions, opinions, values, and beliefs on a particular topic or phenomenon (Arghode, 2012). To find more detailed information regarding the differences between qualitative and quantitative research, please reference Sorin-Peters (2004). Both qualitative and quantitative research take time and voluminous data collection and analysis.

Regardless of whether a research study uses qualitative or quantitative technique, there is no such thing as a perfect study, and there is a hierarchy of research designs because not all research is created equal. Some researchers seek to determine cause-and-effect relationships (experimental research) while others seek to investigate the benefits or risk of a treatment so informed decision making can take place (outcomes research). Since the early 2000s one-dimensional research pyramids have been proposed that focus on weaker research designs (outcomes research) at the bottom of the pyramid and stronger research designs (experimental research) at the top of the pyramid. In theory, as the pyramid narrows at the top, violations of internal validity (potential bias and confounding variables) are less of a threat. Figure 9.1 depicts the traditional (one-dimensional) view on the hierarchy of research designs.

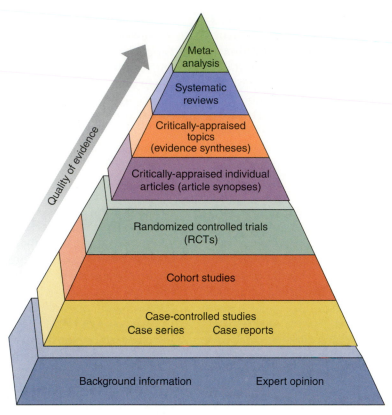

Figure 9.1   The pyramid of research designs. The higher on the pyramid, the better the quality and the less potential for bias.

Often the traditional research pyramid is accompanied by a levels of evidence–ranking scheme that is associated with a particular research design. Levels of evidence is a hierarchy system that dates back to 1979 (The Periodic Health Examination, 1979). Since the 1970s levels of evidence tables have morphed (Howick et al., 2022; OCEBM Levels of Evidence Working Group, 2011; Sackett, 1989), but the one common ground shared by these rankings is that the levels of evidence tables rank studies on the probability of bias (Burns et al., 2011). For example, random control trials (RCTs) are often found at the top tier of the pyramid because they are designed to be unbiased. On the other hand, expert opinion is often found in the lower tiers of pyramids because these publications are biased by one or more authors' experiences or opinions. A physician, for example, likely sees a population of clients that are biased by location, socioeconomic, and other factors. The authors of this chapter suggest referring to the Oxford Centre for Evidence-Based Medicine (OCEBM) Levels of Evidence 2011 table when ranking levels of evidence. This table not only considers bias but the quality of data, which is lacking from earlier versions of levels of evidence–ranking schemes. The OCEBM 2011 table also allows professionals to rank studies on the type of research question that is being investigated.

While the traditional pyramid of research is still well acknowledged across many disciplines, this one-dimensional model has a few problems (Szklo, 1998):

- Qualitative research is not included in the pyramid.
- Not all RCTs do a good job of controlling for bias or have large samples.
- Large population-based cohort outcome studies are automatically placed in the lower tiers of the pyramid despite high levels of external validity.

Regarding the issue of research and statistical validity, it is important to understand that studies with high internal validity do not automatically generate high levels of external validity, because the two are often inversely related. **Internal validity** relates to the design and experimental control that allows for the results to be primarily related to the independent variables being studied. **External validity** relates to the sampling and design issues that allow for the results to be generalized or applied from the study sample to the population of people they represent.

Tomlin and Borgetto (2011) collaborated with a variety of student health care professionals (occupational therapists, physical therapists, and speech pathologists) to develop an alternative model when identifying and considering research hierarchies. Figure 9.2 shows a three-dimensional (3D) research pyramid that acknowledges the dynamics of internal and external validity, includes qualitative research, and "separates, and values at parity, outcome studies, many of which seek to answer real-world questions of professional practice that cannot be investigated through randomized controlled trials" (Tomlin & Borgetto, 2011, p. 192).

- The information presented in pink is the base of the pyramid and results in descriptive-driven studies.
- The yellow and blue sides of the triangle constitute quantitative research through experimental and outcome-driven studies.
- The green side of the triangle represents qualitative research.
- The plus icons on each tip of the pyramid represent the top of the pyramid. Consequently, meta-analyses are placed on the top tiers of each of the pyramidal sides.

- Single-subject study under the experimental research side is on the bottom tier, because each participant is their own control in this type of design.

This 3D pyramid model represents a comprehensive look into the fields of research (qualitative and quantitative), the types of research within those fields, and the interaction between internal and external validity. While qualitative research is important, the majority of exercise science literature focuses on quantitative research. As such, the explanation of these quantitative research designs will be described in the next section of this chapter.

> **KEY POINT**
>
> Depending on the research question, qualitative or quantitative methods can be instituted. Research designs can be viewed hierarchically. The traditional one-dimensional pyramid is well acknowledged but has some flaws that must be recognized. The 3D research pyramid is fluid and recognizes the dynamics between internal and external validity.

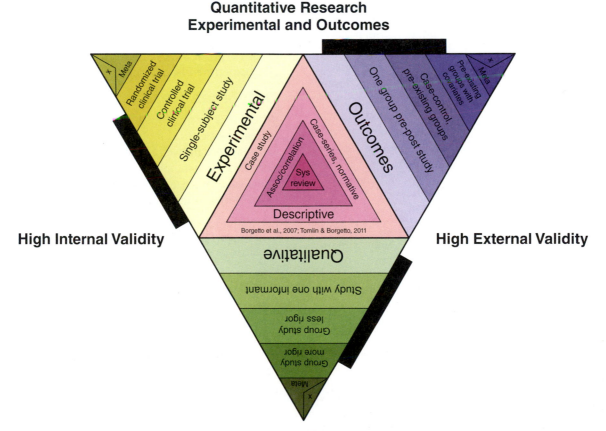

Figure 9.2 3D research pyramid.
Courtesy of University of Puget Sound, Tacoma, WA.

Common Research Designs Used in Exercise Science

Research designs describe the strategy that is used to conduct and answer a question in a study. As mentioned earlier, quantitative research designs often are partitioned into either experimental or outcomes research. **Outcomes research** is also referred to as *observational research*. According to Williams and Kendall (2007), the research design most frequently implemented by sport science researchers is experimental research. Experimental research often is divided into two sections: true experimental (with randomization and control group) and quasiexperimental (no randomization or control group). Before discussing the diverse types of true experimental designs, it is important to note that not every design in the pyramid of research designs will be described in this chapter.

True Experimental Research Designs

True experimental research designs are always prospective in nature, meaning the study is conceived prior to anyone experiencing the outcome. In addition, this type of research helps determine evidence of a cause-and-effect inference or relationship between variables under a controlled environment. Researchers who develop an experimental research design want to determine the efficacy of a new intervention or treatment (independent variables) on certain outcomes (dependent variables). Experimental designs involve a random assignment of participants to either an experimental or control group; thus, these studies are often referred to as *randomized controlled trials* (RCTs), sometimes also called *randomized clinical trials* (Bloomfield & Fisher, 2019). The *Cochrane Handbook* defines RCTs as follows: "If the author(s) state explicitly (usually by some variant of the term 'random' to describe the allocation procedure used) that the groups compared in the trial were established by random allocation, then the trial is classified as a RCT" (Higgins et al., 2019). Many forms of RCT experimental designs are available, but the most common are outlined in table 9.2.

In some cases, a true experimental design cannot be executed, because random assignment of participants or control treatment is not always ethical or possible. Other designs, like quasiexperimental designs, might be more appropriate for when randomization is not feasible. Research on elite (which usually refers to Olympic or international-level) athletes, for example, typically cannot be experimental because it would be unethical to withhold training from a high-level athlete preparing for competition.

Quasiexperimental Research Designs

Like true experimental research designs, quasiexperimental research designs are prospective in nature and also attempt to establish evidence toward a cause-and-effect inference or relationship (Sousa et al., 2007). However, these types of studies do not offer randomization, which compromises the studies' validity and increases the risk for bias (Thompson & Panacek, 2006). The *Cochrane Handbook* categorizes controlled clinical trials (CCTs) as quasiexperimental because "the method of allocation is known but is not considered strictly random" (Higgins et al., 2019). There are four commonly used quasiexperimental designs:

- Nonequivalent (or not random) group pretest–posttest design
- Control group interrupted time series design
- Single group interrupted time series design
- Counterbalanced design

Single-subject study design is also a type of quasiexperimental design but is more often used in human behavior and educational research. This type of study details the behavior of a small number of participants (Sousa et al., 2007). Table 9.3 outlines the common quasiexperimental designs.

Outcomes Research

When true experimental research is not always indicated or ethical to conduct, outcomes or observational research is an alternative and important category of research design. Cohort studies, cross-sectional, and case-control studies are the most common types of outcomes research.

- A **cohort** study can be prospective or retrospective in nature.
  - In a prospective cohort study, a study population is identified, detailed baseline exposure data are collected, and the participants are followed over time until the outcome of interest occurs.
  - A retrospective cohort study is carried out in present time but looks to the past to examine outcomes. Thus, participants are selected for the study based on the exposure of interest, and the outcomes, which were documented

Table 9.2　**Types of True Experimental Designs**

Experimental design	Testing procedure	Example of research question	Synopsis of design
Two arm	R — O — X — O R — O — C — O	Is hot yoga effective at decreasing hypertension in older adults?	The study will compare two groups randomized to either an active intervention (hot yoga) or a control intervention to see if hot yoga influences hypertension.
Three arm	R — O — X^1 — O R — O — X^2— O R — O — C — O	Do taking vitamins and exercising affect obesity in middle-aged adults?	The study will compare three groups randomized to two different interventions (X^1 = vitamins only and X^2 = vitamins with exercise) and a control group to see if including vitamins and exercise in a daily routine influences obesity.
Follow-up	R — O — X — O — O^2 R — O — C — O — O^2	Is hot yoga effective at decreasing hypertension in both the short and long term in older adults?	The study will compare two groups randomized to either an active intervention (hot yoga) or a control intervention to see if hot yoga influences hypertension over two different time periods. Are there short-term effects (O = right after completion of intervention) and long-term effects (O^2 = 6 months after intervention)?
Crossover	R — O — X^E — O — X^C — O R — O — X^C — O — X^E — O	What is the impact of curcumin supplementation on oxidative stress during a bout of physical exertion?	In crossover designs the individuals become their own controls, and participants are given two treatments or interventions. The participants are randomly assigned to two groups. The first group receives the experimental treatment (curcumin) followed by the control treatment (no curcumin). The second group receives the control experiment first and then the experimental treatment. A **wash-out period** is required between successive treatments.

A group is considered an arm in the above examples (two arms means two groups).

R = randomly assigned; O = pre and posttest; O^2 = subsequent follow-up test; X = exposed to an active intervention; X^1, X^2 = series of different treatments; X^E = experimental treatment; X^C = control treatment; C = exposed to a control intervention. A control intervention could be a placebo or standard of care intervention depending on the study.

Table 9.3   **Different Types of Quasiexperimental Research Designs**

Experimental design	Testing procedure	Example of research question	Synopsis of design
Nonequivalent group pretest–posttest design	NR — O — X — O NR — O — C — O	Is hot yoga effective at decreasing hypertension in older adults?	The study will compare two groups to either an active intervention (hot yoga) or a control intervention to see if hot yoga influences hypertension. However, the groups are not randomized.
Control group interrupted time series design	NR — O — O — O — X — O — O — O NR — O — O — O — C — O — O — O	Does taking vitamins affect obesity in middle-aged adults?	The study will compare two groups over multiple time points. There is no random assignment. One group will receive the intervention (vitamins), while the control group will not receive any vitamins.
Single group interrupted time series design	NR — O — O — O — X — O — O — O	Do knee kinematics during a single-leg drop landing task improve following an augmented feedback intervention?	Only one group is measured repeatedly across time both before and after the treatment exposure (augmented feedback).
Counterbalanced design	NR — X^1 — O — X^2 — O — X^3 — O — X^4 — O NR — X^2 — O — X^4 — O — X^1 — O — X^3 — O NR — X^3 — O — X^1 — O — X^4 — O — X^2 — O NR — X^4 — O — X^3 — O — X^2 — O — X^1 — O	What are the effects of four different supplements on oxidate stress during a bout of physical exertion?	This study is like the crossover design except participants are not randomized. The Latin square is the most common counterbalance design in which each group is posttested after each treatment and the number of groups are equal.

NR = not randomly assigned; O = pre and posttest; X = exposed to an active intervention; X^1, X^2, X^3, and X^4 = series of different treatments; C = exposed to a control intervention. A control intervention could be a placebo or standard of care intervention depending on the study.

in the past, are used for analysis. In other words, preexisting data are used to conduct retrospective cohort studies.

- A **cross-sectional** study investigates a group or groups of individuals at a specific time point. This type of study provides a snapshot of information within a given time point.

- **Case-control studies** are retrospective in nature because they start with an outcome and then trace back to investigate an exposure. Once the outcome is identified that outcome would be present within the case group and absent within the control group. The data about exposure to risk factors are then collected. The exposure is known at study inception in a cohort study, while the outcome is known at study inception in a case-control study.

Song and Chung (2010) present helpful figures that illustrate the differences between observational studies in a 2010 review.

Other Research Designs

The **meta-analysis** (MA) and **systematic review** (SR) designs carefully synthesize the results from multiple studies on the same topic for informed decision making. As defined by Moher and colleagues (2015, p. 3), "a SR attempts to collate all relevant evidence that fits pre-specified eligible criteria to answer a specific research question," while a "MA is the use of statistical techniques to combine and summarize the results of multiple studies." The terms should not be used interchangeably, because a MA refers only to the statistical synthesis and the SR is the process used to synthesize the evidence. The Preferred Reporting

Professional Issues in Exercise Science

Critically Appraised Topic

Critically appraised topic (CAT) manuscripts are an excellent resource for practitioners because they provide a brief summary (usually two to three pages) of the latest research for a particular topic. The *Journal of Sport Rehabilitation* regularly posts new CATs with every issue. One great example is from a study published in 2022 investigating the effect of blood flow restriction on muscle hypertrophy and tendon thickness in healthy adults' distal lower extremity. In their study, Post and associates (2022) identified four randomized controlled trials, all investigating the effects of blood flow restriction. Their findings present moderate evidence to support the use of blood flow restriction and low-load resistance training to improve muscle hypertrophy in the gastrocnemius and soleus muscles. One of the major benefits of CATs is that for practitioners with limited time these can be a great way to incorporate research evidence into their practice.

Citation style: APA

Post, D.R., Stackhouse, W.A., Ostrowski, J.L., Bettleyon, J.D., & Payne, E.K. (2022). The effect of blood flow restriction on muscle hypertrophy and tendon thickness in healthy adults' distal lower-extremity: A critically appraised topic. *Journal of Sport Rehabilitation, 31*(5), 635-639. https://doi.org/10.1123/jsr.2021-0176

Items for Systematic Reviews and Meta-Analyses (PRISMA) is a checklist of items that helps authors provide transparent accounts of SRs (Moher et al., 2015) and how readers should evaluate them.

A **critically appraised topic** (CAT) is a standardized summary of research evidence organized around a very specific clinical question (Sadigh et al., 2012). A CAT summarizes between two and five research articles that all answer the same question. The purpose of a CAT is to provide both a critique of the research and a clinical bottom-line statement of the clinical relevance of the results to professional practice. Often CATs are conducted and written by clinical professionals.

A **case report**, sometimes called a *case study*, is the description of events present in one individual. A case study often includes signs and symptoms, interventions, and outcomes following a new or unique exposure and treatment. The term *case* comes from medicine, in which a particular patient is often referred to as a case. Authors sometimes write a report as a case series. A case series reports on the factors, treatment, and outcomes for a small number of similar individual cases seen by the authors (Murad et al., 2018). These studies are valuable because they often describe unique and in-depth client presentations, contextual factors, and interventions to a particular exposure but are at an increased risk of bias due to the small, nonrandomized sample.

Literature reviews, or narrative reviews, provide an overview of current knowledge on a topic to summarize the results and to reveal gaps in the research. These types of reviews are useful when a lot of research on a topic needs to be organized and summarized. Literature reviews are written as part (specific chapter) of a thesis, dissertation, and even grant applications to gain perspective on the topic in order to define and justify a good research question. It is important to know what is known and what is unknown about a topic prior to developing a research project of your own. Narrative reviews do not have standardized structures and procedures, so they have greater potential for author bias in the search, interpretation, and summary of research than meta-analyses and systematic reviews.

Expert opinion papers and editorials might be seen in some journals but lack methodological rigor because the scientific method is not executed in these types of writings. Expert opinion papers often are written by individuals with experience on a certain topic. The individual often reflects on their own clinical perspective, successes, and even challenges associated with that topic. An editorial is a short commentary (usually less than 1,000 words) on a novel perspective or a trending topic in a particular field of interest. White papers and position statements often are produced and sometimes published by professional organizations. A white paper is defined as "a document created by an authoritative group (usually scientists and clinicians) that helps interested parties understand an issue or make a clinical decision" (Roukis, 2015, p. 151). Often white papers are not peer-reviewed for publication but convey recommendations on a particular topic for an organization. Position statements are often written for a scientific or professional group by a team of scientists and clinicians who have an expertise on a particular topic. Often times these statements are peer reviewed and published to give an overview of practice or policy recommendations based on the best available evidence.

Laboratory studies that use animals often study the development, progression, and treatment of disease prior to being conducted on human subjects. These studies often are basic research studies that address biological mechanisms and aid in the development of the more applied research we see in human subject models. Scientific research with humans and animals must meet strict international and national ethical standards. Scientific research with animals is essential in most ground-breaking and Nobel Prize–winning basic research and is performed humanely and with a tiny fraction of the numbers of animals that die from starvation and human neglect.

> **KEY POINT**
>
> A variety of different types of research designs are available to answer a specific research question. It is important to determine if the question is best suited for an experimental or outcome-based design.

Overview of Using Research to Support Evidence-Based Practice

Regardless of your future exercise science–related profession, it is important to engage in practices that are supported by sound evidence. Evidence-based practice is based on three pillars: research evidence, professional experience, and the preferences of your clients. As a young professional your firsthand professional practice experiences might be limited. Meanwhile, an abundance of knowledge and support has been produced and distributed by experts in your field. Reviewing and applying research literature is a strong core foundation of engaging in evidence-based practice. The next few sections of this chapter will discuss principles and strategies related to conducting a systematic literature search, evaluating the quality of the research evidence that you find, and considerations for implementing those findings into practice.

Searching the Literature for Evidence

After you have developed a clearly worded, answerable professional practice question, the next step in the EBP process is to search for the best available evidence. Searching through the available evidence has become a much easier process thanks to advancements in modern computer memory, speed, and the Internet. In the past, exercise science professionals needed to spend countless hours in the library reviewing hard copies of journal articles and microfiche films. Today most research can be found online by searching through specialized bibliometric databases that catalog, and sometimes store, published research. In the next few sections of this chapter, we will discuss a few of the many research databases that are commonly used in exercise science, helpful search strategies that you can use to facilitate and enhance your search (i.e., using specific key words rather than phrases or full questions), how to evaluate critically the quality of the evidence, and suggestions for synthesizing evidence to answer your question.

Databases

When seeking to answer a question, textbooks and scholarly journals often are the first place many people consider searching. Depending on how often a textbook is updated or revised, however, it often cannot provide the best or even most recent evidence. On the other hand, while journals provide up-to-date data and knowledge, hundreds of journals are associated with each subdiscipline of exercise science. Trying to read through all these journals would not be an effective or practical endeavor. A more time-efficient strategy is to conduct a literature search using online academic search engines and bibliographic databases to find the best evidence. Academic search engines and bibliographic databases are structured collections of a variety of sources of knowledge such as periodicals, scholarly research articles, books, graphics, and multimedia that can be searched to answer a question. Several different sources are available to search. Some journals are included in multiple databases, but no single database contains all available research. Some databases are owned by publishers who provide more indexing of their journals than other publishers' journals.

It is important to have a basic understanding of the different databases so that you can determine which databases would be the most appropriate to find the best evidence for your question. Examples of databases commonly used in exercise science include Google Scholar, PubMed/MEDLINE, PsycINFO, and SPORTDiscus. It is important to remember that these examples are just a few of numerous databases; however, having a basic understanding of these databases can help you decide which databases are best to include in your search. Ask your university librarian for assistance to access other databases that might be relevant to your research. The librarian also can assist you with search strategies that take advantage of computing power, the search engine of the database, and Boolean search terms to customize your search.

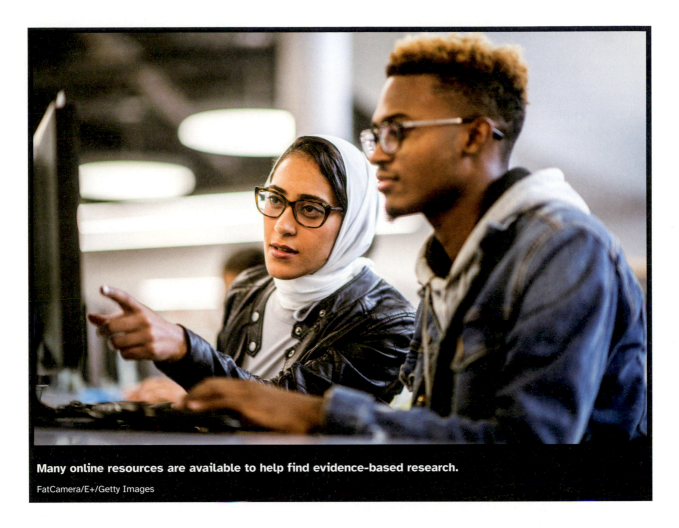

Many online resources are available to help find evidence-based research.

FatCamera/E+/Getty Images

Google Scholar   Google Scholar is a search engine developed by Google that allows you to search broadly for scholarly literature that includes a wide variety of published articles, theses, books, abstracts, patents, and court opinions from academic publishers, professional societies, online repositories, universities, and other websites. Due to its ease of use, Google Scholar has been identified as one of the most used information sources in academia. As of January 2018, the search engine included over 389 million records, making it one of the largest search engines available (Gusenbauer, 2019). This number continues to rise as more research records are identified. The academic search engine uses automated software known as *robots* or *crawlers* to find reference materials, with the goal of finding all scholarly information and storing it within a single database.

Despite the large scope of its database, Google Scholar has several limitations that must be considered. First, many researchers are frustrated by the lack of transparency with almost every aspect of the search platform. For example, when conducting a search, the default option is to sort items "by relevance."

Google Scholar reports that the relevance of articles is determined the same way researchers do but offers no documentation on how these articles are sorted by their page-rank algorithm. As an alternate option you can choose to sort the records based on publication date, or you can examine publications only within a specific range of dates if you are looking for specific studies. One important concern with the automated software for identifying new records is that because the records are only added as they are found by the crawlers or robots, some of the more recently published information might not have been discovered by the search robots. Secondly, no criteria are provided for what qualifies as "scholarly"; the database has been described as "getting bigger and bigger but in the wrong way, through hoarding giga collections of irrelevant and/or non-scholarly content" (Jacsó, 2012).

The crawler-based search engine includes many reputable resources but also includes what is referred to as *gray literature*. **Gray literature** are materials and research, including white papers, government documents, working papers, or even conference abstracts (short summaries) and procedures, that have been produced by groups and organizations outside of traditional academic publishing groups. Another example of gray literature is deceptive, or predatory, journals. The laudable effort to provide open (free) access to scientific research that is not behind a publisher's subscription pay wall has resulted in for-profit fake journals with no real peer review. While some gray literature might be reputable or help answer your question, the quality and academic rigor of some of these items might be questionable. As with any resource, it is important to evaluate critically the quality of each individual research report. The process for critically appraising research evidence is discussed later in this chapter.

MEDLINE, PubMed, and PubMed Central

MEDLINE is the National Library of Medicine's (NLM) journal citation electronic database that started in the 1960s (following the old print Index Medicus) and includes more than 28 million references from biomedical and life sciences journals. The MEDLINE database is accessible through the NLM as a section of the PubMed database and through many other search databases such as EBSCOhost and OVID, thanks to the NLM's Data Distribution Program. A distinguishing feature of the MEDLINE database is the use of the NLM-controlled vocabulary, Medical Subject Headings (MeSH), to index all its citations. The MeSH is a hierarchically organized list of vocabulary terms that can provide synonyms and related concepts for specific search terms. Using the MeSH library can be helpful for identifying relevant articles even though an author or publisher might not have used the same terminology. This is important because terminology sometimes changes with time or varies across the world.

The PubMed (PMC) database has been available since 1996 and includes more than 32 million references (National Library of Medicine, 2020). In addition to the entire MEDLINE database, the PubMed database also includes citations from studies supported by grant funding from the National Institutes of Health (NIH). Citations from before 1966 that have not been updated with the current MeSH, citations that predate the inclusion of a journal within the MEDLINE index, citations for the majority of books available on the bookshelves of the National Center for Biotechnology Information, and ahead-of-print and in-progress citations when available are also included. Both PubMed and MEDLINE citations might contain links to full-text articles or manuscripts that can be accessed through PubMed Central (PMC), which is a free-to-access electronic database that serves as a digital counterpart to the NLM print journal collection. Currently, PMC is one of the most used databases among health care professionals because it provides free access to the database and includes MeSH, which can enhance your search efforts.

PsycINFO   PsycINFO is a useful database for those searching for answers to questions that are more psychological in nature. For example, you might be interested in learning more about the role that motivation and confidence play in athletic performance or recovery from injury. The PsycINFO database contains over 5.4 million records of scholarly literature from journals published by the American Psychological Association (APA) related to behavioral and social sciences (APA, 2021). Like PMC, the PsycINFO database can be accessed electronically and includes access to full-text articles, book chapters, and other scholarly works; however, unlike PMC, it requires a subscription to access. Thankfully, most institutions provide access to students through the university library.

SPORTDiscus   SPORTDiscus is a bibliographic database that contains scholarly literature related to coaching, sport administration, sport science, physical fitness, exercise, health and physical education, sports medicine, nutrition, physical therapy, occupational health and therapy, exercise physiology, and many other exercise science–related fields. In the early 1970s sport scientists often relied on searching MEDLINE or ERIC (Educational Resources Information Center) databases, which often had too few relevant resources (Chiasson, 1997). The database has been described as "the leading bibliographic database for sports and sports medicine research" and is one of several databases that can be accessed through EBSCOhost (*SPORTDiscus*, n.d.). The broad range of topics covered by the SPORTDiscus database makes it a popular option among exercise science students and professionals. The database includes access to more than 1.7 million records from 459 active, peer-reviewed, indexed, and abstracted journals from 1892 to the present. This includes international references in nearly 60 languages.

KEY POINT

Several different databases can be searched to help find answers to a particular question. Deciding which databases to use depends on the type and context of the question you are asking.

Conducting a Literature Search

Once you have selected the databases that you feel are most important to include in your search, the next step is to begin searching through the available literature. Depending on which databases you include, there are potentially millions of research articles to sift through to find the answers to your questions. Although you might have spent a considerable amount of time browsing the Internet throughout your lifetime, the skills required for conducting a systematic literature search are different from those that you might use when you ask Siri to find something on the Internet or while searching Google. For instance, typing entire sentences or questions into the search bar most often leads to disappointing or, in some cases, no results. This is because most search engines are not capable of interpreting phrases and instead use key words to search and identify relevant articles.

Consider the following example question: "Does soda consumption have an effect on obesity rates in children?" If you type the full phrase directly into the search bar, you will get zero results in the SPORTDiscus With Full Text database. Performing the same search using PubMed Central yields three results. This is because with both databases the question has too many terms to search, which causes the search algorithm to get confused. To understand why this occurred we must think about how the search algorithm works. For most search engines and databases all the words included in the search bar are treated as separate key words. During the record retrieval process records are selected for inclusion only if they include all the listed key words. To avoid confusing the search algorithm it is recommended that you use a systematic (and sequential) process to identify the relevant records for each component of your question. Also, it is important to mention that by default, stop words (e.g., *does, have, an, on, in*) are excluded from the search. Inclusion (or exclusion) of these words adds no benefits to your search and can potentially complicate or confuse the search algorithm; therefore, it is better to not include these words in your search. Reducing your search to include only the following key words—*soda obesity children*—provides much better results, with 42 and 174 records for SPORTDiscus and PubMed, respectively.

Although using key words is beneficial, search engines and databases will only include those exact words that were listed. If a synonym or alternative word choice was used, that study might not be included in the search results. For instance, children might be referred to as *juveniles, youths,* or *adolescents*. Instead of *soda*, the authors might have used the terms *soft drink, Coke,* or *pop*. Identifying and including appropriate synonyms and alternative key words is critical to ensure that important studies are not excluded from your search results. Recall, however, that when conducting a search, the algorithm attempts to find records that contain all the listed key words. To enable searching of multiple key words, we must use Boolean operators.

Using Boolean Operators   Boolean operators are simple words that, when added to a search bar, can help refine or expand your search depending on the terms that were used. In general, most of the Boolean operators function similarly regardless of which database is being searched. Some search engines such as Google Scholar might include unique operators that only function within that search engine or database. For example, the Boolean operator "NOT" is commonly used when you want to exclude a specific word or term. You might want to investigate diabetes, but you want to exclude patients with gestational diabetes. In this case you would include "diabetes NOT gestational" as part of your search. Also, when including multiple key terms or Boolean operators, it is helpful to include "parentheses" around specific phrases to ensure that the search engine or database interprets the operator in the way it is intended. With Google Scholar the Boolean operator "NOT" is implemented using "-" instead of with the word *NOT*. Other common Boolean operators include "AND," "OR," and "AND NOT." The operator *AND* is used when you are interested in finding studies that include both terms. By default, many search engines automatically insert "AND" between all listed key words when no operators are included. The operator *OR* is used when you are looking for studies that contain one or the other word.

Exploding and Truncating Key Words
Another useful tool that can be implemented when searching databases is exploding or truncating key terms. Exploding key terms can be used in any database that includes a controlled vocabulary or thesaurus (e.g., MEDLINE and PsycINFO). An example of MeSH for the key word *exercise* is listed in figure 9.3. As you can see, *exercise* is a specific type of motor activity. Running, swimming, and walking are all included as specific types of exercise, while jogging and marathon running are listed as types of running. If you explode the term *exercise*, the word *exercise* as well as all the different types of exercise (as listed in figure 9.3) will be included in your search as well. With PMC, key words that are listed within MeSH are exploded automatically. This is not the case for all

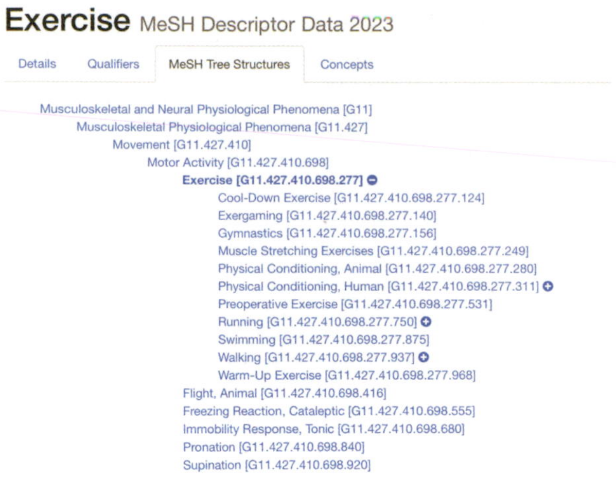

Exercise MeSH Descriptor Data 2023

Details   Qualifiers   **MeSH Tree Structures**   Concepts

Musculoskeletal and Neural Physiological Phenomena [G11]
Musculoskeletal Physiological Phenomena [G11.427]
Movement [G11.427.410]
Motor Activity [G11.427.410.698]
Exercise [G11.427.410.698.277] ⊖
Cool-Down Exercise [G11.427.410.698.277.124]
Exergaming [G11.427.410.698.277.140]
Gymnastics [G11.427.410.698.277.156]
Muscle Stretching Exercises [G11.427.410.698.277.249]
Physical Conditioning, Animal [G11.427.410.698.277.280]
Physical Conditioning, Human [G11.427.410.698.277.311] ⊕
Preoperative Exercise [G11.427.410.698.277.531]
Running [G11.427.410.698.277.750] ⊕
Swimming [G11.427.410.698.277.875]
Walking [G11.427.410.698.277.937] ⊕
Warm-Up Exercise [G11.427.410.698.277.968]
Flight, Animal [G11.427.410.698.416]
Freezing Reaction, Cataleptic [G11.427.410.698.555]
Immobility Response, Tonic [G11.427.410.698.680]
Pronation [G11.427.410.698.840]
Supination [G11.427.410.698.920]

Figure 9.3   MeSH heading example using PubMed Central Database.

databases, and each database uses a unique method for exploding terms. Be certain to check the specific criteria for including MeSH terms for the database you have selected.

In general, using exploded key words will provide you with a greater number of records retrieved during your search. In most cases, this is beneficial and will help you find additional sources related to your research question. In some cases, however, you might find that some of the included terms are not related to your specific question. For example, if you explode "exercise," you might find several articles discussing the benefits of stair climbers. To make sure you are including only the most relevant records, be sure to review the key words that are included with exploded terms. In PMC, which enables exploded terms by default, you can exclude specific key words using Boolean operators, or turn off exploding terms using the advanced search function or the code [mh:noexp].

As mentioned previously, most databases only retrieve records that include the exact key words as specified. This means that spelling counts. The key word *child* will only retrieve records that use the key word *child*; records using the key word *children* likely will not be included in the search. You could include all the possible variants of "child" using the Boolean operator *OR*; however, even using this approach you might miss an uncommon variation of the word. In addition, this can make your search entries longer, leading to potential confusion for search algorithms. An easier approach is to use truncation. Truncation, which is also sometimes called *wildcard search*, allows you to search for a specific key word and all spelling variations of that term. To truncate a search term, remove the ending of the word and add an asterisk (*) to the end of the word. For example, if you are

searching the word *child* you would use the search term *child**. This enables some database search engines to include all records that begin with the letters you entered.

> **KEY POINT**
>
> The skills for conducting a literature search are different from the skills you would use to perform a Google search. Using targeted key word searches and incorporating MeSH, Boolean operators, and exploding or truncated terms can enhance the quality of your search greatly.

Critically Reviewing Research Literature

Once you have finished conducting your literature search the next step in the EBP process is to evaluate the quality and relevance of the articles you found. Prior to being published in a traditional print or electronic journal, studies undergo a rigorous peer review process. Journal editors select experts to review studies anonymously that are submitted for publication. The reviewers read and provide suggestions for improving the submitted works before they are finally accepted (or in some cases rejected) for publication. Despite this rigorous review process, which often takes months to complete, not all published research manuscripts are high quality or free from biases. One example of potential bias is related to the interpretation of results based on statistical tests. The statistics that are used to evaluate the effectiveness of interventions or strength of associations between variables are based on probabilities.

Statistically significant findings usually mean the effect is not likely zero, given the data variability and probability theory. Good research follows up statistically significant results with other statistics on the size or meaningfulness of the effects reported or confidence intervals. Common examples of effect sizes used in exercise science include Cohen's d, odds ratios (OR), relative risk or risk ratios (RR), and the coefficient of determination (r^2). Effect sizes are usually classified as small, medium, or large (Sullivan & Feinn, 2012). Interpreting effect sizes still can be a bit subjective; however, with large effect sizes the difference between groups or strength of association is easily detectable. A medium effect size offers a meaningful difference; however, it might be more difficult to identify. Finally, small effect sizes are associated with negligible differences or weak associations. When a small effect size is present, even with statistical significance, the results might not be all that meaningful.

Although most researchers have the best of intentions, it can be difficult to recognize or appreciate all potential sources of bias. A meta-analysis examined the amount of publication bias present across 5,014 separate meta-analyses (Furuya-Kanamori et al., 2020). (Recall that meta-analyses provide the highest grade of research evidence.) In their review, the authors concluded that publication bias was significantly underestimated across studies. Publication bias is one of many distinct types of bias present in most research. This specific type of bias suggests that authors are less likely to submit or be able to publish the results of a study with nonsignificant findings. In general, authors are more likely to be able to publish reports with statistically significant results than nonsignificant results. It is important to keep this in mind when reviewing published research. Additionally, the strength or grade of evidence also should be considered. Recall that studies such as systematic reviews, meta-analyses, and RCTs provide stronger evidence than cohort or case-control studies. Critically assessing the quality of research is critical to ensure that bias is minimized and that convincing evidence is provided to support the answer to your question.

Much like the literature search process, the critical review process should be a systematic process used to evaluate carefully each piece of evidence to determine the quality of the findings, the relevance of the findings to your question, and the strength of those findings. The goal is to figure out whether evidence is from a viable source that can be translated into professional practice to answer your question. In this section we will focus primarily on how to apply the critical appraisal process to research literature, but the same steps can be used to evaluate information from colleagues and other professionals as well. Remember that research literature is only one aspect of EBP. In addition to research evidence, clinical expertise (which can come from other professionals in the field, professors, or your own evidence) and patient values are cornerstones of EBP.

To assist with the critical appraisal process, several appraisal scales have been developed. Each scale contains a variety of questions that help determine the validity, reliability, and relevance of the results. In addition, many of these scales also help identify potential sources of bias. The various scales that are commonly used vary depending on the type of study that is being evaluated. For instance, the PEDro scale was developed to assist with assessing the methodological quality of randomized controlled trials in physical therapy, or physiotherapy. The scale contains a series of 11 yes-no criteria that are evaluated to provide a score for the study. The 11 criteria assess various aspects of the study such as participant eligibility criteria, participant group assignment, the blinding process, the statistics used, and retention of participants throughout the clinical trial.

Although no set cut point exists for what is considered a "good" study, the score from these criteria can be compared across multiple randomized controlled trials to provide a relative evaluation of methodological quality. It is also important to mention that not all critical appraisal checklists are appropriate for all study designs. Some scales might include questions or items that are not relevant for specific types of studies. Furthermore, while many scales have scoring systems included as part of the scale, not all scales do. Examples of critical appraisal scales that are commonly used in exercise science and kinesiology are listed in table 9.4. Regardless of which scale is used, the decision about the quality of the study should be multifactorial. Using critical appraisal scales can help facilitate that decision; however, these scales are only a tool to assist with your own decision making and should not be relied on exclusively.

KEY POINT

It is critically important to consider the quality of the evidence that you find when conducting your literature search. Studies with lower levels or grades of research evidence should be less influential than those with higher levels or grades of research evidence, especially when studies are published with opposing conclusions.

Avoiding Pitfalls of Research

Being a consumer of academic research and implementing the findings of those studies into practice is a cornerstone of EBP. As a student you will learn about many different concepts and topics related to exercise science over the next few years. Once you become a professional in the field much of your learning will come through personal experience, from peers and colleagues, or through your own review of published research. As you begin reviewing research studies and attempting to answer questions you have developed, you should be aware of some common pitfalls that many students and young professionals often make.

Table 9.4   Examples of Commonly Used Critical Appraisal Tools in Exercise Science Research

Critical appraisal scale	Study designs	Item number	Score range
Jadad Scale (Jadad et al., 1996)	Randomized controlled trials	5	0-5 points
Downs and Black Checklist (Downs & Black, 1998)	Randomized controlled and nonrandomized trials	27	0-28 points
Tool for the assessment of study quality and reporting in exercise (TESTEX) scale (Smart et al., 2015)	Exercise training trials	12	0-15 points
Newcastle-Ottawa Scale (NOS) (Wells et al., 2011)	Cohort or case-control studies	8	0-9 stars
Strengthening the Reporting of Observational Studies in Epidemiology (STROBE) (von Elm et al., 2008)	Cohort, case-control, and cross-sectional studies	22	n/a
Standards for the Reporting of Diagnostic accuracy studies (STARD)	Diagnostic accuracy studies	25	n/a
MetaQAT (Rosella et al., 2016)	Nonspecific tool for public health research evidence	9	n/a
Joanna Briggs Institute (JBI) checklists (Joanna Briggs Institute, 2021)	Cross-sectional, case-control, case reports, cohort studies, diagnostic accuracy studies, economic evaluations, prevalence studies, qualitative research, quasiexperimental studies, randomized controlled trials, systematic reviews, and text and opinion	6-13*	n/a
CanChild Critical Review Form for Qualitative Studies (Letts et al., 2007)	Qualitative reviews	20	n/a

*JBI critical appraisal tools include different scales for each type of study design. The number of questions varies between the different scales.

One major mistake that often occurs is something we have already discussed briefly earlier in this chapter. When reviewing research evidence to answer a question, you should not make inferences or interpretations about a question from a single study. Even with new emerging research (e.g., COVID-19 research in 2021), it is rare that only a single study is published on a particular topic. Making inferences based on a single study is cherry-picking and can lead to exception fallacy. This is equivalent to making a judgment about a group of people based on evidence from a single person.

Another common mistake made among professionals is mistaking statistical significance for importance or relevance. Statistical significance is just the cue to pay attention to differences or associations in the sample data. The effects of an intervention might indicate that those who received the treatment effect had a significantly higher score on the outcome; however, the magnitude of the effect is so small that it is meaningless in clinical practice. The probability, or p-value threshold (often $p < .05$), for what is an acceptable risk of a false effect being detected (type I statistical error) that is selected in a study is an arbitrary value that has become customary. Probabilities range from 0 (will not occur) to 1 (certain to occur), so $p < 0.05$ can be expected in about 1 of 20 similar situations. One example of this is seen in a study including 22,000 adults that found a significant association between people taking aspirin and a reduction in heart attacks. The difference in heart attack probability for those taking aspirin compared with those not taking aspirin was less than 1% (Bartolucci et al., 2011). It is, unfortunately, not rare for authors and readers of scientific articles to overemphasize statistically significant results. Studies that use correlation analyses also are prone to finding statistical significance with exceptionally low and practically meaningless correlation coefficients, especially when sample sizes are large. Correlations that account for 5% of the variance of a variable means that you have no evidence about the other 95% of what is going on with the variable.

Evaluating effect sizes instead of p-values and statistical significance is frequently recommended as a solution to help reduce the risk of mixing statistical significance with importance. A p-value indicates the likelihood that you get the results you did by random chance if the null (no difference or association) hypothesis were true. Effect sizes, on the other hand, attempt to provide a measure of the magnitude of the effect that was observed in the study. Although using effect sizes might be an improvement over using p-values, they are not a perfect solution. Even effect sizes can be misleading if not examined in the context of the situation. Consider for a moment that a new treatment method was identified that would reduce the risk of developing a health condition by 50%. Initially this might seem like a great treatment. Yet if that health condition only has a lifetime risk of 0.002%, reducing the risk by 50% would not make a meaningful change in overall risk. A statistic commonly used to investigate the effectiveness of medical treatment interventions, called the *numbers needed to treat (NNT) analysis*, indicates that the new treatment method in the previous example would need to be administered to 100,000 people to reduce the number of developed cases by one. Even though there is a 50% reduction in risk, because the initial risk is already so low the new treatment offers minimal benefit, unless this is a serious condition and you are the lucky patient.

Confirmation bias is another research pitfall that affects many people, particularly students and young professionals. People are often far more likely to make interpretations that support their existing beliefs, even when the evidence supports an opposing view (Kahan et al., 2017). A disturbing example of this phenomenon is demonstrated in a study published from data collected during the 2012 American National Election Study. Results from the study found that Hispanic respondents with the lightest skin were several times more likely to be perceived as having high intelligence compared with those having the darkest skin (Hannon, 2014). Numerous other examples of confirmation bias are present throughout the research literature. As a student, one of your goals upon entering the professional field will be to avoid confirmation bias. Seeking out information from a range of sources, without relying on a sole source of evidence, is one of the many ways to help avoid confirmation bias. Also, using strategies such as the **Six Thinking Hats technique**, which provides guidance for examining a question from six different perspectives (i.e., hats), can help identify and avoid potential confirmation bias.

More than anything else, it is important to remember that research is conducted by humans. No one is perfect, and this applies to research evidence as well. Many practices and standards have been put in place to reduce the risk of bias and to identify potential errors during the research process, such as the peer review process that studies undergo prior to publication. However, despite best intentions, it is possible that errors have occurred. Remember that research is only one piece of the EBP process. Discuss important findings with your colleagues and peers. Find out what works best for your practice, and document evidence to support your conclusions. Do the best that you can, and do not hesitate to reach out to experts for additional advice.

KEY POINT

As you begin to navigate the complex research process it is important to remember and avoid common research pitfalls such as cherry-picking studies, using search practices that lead to confirmation bias, or confusing statistical and clinical significance. You will learn more about tips and strategies to improve your search techniques and evaluation of exercise science research throughout your coursework.

Applying Research Evidence to Practice

Despite great progress toward the standardization of EBP, implementation remains one of the most challenging and difficult aspects of the entire process. The adoption of EBP requires a multifaceted approach that includes educating professionals on the concepts of evidence-based practice and the development of effective strategies for implementation within professional practice. In fact, lack of sufficient knowledge of the EBP process is one of the most commonly reported barriers (Manspeaker & Van Lunen, 2011; Welch et al., 2014). An investigation of EBP implementation across a variety of allied

EXERCISE SCIENCE COLLEAGUES

Sophia Hooks

Sophia Hooks earned both her bachelor's and master's degrees in exercise sports science. Her education presented job opportunities as an exercise physiologist in a cardiovascular unit and as a product specialist with Stryker, a medical technology company. Both of these positions required Sophia to integrate her own clinical skills into the research to provide patient-centered care. On the cardiovascular floor it was essential to stay current on best clinical and research practices in order to provide her patients with quality care. Her current position with Stryker involves the integration of Mako Robotic-Arm assisted surgery into patient care. The device has been studied in clinical trials. Consequently, Sophia's understanding of this device is rooted in research, which allows her to collaborate with surgeons to provide evidence-based medicine to the patients with whom she interacts.

Courtesy of Sophia Hooks.

health care professions indicated that educational training (e.g., attending an EBP course or reviewing a textbook discussing EBP) and having an academic degree were all significantly associated with greater awareness and adoption of EBP (Weng et al., 2013).

As a student reading this textbook chapter, you already are taking the first steps toward improving your knowledge of EBP. As with any new change, however, the key to success is to plan ahead and be proactive rather than reactive. The decision to adopt EBP services should be a deliberate action. Kurt Lewin, a highly recognized psychologist who specializes in organizational change, proposed four principles necessary to implement change (Lewin, 1951; McCluskey & Cusick, 2002):

1. Change should be implemented only for good reason.
2. Change should be gradual.
3. Change should be planned, not sudden.
4. All individuals who might be affected should participate in planning the change.

Regardless of your chosen profession or work setting, making the deliberate choice to implement EBP and planning can enhance your success greatly. Other successful strategies that have helped some professions include using a team approach, motivation for change, and the use of quick and easy tools to help facilitate client management (Lu et al., 2021).

Wrap-Up

Evidence-based practice is a process that is used to search for, appraise, and use research findings to guide professional practice. When most think about research they focus on the process of doing the research; however, EBP is more about the act of using the research rather than doing research. Understanding the fundamentals of the research process, knowing how to identify and critically review research evidence, and incorporating that evidence into practice is the cornerstone of EBP. In many exercise science professions, the benefits of adopting EBP have been well documented. Developing and regularly engaging in EBP will help develop and improve your professional skills.

More Information on Research and Evidence-Based Practice

Organizations

Centre for Evidence-Based Medicine

Cochrane

Committee on Publication Ethics

International Committee of Medical Journal Editors

Journals

Cochrane Database of Systematic Reviews

Exercise and Sport Sciences Reviews

Journal of Athletic Training

Measurement in Physical Education and Exercise Science

Medicine & Science in Sports & Exercise

Qualitative Research in Sport, Exercise and Health

Research Quarterly for Exercise and Sport

Sports Medicine

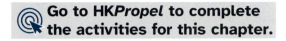

Go to HK*Propel* to complete the activities for this chapter.

Review Questions

1. What are the five steps of evidence-based practice?

2. What are some of the common types of research designs used in exercise science?

3. What are some of the major differences between Google Scholar and PubMed Central databases?

4. Which of the following—effect sizes or p-values—is more useful for making practical decisions regarding the significance of findings from a research study? Why?

Suggested Readings

Bishop, D., Burnett, A., Farrow, D., Gabbett, T., & Newton, R. (2006). Sports-science roundtable: Does sports-science research influence practice? *International Journal of Sports Physiology and Performance, 1*(2), 161-168.

Johnson, C. (2008). Evidence-based practice in 5 simple steps. *Journal of Manipulative and Physiological Therapeutics, 31*(3), 169-170.

McCluskey, A., & Cusick, A. (2002). Strategies for introducing evidence-based practice and changing clinician behaviour: A manager's toolbox. *Australian Occupational Therapy Journal, 49*(2), 63-70.

Raab, S., & Craig, D.I. (2016). *Evidence-Based Practice in Athletic Training.* Human Kinetics.

Sullivan, G.M., & Feinn, R. (2012). Using effect size—or why the p-value is not enough. *Journal of Graduate Medical Education, 4*(3), 279-282.

Thomas, J.R., Silverman, S.J., Martin, P.E., & Etnier, J.L. (2023). *Research methods in physical activity* (8th ed.). Human Kinetics.

References

American Psychological Association (APA). (2021). *APA PsycInfo.* American Psychological Association. www.apa.org/pubs/databases/psycinfo

Arghode, V. (2012). Qualitative and quantitative research: Paradigmatic differences. *Global Education Journal, 2012*(4), 155-163.

Aziz, H.A. (2017). Comparison between field research and controlled laboratory research. *Archives of Clinical and Biomedical Research, 1*(2), 101-104.

Bartolucci, A.A., Tendera, M., & Howard, G. (2011). Meta-analysis of multiple primary prevention trials of cardiovascular events using aspirin. *American Journal of Cardiology, 107*(12), 1796-1801.

Bishop, D., Burnett, A., Farrow, D., Gabbett, T., & Newton, R. (2006). Sports-science roundtable: Does sports-science research influence practice? *International Journal of Sports Physiology and Performance, 1*(2), 161-168.

Bloomfield, J., & Fisher, M.J. (2019). Quantitative research design. *Journal of the Australian Rehabilitation Nurses Association, 22*(2), 27-30.

Burns, P.B., Rohrich, R.J., & Chung, K.C. (2011). The levels of evidence and their role in evidence-based medicine. *Plastic and Reconstructive Surgery, 128*(1), 305-310.

Chiasson, G. (1997). SPORTDiscus and Information Utilization. *Quest, 49*(3), 322-326.

Dawes, M., Summerskill, W., Glasziou, P., Cartabellotta, A., Martin, J., Hopayian, K., Porzsolt, F., Burls, A., & Osborne, J. (2005). Sicily statement on evidence-based practice. *BMC Medical Education, 5*(1). https://doi.org/10.1186/1472-6920-5-1

Downs, S.H., & Black, N. (1998). The feasibility of creating a checklist for the assessment of the methodological quality both of randomised and non-randomised studies of health care interventions. *Journal of Epidemiology and Community Health, 52*(6), 377-384.

Flexner A. (1910). *Medical education in the United States and Canada.* Washington, DC: Science and Health Publications.

Furuya-Kanamori, L., Xu, C., Lin, L., Doan, T., Chu, H., Thalib, L., & Doi, S.A.R. (2020). P value-driven methods were underpowered to detect publication bias: Analysis of Cochrane review meta-analyses. *Journal of Clinical Epidemiology, 118,* 86-92.

Gusenbauer, M. (2019). Google Scholar to overshadow them all? Comparing the sizes of 12 academic search engines and bibliographic databases. *Scientometrics, 118*(1), 177-214.

Hannon, L. (2014). Hispanic respondent intelligence level and skin tone: Interviewer perceptions from the American national election study. *Hispanic Journal of Behavioral Sciences, 36*(3), 265-283.

Henard, D.H., & McFadyen, M.A. (2005). The complementary roles of applied and basic research: A knowledge-based perspective. *Journal of Product Innovation Management, 22*(6), 503-514.

Higgins, J.P.T., Thomas, J., Chandler, J., Cumpston, M., Li, T., Page, M.J., & Welch, V.A. (Eds.). (2019). *Cochrane handbook for systematic reviews of interventions* (2nd ed.). John Wiley & Sons.

Howick, J., Chalmers, I., Glasziou, P., Greenhalgh, T., Heneghan, C., Liberati, A., Moschetti, I., Phillips, B., & Thornton, H. (2022). *Explanation of the 2011 Oxford Centre for Evidence-Based Medicine (OCEBM) levels of evidence (background document).* Oxford Centre for Evidence-Based Medicine. www.cebm.ox.ac.uk/resources/levels-of-evidence/explanation-of-the-2011-ocebm-levels-of-evidence

Jacsó, P. (2012). Using Google Scholar for journal impact factors and the h-index in nationwide publishing assessments in academia—Siren songs and air-raid sirens. *Online Information Review, 36*(3), 462-478.

Jadad, A.R., Moore, R.A., Carroll, D., Jenkinson, C., Reynolds, D.J., Gavaghan, D.J., & McQuay, H.J. (1996). Assessing the quality of reports of randomized clinical trials: Is blinding necessary? *Controlled Clinical Trials, 17*(1), 1-12.

Joanna Briggs Institute. (2021). *Critical appraisal tools.* Faculty of Health and Medical Sciences at the University of Adelaide. https://jbi.global/critical-appraisal-tools

Johnson, C. (2008). Evidence-based practice in 5 simple steps. *Journal of Manipulative and Physiological Therapeutics, 31*(3), 169-170.

Kahan, D., Peters, E., Dawson, E., & Slovic, P. (2017). Motivated Numeracy and Enlightened Self-Government. *Behavioural Public Policy, 1,* 54-86.

Letts, L., Wilkins, S., Law, M., Stewart, D., Bosch, J., & Westmorland, M. (2007). *Guidelines for critical review form: Qualitative studies (Version 2.0).* McMaster University occupational therapy evidence-based practice research group. https://www.canchild.ca/system/tenon/assets/attachments/000/000/360/original/qualguide.pdf

Lewin, K. (1951). *Field theory in social sciences.* Harper & Row.

Lin, S.H., Murphy, S.L., & Robinson, J.C. (2010). Facilitating evidence-based practice: Process, strategies, and resources. *American Journal of Occupational Therapy, 64*(1), 164-171.

Lu, K.D., Cooper, D., Dubrowski, R., Barwick, M., & Radom-Aizik, S. (2021). Exploration of barriers and facilitators to implementing best practice in exercise medicine in primary pediatric care—Pediatrician perspectives. *Pediatric Exercise Science, 1.* https://doi.org/10.1123/pes.2020-0214

Manspeaker, S., & Van Lunen, B. (2011). Overcoming barriers to implementation of evidence-based practice concepts in athletic training education: Perceptions of select educators. *Journal of Athletic Training, 46*(5), 514-522.

McCluskey, A., & Cusick, A. (2002). Strategies for introducing evidence-based practice and changing clinician behaviour: A manager's toolbox. *Australian Occupational Therapy Journal, 49*(2), 63-70.

Melnyk, B.M., Fineout-Overholt, E., Gallagher-Ford, L., & Stillwell, S.B. (2011). Evidence-based practice, step by step: Sustaining evidence-based practice through organizational policies and an innovative model. *American Journal of Nursing, 111*(9), 57-60.

Moher, D., Shamseer, L., Clarke, M., Ghersi, D., Liberati, A., Petticrew, M., Shekelle, P., & Stewart, L.A. (2015). Preferred reporting items for systematic review and meta-analysis protocols (PRISMA-P) 2015 statement. *Systematic Reviews, 4*(1), 1.

Murad, M.H., Sultan, S., Haffar, S., & Bazerbachi, F. (2018). Methodological quality and synthesis of case series and case reports. *BMJ Evidence-Based Medicine, 23*(2), 60-63.

National Library of Medicine. (2020, September 11). *MEDLINE, PubMed, and PMC (PubMed Central): How are they different?* Retrieved July 13, 2021, from www.nlm.nih.gov/bsd/difference.html

OCEBM Levels of Evidence Working Group. (2011). *The Oxford 2011 levels of evidence.* www.cebm.net/index.aspx?o=5653

Post, D.R., Stackhouse, W.A., Ostrowski, J.L., Bettleyon, J.D., & Payne, E.K. (2022). The effect of blood flow restriction on muscle hypertrophy and tendon thickness in healthy adults' distal lower-extremity: A critically appraised topic. *Journal of Sport Rehabilitation, 31*(5), 635-639.

Rosella, L., Bowman, C., Pach, B., Morgan, S., Fitzpatrick, T., & Goel, V. (2016). The development and validation of a meta-tool for quality appraisal of public health evidence: Meta Quality Appraisal Tool (MetaQAT). *Public Health, 136,* 57-65.

Roukis, T.S. (2015). White papers, position papers, clinical consensus statements, and clinical practice guidelines: Future directions for ACFAS. [Editorial]. *The Journal of Foot & Ankle Surgery, 54*(2015), 151-152.

Sackett, D.L. (1989). Rules of evidence and clinical recommendations on the use of antithrombotic agents. *Chest, 95*(2 Suppl), 2s-4s.

Sackett, D.L., Rosenberg, W.M.C., Gray, J.A.M., Haynes, R.B., & Richardson, W.S. (1996). Evidence based medicine: What it is and what it isn't. *British Medical Journal, 312*(7023), 71-72.

Sackett, D.L., Strauss, S., Richardson, W., Rosenberg, W., & Hayes, R.B. (2000). *Evidence-based medicine: How to practice and teach EBM* (2nd ed.). Churchill Livingstone.

Sadigh, G., Parker, R., Kelly, A.M., & Cronin, P. (2012). How to write a critically appraised topic (CAT). *Academic Radiology, 19*(7), 872-888.

Smart, N.A., Waldron, M., Ismail, H., Giallauria, F., Vigorito, C., Cornelissen, V., & Dieberg, G. (2015). Validation of a new tool for the assessment of study quality and reporting in exercise training studies: TESTEX. *JBI Evidence Implementation, 13*(1), 9-18.

Song, J.W., & Chung, K.C. (2010). Observational studies: Cohort and case-control studies. *Plastic and Reconstructive Surgery, 126*(6), 2234-2242.

Sorin-Peters, R. (2004). Discussion: The case for qualitative case study methodology in aphasia: An introduction. *Aphasiology, 18*(10), 937-949.

Sousa, V.D., Driessnack, M., & Mendes, I.A. (2007). An overview of research designs relevant to nursing: Part 1: Quantitative research designs. *Revista Latino-Americana de Enfermagem, 15*(3), 502-507.

SPORTDiscus. (n.d.) EBSCO. Retrieved July 2, 2021, from www.ebsco.com/products/research-databases/sportdiscus

Sullivan, G.M., & Feinn, R. (2012). Using effect size—or why the p value is not enough. *Journal of Graduate Medical Education, 4*(3), 279-282.

Szklo, M. (1998). Population-based cohort studies. *Epidemiologic Reviews, 20*(1), 81-90.

The Periodic Health Examination. Canadian Task Force on the Periodic Health Examination. (1979). *Canadian Medical Association Journal, 121*(9), 1193-1254.

Thomas, J.R., Nelson, J.K., & Silverman, S.J. (2015). *Research methods in physical activity.* Human Kinetics.

Thompson, C.B., & Panacek, E.A. (2006). Research study designs: Experimental and quasi-experimental. *Air Medical Journal, 25*(6), 242-246.

Tomlin, G., & Borgetto, B. (2011). Research pyramid: A new evidence-based practice model for occupational therapy. *American Journal of Occupational Therapy, 65*(2), 189-196.

von Elm, E., Altman, D.G., Egger, M., Pocock, S.J., Gøtzsche, P.C., & Vandenbroucke, J.P. (2008). The Strengthening the Reporting of Observational Studies in Epidemiology (STROBE) statement: Guidelines for reporting observational studies. *Journal of Clinical Epidemiology, 61*(4), 344-349.

Welch, C.E., Hankemeier, D.A., Wyant, A.L., Hays, D.G., Pitney, W.A., & Van Lunen, B.L. (2014). Future directions of evidence-based practice in athletic training: Perceived strategies to enhance the use of evidence-based practice. *Journal of Athletic Training, 49*(2), 234-244.

Wells, G.A., Shea, B., O'Connell, D., Peterson, J., Welch, V., & Losos, M. (2011). The Newcastle-Ottawa Scale (NOS) for assessing the quality of nonrandomized studies in meta-analyses. *Ottawa: Ottawa Hospital Research Institute,* 1-12. https://www.ohri.ca/programs/clinical_epidemiology/oxford.asp

Weng, Y.-H., Kuo, K.N., Yang, C.-Y., Lo, H.-L., Chen, C., & Chiu, Y.-W. (2013). Implementation of evidence-based practice across medical, nursing, pharmacological and allied healthcare professionals: A questionnaire survey in nationwide hospital settings. *Implementation Science, 8*(1), 112.

Williams & Kendall. (2007). A profile of sports science research (1983-2003). *Journal of Science and Medicine in Sport, 10*(4), 193-200.

CHAPTER 10

Sport Performance: Strength and Conditioning, Nutrition, and Sport Science

Broderick L. Dickerson, Drew E. Gonzalez, Scott M. Battley, and Richard B. Kreider

CHAPTER OBJECTIVES

In this chapter, we will

- outline the importance and history of sport performance in regard to strength and conditioning, sport nutrition, and sport science;
- provide an overview of common research methods in various contexts for respective subdisciplines in exercise science;
- provide recommendations and strategies from evidence-based research focused on athletes and the general population; and
- highlight the contribution of specific sports' performance areas on athletes.

Jordan is a collegiate football player and is currently in the precompetitive season. Every morning he undergoes structured training focused on specific performance outcomes (e.g., speed and agility, strength, power) organized by the certified strength and conditioning coach who oversees the preseason training program. Jordan and his teammates perform supervised skill-specific drills on the field during football practice to prepare for the upcoming season. Additionally, Jordan and his teammates perform sprints and field drills after practice, which are also coordinated and supervised by a certified strength and conditioning coach who oversees all training related to physical development. Throughout the day, Jordan follows specific dietary guidelines recommended by the team's dietitian (in collaboration with sport nutrition specialists) to optimize performance outcomes and recovery. Pre- and postgame or practice meals and snacks are provided by the team's nutrition staff to keep him and his teammates fueled and well hydrated.

Sport scientists use wearable technology, video, and various software to monitor workload, exercise intensity, and recovery, as well as to attenuate risk of injury from repeated training sessions. Jordan and his teammates can be tracked through global positioning systems (GPS) and video analysis software that give detailed information and feedback regarding performance metrics during training sessions, practices, scrimmages, and games. Strength and conditioning coaches and nutrition professionals collaborate with the sport scientists and sport-specific coaches on the usage of analytics data to provide feedback and guidance to Jordan and his teammates in an effort to get a leg up on the competition. Jordan benefits from the integrated and collaborative knowledge of the sport performance team to optimize every facet of Jordan's game to ensure success and a high level of performance during the season.

The sport performance team works with athletes to enhance performance and reduce the risk of injury. A variety of exercise professionals serve on the sport performance team depending on the sport and the level (i.e., high school age through the elite or professional level) of competition. Athletes rely on the expertise of multiple subdisciplines of exercise and sport science to build a solid physical and mental foundation that allows for the most effective training, peak performance during competition, and recovery. The subdisciplines that particularly contribute include strength and conditioning, dietetics and sport nutrition, sport science, biomechanics, exercise psychology, and sports medicine. All these areas of exercise science should be integrated to optimize training in the competitive season and off-season. Students should be familiar with the collaborative effort required among the various subdisciplinary experts of the sport performance team such as the specialized knowledge, duties, and responsibilities of strength and conditioning specialists, sports nutritionists, registered dietitians, and sport scientists.

The sport performance team aims to

- educate and train athletes with appropriate strength and conditioning training programs targeted to optimize performance while reducing risk of injury;
- recommend and implement the most up-to-date, evidence-based nutrition strategies that support training and performance outcomes;
- promote and optimize recovery;
- use real-time data in order to augment recovery and enhance performance;
- strengthen mental skills of athletes to accommodate rigorous training and competition; and
- provide rehabilitation and training programs that decrease a chance of reinjury to injured athletes.

If all of these aims are met, the athletes will be provided the opportunity to optimize their performance and increase their potential of in-game success.

This chapter provides an overview and discussion of the general roles and responsibilities, history, and knowledge of the main subdisciplines of sport performance: strength and conditioning, sport nutrition, and sport science. The latter has emerged as a relatively new subdiscipline that focuses on data analytics and athletic performance. The reader should be able to apply fundamental theories of the main subdisciplines to sport performance with the information outlined in this chapter.

What Do Sport Performance Professionals Do?

Sport performance has emerged into an interdisciplinary field encompassing several exercise science–related subdisciplines. A sport performance professional is someone who concentrates their expertise on a specific focus or practice with respect to the aforementioned subdisciplines. Professional and intercollegiate sports, sport organizations, and private companies are embracing scientific approaches to enhance performance. Most competitive teams and programs employ several key sport performance professionals who work closely with the coaches and teams to prepare athletes for peak performance and succeed in competition. These specialists include strength and conditioning coaches, sport dietetics and nutritionists, athletic trainers, physical therapists, sports medicine and specialty physicians, biomechanists, sport scientists, and sport psychologists (figure 10.1) (Dijkstra et al., 2014).

KEY POINT

Sport performance is a multifaceted field that has several key subdisciplines: strength and conditioning, sport nutrition, sport science, and sport psychology and coaching.

The general hierarchy of improving sporting performance should be considered by all members of the sport performance team. It is essential to provide

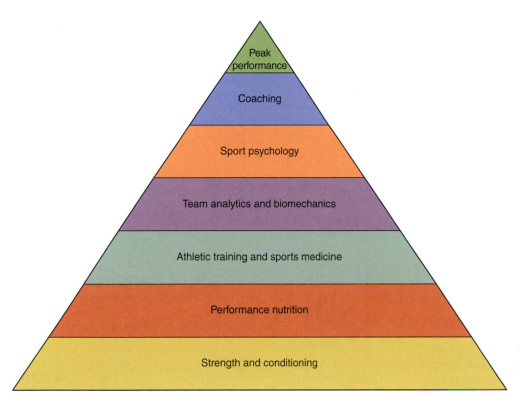

Figure 10.1 Hierarchy of improving sport performance.

athletes and sport teams with the most appropriate training programs and dietary guidelines to accommodate training and performance needs and to minimize the risk of injury. Therefore, strength and conditioning and nutrition are the foundations of peak athletic performance. Strength and conditioning programs are implemented to improve skill and health-related fitness components in addition to the goal of improving an athlete's athletic profile. Moreover, the ability to perform vigorous and repetitive muscular contractions is dependent on energy availability from energy-yielding nutrients consumed through the diet. While the sport performance team has many critical components, these foundational subdisciplines lay the groundwork on which athletic success is built.

Athletic training and sports medicine are essential in the rehabilitation process for injured athletes who will then return to training, which is conducted in ways that minimize the risk of reinjury. Sport scientists use analytics to fine-tune performance and provide feedback to the entire sport performance team regarding training adaptations, competition performance outcomes, and athlete recovery. With the advancements in technology, new databases and analytic devices and software can be used to provide real-time feedback to the sport performance staff

and athletes to better assess sport-specific outcomes. Biomechanists can be consulted on appropriate movement technique that confers the best chances for success. Sport psychologists evaluate, educate, and train athletes to prepare them for competition in ways that bolster mental toughness, perceptual skills, and player attitudes. Improving the mentality and cognition of the athlete is critical considering the intense training schedules and requirements, as well as having to perform in high-pressure competition.

Athletic performance is the product of collaboration between various individuals that make up the entire sport performance team. Athletic administrators, facilities and equipment managers, travel and logistics personnel, videographers, researchers and academic support, and communications and marketing personnel all play a role in the functioning of the entire sports performance team (Smith & Smolianov, 2016). It is important to note that some sports might require more contribution from specific specialists to optimize athletic performance while minimizing the risk of injury. For example, technique-centric sports (e.g., gymnastics or golf) might use the expertise of biomechanics professionals and sport scientists to achieve success in competitive events. On the other hand, football teams might require more attention from athletic trainers and sports medicine personnel given the nature of the contact sport. Therefore, it is likely that most programs have developed an interdisciplinary approach that considers the nature of the sport and athlete and the best practices and strategies to ensure optimal performance outcomes.

Sport performance professionals are carving a new path that brings together researchers, practitioners, and analysts to continue expanding on the application of knowledge from exercise science, nutrition, and athletic performance fields. In March 2018, the National Strength and Conditioning Association (NSCA) initiated an official scope of practice for the sport performance profession. This chapter will discuss details later regarding the new NSCA certification as well as other certifications that sport performance professionals can obtain. All of the sport performance subdisciplines have worked together to shape the sport performance profession in parallel to the NSCA mission, which has led to the development of its new certification.

Research within sport performance has focused on ergogenic aids, noninvasive wearable and field-based technologies, and various exercise and strength training protocols to improve performance and recovery outcomes. Given the growth in the importance of sport dietetics, along with the growth of sport nutrition organizations such as the International Society of Sports Nutrition since 2000, it appears that those in sport nutrition will continue to advance the knowledge base of supplementation practices and nutritional interventions that result in better on-the-field performance outcomes. Additionally, those in the strength and conditioning field are continuously expanding their base of knowledge with the NSCA's *Journal of Strength and Conditioning Research*, which highlights the latest findings in the strength and conditioning field from some of the top exercise and sport scientists across the world. Lastly, it is important to note that the Collegiate Strength and Conditioning Coaches Association (CSCCa) offers gold standard certifications in strength and conditioning for the athletic population. The CSCCa has certified Master Strength and Conditioning Coaches (MSCC) who serve as mentors to the future Strength and Conditioning Coach Certified (SCCC) personnel. All of these professional organizations and associations serve a critical role in the progression of the sport performance profession. While a large collaborative group of professionals already is working within their respective fields to optimize athletic performance outcomes, the future of sport performance will continue to foster an evidence-based and data-driven approach to advance the field as newer data analytics software and technology become available.

Additionally, the need for exercise scientists and researchers dedicated to helping the field grow through ethical and sound research will always be needed. Professionals and practitioners support the field by collaborating with researchers; using the findings of well-designed, peer-reviewed research studies; and applying the findings into practice. It is because of the scientists and professors conducting important research that the sport performance team can implement evidence-based nutritional and exercise practices to enhance athletic performance and competitive readiness while minimizing the risk of injury.

As sport evolves, so do the coaches and personnel working behind the scenes of athletic success. We tend to marvel at athletes breaking world records or having career-defining performances. A lot of what sport performance professionals and scientists do is not always noticed; however, their contribution to the profession has a tremendous effect in helping athletes and teams reach peak levels of performance. Professional sports and sporting events are gaining more popularity. Therefore, the need is growing for more scientists and specialists in the sport performance field.

History of Sport Performance

Sport performance originated in the academic setting as faculty in the field of exercise science, kinesiology, and sport science throughout the world published research on human performance. Both basic and applied science of muscle actions and bioenergetics contribute to our knowledge of sport performance. Some of the earliest, targeted sport science and high-performance research was conducted by scientists working in national Olympic training centers. Academic biomechanists Dr. Gideon Ariel and Dr. Charles Dillman were the first two leaders of the sport science program for the United States Olympic Committee. The late Dr. Paavo Komi from the University of Jyväskylä was a leader in the integration of basic biomechanics research and research in sport through his work with the Finnish Olympic Committee. Numerous exercise and sport scientists have assisted Olympic athletes and coaches in many national and state or province Olympic training centers. German researcher Dr. Albert Gollhofer also served as a seminal biomechanist and helped establish sport science in Europe during the late 1980s. More recently, New Zealand researcher and coach Mike McGuigan has set a foundation on how to test athletes properly, allowing for translatable strength training interventions and ensuring optimal success on the competitive grounds. These are just some of many scientists from all over the world whose research is being used to improve athletic performance and remediate injuries.

The successes of sport science and performance services at the Olympic level have been slow to be adopted by most professional sport teams in America. The resistance to sport analytics in baseball was highlighted for the general public by the book and movie *Moneyball*. Exercise and sport science researchers from Europe, Asia, Australia, and New Zealand, however, found high-level coaches who were trained in this discipline and were ready to collaborate with sport performance professionals and programs.

More recently, the for-profit professional sport teams in the United States have followed the lead of the professional sport teams in Asia, Europe, Australia, and New Zealand. Since 2010 almost every major professional sport team in the United States has expanded beyond strength and conditioning and sports medicine services to providing an integrated team of sport performance professionals. To meet the needs of the growing sport performance profession, the NSCA has developed the Certified Performance and Sports Scientist (CPSS) certification, which emphasizes a scientific approach to improving individual and team athletic performance and reducing injury risk. The CPSS certification highlights the growing interest and importance of the sport science and performance discipline within exercise science.

The rest of this section will provide a few more examples specific to the primary components of sport performance professional career paths and services. It should be noted that the time frame to establish and develop each subdiscipline varied, either occurring rapidly over the course of a few decades or for centuries building up to a bona fide practice. As each field grew and matured in its own regards, the efforts set forth by headstrong and seminal researchers allowed an opportunity for strength and conditioning, sport nutrition, and sport science to evolve into a common practice of sport performance. Table 10.1 lists the numerous personnel and their duties on the sport performance team.

History of Strength and Conditioning

Strength training was at the center of what would evolve into sport performance and sport science. As early as 3600 BC, Chinese emperors enforced exercise testing and weight-lifting assessments for individuals to gain entry to the military. Additionally, weight training was a component of life for ancient Egypt, Greece, and India (Fry & Newton, 2000). In the late 1960s, despite several decades of research on the benefits of strength training, many sport coaches in the United States advocated against strength training due to fear that it might adversely affect athletic performance. Only a few strength coaches were working or volunteering with sport teams at that time. In 1969 the University of Nebraska hired Boyd Epley as the first full-time strength coach, and his success quickly was recognized. Not long after the University of Nebraska won the national championship in football in 1970 and 1971, several other universities and a couple of professional organizations hired strength and conditioning coaches (Shurley & Todd, 2012).

Boyd Epley and 76 strength coaches saw a need to share their experiences and knowledge to expand the profession of strength and conditioning. In 1978 the NSCA was founded with the mission and vision to serve as a worldwide authority on strength and conditioning, disseminate research-based knowledge, and provide practical application to improve athletic performance and fitness. Today, the NSCA has over 30,000 members across 72 countries as well as other national strength and conditioning associations. By 2000 the CSCCa was founded to represent and promote collegiate strength and conditioning coaches. Most full-time strength and conditioning coaches belong to the NSCA or CSCCa (or both) and are certified as a Certified Strength and Conditioning Specialist (NSCA: CSCS), a Strength and Conditioning Coach Certified (CSCCa: SCCC), or a Master Strength and Conditioning Coach (CSCCa: MSCC). Unlike some exercise training certifications, these specific credentials are rigorous, requiring a college degree in exercise science or kinesiology to sit for these accredited exams. Holding a certification with such merit can provide aspiring practitioners the initial credentials to practice as a strength and conditioning specialist with athletes.

The mission and visions of organizations such as the NSCA and CSCCa serve as guardians of the knowledge and evidence-based practice of strength and conditioning to enhance performance outcomes while reducing injury risk. Additionally, with advancements in miniature sensors, computers, and networking technology, traditional lab-based exercise science assessments now can be performed on the field or court with acceptable accuracy and ease of use. These advancements have led the NSCA to develop a new certification focused on sport performance and science. In 2021 the CPSS textbook, *NSCA's Essentials of Sport Science*, was published and provides an overview of the knowledge, expertise, and principles of sport science knowledge.

Table 10.1 **Roles and Responsibilities of the Sport Performance Team**

Sport performance disciplines	Careers or certifications	Duties
Strength and conditioning	Certified Strength and Conditioning Specialist (CSCS) Registered Strength and Conditioning Coach (RSCC) Strength and Conditioning Coach Certified (SCCC) Master Strength and Conditioning Coach (MSCC)*	Improve performance of athletes Design, implement, and supervise training programs Oversee organization and administration of strength and conditioning facilities *Can mentor SCCC personnel
	Tactical Strength and Conditioning Facilitator (TSAC-F)	Improve performance of tactical populations Design, implement, and supervise training programs Oversee organization and administration of on-site strength and conditioning facilities
Personal training	NSCA-CPT[1] ACSM-CPT[2] ACSM-GEI[3]	Train general population clients one on one[1,2] Develop exercise testing[1,2] Conduct field health assessments[1,2,3] Lead fitness classes[3]
Sports medicine and health and rehabilitation	Physical therapist	Rehabilitate clients through injury by providing treatment exercises for injury Diagnose movement dysfunction Educate clients on assisted mobilizers Educate proper exercise equipment usage
	Athletic trainer	Diagnose and treat sport-related injury Educate on injury prevention Monitor athletic competition Provide immediate health care to injured athletes Perform musculoskeletal evaluations
	ACSM-EP[4] ACSM-CEP[5]	Develop exercise prescriptions for healthy or medically controlled diseased populations[4,5] Develop exercise plans for the chronically diseased[5] Administer fitness tests[4,5] Maintain records of exercise tests performed[4,5]
Nutrition	Registered dietitian (RD)	Create nutrition plans to cater to client needs Medical nutrition therapy Analyze nutrient intake Analyze client health status and goals Provide nutrition counseling Educate on proper nutritional behaviors and foods

Sport performance disciplines	Careers or certifications	Duties
	Dietitian assistant	Provide support for RDs Provide education to clients Assist in counseling clients on nutrition
	Board Certified Specialist in Sports Dietetics (CSSD) for RDs	Provide education and meal plans to collegiate and professional athletes Consult with military members on meal plans and monitoring nutrition
	Certified Nutrition Specialist (CNS)	Alternative to RD Customize meal plans Provide nutrition education to clients Provide counseling on nutrition
	Certified Sports Nutritionist through the International Society of Sports Nutrition (CISSN)[6] ISSN–Sports Nutrition Specialist (ISSN-SNS)[7]	Consult with athletes on nutrition[6,7] Give advice on supplements and diets[6,7] Cannot work with patients with medical nutrition therapy[6,7]
	Precision Nutrition Level 1 Certification[8] Precision Nutrition Level 2 Certification Master Class[9]	Provide nutritional coaching to clients and athletes[8,9] Implement behavior-modifying nutrition coaching[8,9] One-year mentorship under trained nutrition professional[9]
	NASM-CNC	Offer nutrition coaching to clients
Analytics	Certified Performance and Sports Scientist (CPSS)	Assist in implementing training programs for athletes Operate training equipment and software Consult with coaches on training programs and recovery methods
	Sport data analyst	Collect and store data on performance variables Provide competition data to television companies
All sport performance disciplines	Researcher or professor	Determine research topics Collect, store, and analyze data Submit manuscripts for publication Present research at conferences Submit proposals for funding Teach university-level courses

1 = NSCA-CPT; 2 = ACSM-CPT; 3 = ACSM-GEI; 4 = ACSM-EP; 5 = ACSM-CEP; 6 = Certified Sports Nutritionist through the International Society of Sports Nutrition; 7 = ISSN Sports Nutrition Specialist; 8 = Precision Nutrition Level 1 Certification; 9 = Precision Nutrition Level 2 Certification Master Class

Strength and Conditioning Pioneers

Dr. Mike Stone, Meg Stone, and the Mission and Goals of the Center of Excellence for Sport Science and Coach Education

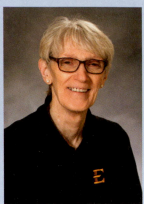

Courtesy of Mike and Meg Stone.

Mike Stone is a researcher, sport scientist, and professor in exercise physiology and biomechanics. He has a particular interest in Olympic weightlifting and is well known for his work with respect to skeletal muscle adaptations via strength and power training, endocrine adaptations, and athletic performance enhancement. Mike has been published extensively, with approximately 840 articles, book chapters, and books combined. His contributions to sport performance have been critical in the shaping and evolution of the field. Mike is a recipient of several prestigious awards such as the NSCA Sports Scientists of the Year (1991), the NSCA Lifetime Achievement Award (2000), the American Society of Exercise Physiologist Scholar Award (2003), and the Doc Councilman for Sports Science Award in Weightlifting (2010). Currently, Mike is a graduate program coordinator and lab director at East Tennessee State University.

Meg E. Stone is well known as a two-time Olympian who competed in the discus for Great Britain. She holds the NCAA outdoor record in discus (221'3"), which was set during her time at the University of Arizona (UA). By 1994 Meg had become a head strength and conditioning coach at UA, where she worked with all NCAA sports, especially football. In fact, Meg was the first woman to ever hold a head strength and conditioning coaching position at the university level. Soon after her time at UA, Meg moved to Texas Tech, holding the same head strength and conditioning coach position. In 1996 Meg became the associate head coach at Appalachian State University, where she focused full time on track and field. Meg made a return to her homeland, Scotland, as the national track and field coach, the first woman to hold this position. Over the course of Meg's career, she has coached at the highest levels, overseeing several Olympians as well as several NBA, NFL, and MLB athletes. In addition to her work with Olympians and high-level athletes, Meg has worked with road cycling and Paralympic groups through the Carmichael Training Systems in Colorado Springs, Colorado. Meg has published on the topic of principles and practice of resistance training with her husband, Mike Stone, and has been honored by the College Strength Coaches Association with the Legends in the Field Award and by the NSCA as a Fellow of the Association (FNSCA).

Mike and Meg have worked to develop the Center of Excellence for Sport Science and Coach Education (CESSCE), which is an example of how academic programs are aided in the development of professionals in the sport science and coaching arena. The mission of the CESSCE is twofold. First, the mission is to develop young sport scientists and coaches and to provide a cutting-edge, evidence-based training process for competitive athletes. This process includes developing and monitoring a training program integrating sport scientists and sports medicine personnel into the process. The second aspect of the mission is to provide educational and practical application experiences for undergraduate and graduate students interested in sport coaching. CESSCE provides an educational process for the development of young coaches (particularly for strength and conditioning) and sport scientists. This process involves a close working relationship with the athletic department and sports medicine personnel. Doctoral and master's students entering the program are allocated a team in the athletic department, and their responsibility is to develop an evidence-based annual training plan and athlete-monitoring process based on their learning experience in the classroom. They are expected to implement their plan, with supervision, in the practical situation with their team for the period of their academic studies.

The CESSCE has developed as a multifaceted educational, research, and service entity. With these concepts of education, research, and service in mind, the CESSCE has engaged in a diverse array of projects including practical service-oriented activities such as monitoring and testing local high school athletes, East Tennessee State University athletes, and Olympians and Olympic hopefuls; educating coaches and sports scientists; and conducting applied and translational research projects. Additionally, through a cooperative program with the local school system and community entities, service programs have been delivered for the surrounding community. CESSCE was established in 2012 as a truly unique multidisciplinary program. Several universities are making efforts to implement a similar program, realizing the benefits of enhancing the education of sport coaches, sport scientists, and strength and conditioning practitioners.

KEY POINT

Various college-level courses and exercise science degree programs are offered at universities to students who want to pursue careers in a sport performance–related field. Exercise graduates interested in strength and conditioning careers then must seek advanced certifications from the NSCA or the CSCCa.

History of Sport Nutrition

Sport nutrition, a relatively new field within exercise science, was established in the late 1970s and encapsulates the nutritional practices of all kinds of athletes. Growing evidence on athletes' use of dietary supplements and practice with nutritional strategies to enhance performance outcomes have led to tremendous interest in the field of sport nutrition. Historically, athletes dating back to ancient Greece and Rome implemented dietary strategies such as ingestion of whole grains, fruits, cheeses, wines, and meats and fish to optimize performance outcomes (Applegate & Grivetti, 1997; Harris, 1966; Juzwiak, 2016). Spartan athletes such as Charmis and distance runners Dromeus of Stymphalos purportedly ingested dried figs and meats, respectively, prior to competition (Juzwiak, 2016). Additionally, the concept of creatine-loading protocols through ingestion of meats before competition was unknowingly practiced by athletes (Butts et al., 2018).

As the world of sport has evolved, so have dietary supplement and nutritional strategies to enhance performance and recovery outcomes. In the mid-1800s, research on nutrition and athletic performance started to gain traction. In 1832 creatine was discovered in muscle by Michel Eugène Chevreul (Heffernan, 2015). In 1842 Justus von Liebig suggested protein was a primary source of fuel during exercise, and later he proposed that meat extract might increase strength, performance, and health (Büttner, 2000). In the early 1900s the first studies demonstrating that creatine ingestion (with and without carbohydrate) can increase muscle creatine, weight gain, and exercise capacity and performance were published (Brown & Cathcart, 1909; Paton & Mackie, 1912). In the 1920s and 1930s studies indicated that diet can influence carbohydrate and fat usage at rest and during exercise (Christensen & Hansen, 1939). By the 1940s to the 1960s research began to assess the role of vitamins and minerals as they related to health (Applegate & Grivetti, 1997).

The early work on nutrition set a foundation for future work investigating the role of supplementation, metabolism, and specific dietary compositions as they relate to health and exercise performance. Certain foundational studies conducted by pioneering researchers in the 1960s and 1970s helped propel sport nutrition to a legitimate focus of study in exercise science. During the 1960s and 1970s preeminent researchers in Europe such as Dr. Bengt Saltin, Dr. Jonas Bergström, and Dr. Eric Hultman conducted seminal research on muscle glycogen stores and the effects of manipulating carbohydrate intake on muscle glycogen availability prior to exercise (Bergström & Hultman, 1972; Hermansen et al., 1967; Karlsson & Saltin, 1971). Bergström and Hultman published milestone articles on the first studies to assess human exercise metabolism, which denoted the importance of macronutrient consumption to provide energy for exercise and for recovery purposes. Dr. Roger Harris, along with Bergström and Hultman, conducted studies using novel muscle biopsy techniques to observe metabolic changes in muscle after exercise or carbohydrate loading. These pioneering methodologies influenced the administration of the common practice of nutritional loading protocols used when supplementing with certain substances such as creatine monohydrate and beta alanine.

Around the same time, researchers in the United States such as Dr. David Costill and Dr. Edward Coyle were making waves in the field with their own research methods and findings. Costill and Coyle coordinated studies examining muscle glycogen availability after strenuous exercise (Costill et al., 1970), how proper hydration habits can influence exercise or sport performance positively (Costill et al., 1970), and the effects of various athletic drinks on gastric emptying (Coyle et al., 1978). Dr. Robert Cade and colleagues created Gatorade in the 1960s and demonstrated that ingestion of glucose electrolyte solutions can help prevent dehydration and improve athletic and exercise performance during prolonged events (Cade et al., 1972). Numerous studies were then conducted to dive further into the effects of glucose electrolyte solutions on hydration and athletic performance. In 1976 Dr. Mel Williams published a book on the nutritional components of physical athletic performance that broadly synthesized the literature available at the time (Williams, 1976). This book was used as a graduate textbook in many postbaccalaureate programs and helped solidify sport nutrition as a scholarly area of research and academic focus within kinesiology and exercise science. Many academic programs then added sport nutrition to their curriculum during the 1980s. The courses added during the 1980s focused on how nutrition can affect performance, body composition, and general health and well-being.

Since the 1970s the role of nutrition in exercise, recovery, athletic performance, and body composition exploded into the newly founded field of sport nutrition. In 1991 the *International Journal of Sport Nutrition* was created and was renamed the *International Journal of Sports Nutrition and Exercise Metabolism* in 2001. In 2003 the International Society of Sports Nutrition (ISSN) was founded, dedicated to promoting science and application based on evidence to advance our understanding of sport nutrition and supplementation. The *Journal of the International Society of Sports Nutrition* was founded in 2004, and since then, numerous position stands have been published to help scientists and practitioners understand, synthesize, and apply sport nutrition research within the exercise science and sport performance field (Aragon et al., 2017; Buford et al., 2007; Campbell et al., 2007; Campbell et al., 2013; Goldstein et al., 2010; Gonzalez et al., 2022; Grgic, Pedisic, Saunders, et al., 2021; Guest et al., 2021; Jäger et al., 2017; Jäger et al., 2019; Kerksick et al., 2008; Kerksick et al., 2017; Kerksick et al., 2018; Kreider et al., 2017; Kreider et al., 2010; La Bounty et al., 2011; Schoenfeld et al., 2017; Tiller et al., 2019; Trexler et al., 2015; Wilson et al., 2013). Additionally, the Academy of Nutrition and Dietetics, Dietitians of Canada, and ACSM also have published position stands on nutrition and athletic performance (Thomas et al., 2016).

Sports nutrition has grown tremendously in the 2000s: many universities have created master's and PhD degree programs focused on sport nutrition, and certifications have been developed by organizations such as the ISSN. The Certified Sports Nutritionist certification by the ISSN is considered the premier certification in the field of sport nutrition and its application. Additionally, the Board-Certified Specialist in Sports Dietetics (CSSD) was developed in 2005 and is the first and only sport nutrition certification that is accredited by the National Commission for Certifying Agencies (NCCA). Finally, in 2010 the Collegiate and Professional Sports Dietitians Association was created and is dedicated to helping meet the needs of sport dietitians working in the field.

Although the role of the nutritionist and registered dietitian might sound similar, the qualifications for each differ. Registered dietitians (RDs or RDNs) have to obtain at least a bachelor's degree in nutrition, accumulate hours of internship through a university-accredited host (e.g., hospital or clinics), and pass national board exams. Clinical populations might seek the help of RDs to help treat disease or to treat those with comorbidities. Additionally, RDs can provide meal plans and nutritional guidelines to athletes of team or individual sports if they are hired by an athletic organization or university. In the United States RDs must become certified through the Academy of Nutrition and Dietetics to be able to practice. If an aspiring RDN wanted to focus on athletic dietetics, they would certify through the CSSD. The term *nutritionist* previously had little legal meaning, much like *personal trainer*, so anyone could call themselves a nutritionist. The Academy of Nutrition and Dietetics has moved to protect their professional scope of practice by changing the name of their organization and legal credential (registered dietitian nutritionist, or RDN).

KEY POINT

Sport nutrition has become a vibrant field of study in exercise science. Universities offer courses and programs that are dedicated to teaching students about sport nutrition and supplementation. Sport nutrition scientists' work complements peer scholars and professionals in the Academy of Nutrition and Dietetics.

History of Sport Science

Sport science developed as a distinct discipline, similar to exercise science and kinesiology, from the broad, preventive health beginnings of physical education in higher education (Borms, 2008; Massengale & Swanson, 1997). University faculties in Europe and Asia developed a somewhat greater emphasis on sport than physical activity (kinesiology) or exercise (exercise science) that is more common in the United States. Today, many countries have their own sport science societies, journals, and certifications in sport science and sport performance. The British Association of Sport and Exercise Sciences (BASES), for example, publishes the *Journal of Sports Sciences* and offers High Performance Sport Accreditation to professionals. The journal has a section on sport performance featuring sport science research in metrics and analytics. It is important to note that sport science is a true theoretical, academic discipline and is not synonymous with coaching. Research and theory of sport coaching is not as well developed as exercise science, although, like coaching, some sport science research has been criticized for statistical errors and low-level study designs (McLean et al., 2021). Examples of journals dedicated to reporting coaching education and research include *Coaching: An International Journal of Theory, Research, and Practice*, the *Journal of Intercollegiate Sport*, and the *International Sport Coaching Journal*.

Additionally, the International Society of Performance Analysis of Sport (ISPAS) is dedicated to providing education and training on how to perform analysis optimally on performance and correctly interpret findings along with applying the findings to training. The organization founded its accompanying journal in 2001, *International Journal of Performance Analysis in Sport*, with the hopes of disseminating literature on performance analysis. Through this organization one can become an accredited performance analyst. Other journals also exist to provide education and literature on sport science such as *Journal of Sports Sciences*, *European Journal of Sports Science*, and one dedicated to data analysis called the *International Journal of Computer Science in Sport*.

Sport science has continued to evolve and branch out to other subdisciplines. In fact, interest has increased in sport neuroscience and how this subdiscipline can aid athletes to have a competitive edge. The Society for NeuroSports, an organization established in 2019, is dedicated to the integration of exercise and sport science with neuroscience. In 2020 the organization established the *Journal of the Society of NeuroSports*, which aims to further evidence-based research on athletic performance and neuroscience. The Society for NeuroSports also offers a certification titled the Certified by NeuroSports (CNSP) as well as an annual conference. This organization is led by a group of sport scientists looking to further the advancement of research on how exercise and nutrition can affect the brain.

KEY POINT

A career in sport science offers practitioners an opportunity to work with high-level athletes in collegiate, professional, and Olympic settings. A sport scientist is responsible for assimilating training and performance or competition data on athletes to establish training protocols that could improve performance or time to recovery.

Research Methods in Sport Performance

Researchers in strength and conditioning, sport nutrition, and sport science and analytics employ various methodologies to answer important questions related to how exercise, training, and nutrition can optimize performance, training adaptations, and health. The following sections discuss the main types of research used in the subtypes of sport performance. The advancements in technology, and usage of different study designs, help ensure optimal data collection methods. Thus, the study results are more likely to be accurate and potentially can offer practitioners more tools in the toolbox. Strength and conditioning research will be discussed in two main research areas, which will highlight a progression of certain research ideas from basic to applied research. Further, sport nutrition research is conducted with various between- and within-subject designs, and the data can be analyzed in numerous ways to show whether performance outcomes have resulted. Lastly, sport science and analytics research relies on technology

Career Opportunity

Sport Scientist

The need for sport and performance scientists is steadily growing as technological advancements are making monitoring performance, fatigue, and recovery in athletes more convenient than ever before. New methodologies to assess athletic performance in conjunction with the ability to store, synthesize, and view large amounts of performance data (Big Data and Analytics) are more common in research and sport performance services. Sport science is an emerging career option within the realm of sport performance and allows prospective researchers or qualified exercise science graduates the opportunity to be involved with athlete development. Sport scientists are expected to work with athletes directly by assessing training methods, exercise intensity, and recovery; implementing analytical ways to monitor performance; evaluating characteristics of key performance indicators; and handling, evaluating, and disseminating information pertinent to performance. Students interested in sport science can expect to conduct research for a university and work directly with a sporting organization or be hired by the organization as a professional sport scientist. Expected research foci within sport science typically lie within the biomechanics, exercise physiology, and data analytics subdisciplines.

(e.g., wearable devices, GPS software, trackers) for data collection and can be used to give athletes real-time performance feedback and metrics to boost their competitive edge.

Strength and Conditioning Research

Applied research is critical to the strength and conditioning and sport performance fields. Practical observations and direct, prospective interventions with athletes are considered some of the best research methods used in sport settings due to the direct assessment of performance. **Applied research** assesses the practicality of basic research results by addressing pragmatic questions posed by coaches and sport scientists (Acs, 2015). With applied research methods in mind, the strength and conditioning specialist can assess how different training programs affect performance outcomes, physiological adaptations, and the recovery process in direct, ecologically valid conditions.

Sport performance research strives to select the most relevant performance-based outcomes (dependent variables) connected to the success of the athlete. **Primary performance outcomes** are any type of metric that directly translates to competitive events (e.g., whether the athlete is getting bigger, faster, stronger, more accurate). These outcomes determine the amount of direct efficacy of certain interventions regarding the definitive enhancement of the athlete's performance. Examples of these outcomes include increased bench press or deadlift 1RM load, decreased 40-yard (37 m) sprint time, or improved vertical jump height. Often sport performance researchers use biomechanics principles and exercise physiology-based measures to assess variables that are major factors to successful sport performance outcomes (e.g., acceleration, forces, energy consumption, fatigue, impulse, mechanical power, moments of force). Laboratory-based exercise physiology procedures are a critical component to testing research hypotheses regarding primary performance outcomes. It is important to note that during field-based testing, certain manipulations and considerations must be given to the testing procedures employed. In-field testing offers less control on the environment and other influences; however, it might be more practical to assess performance outcomes on the actual event or sport-specific skill toward which the focus is directed. While applied research has a more direct relationship to on-field performance, all members of the sport performance team should be familiar with and respect basic research.

Basic science research in strength and conditioning is designed with strict controls to establish a better understanding of mechanistic explanations, theories, or principles of an athlete's responses to training (Acs, 2015). Basic research provides the theoretical framework and knowledge that can be future studied with applied research methodologies. Exercise physiologists can take a basic science (or fundamental or mechanistic) approach to research questions aimed at uncovering how various biological and molecular processes change due to training, as well as the subsequent changes in the muscle's architecture or alterations in metabolism. From the findings of basic research, which are often performed on animal models with an in vitro (cell or tissue is removed from the organism) or in situ (tissue is in its original position) method, exercise physiologists and other exercise scientists can make inferences about the findings that can contribute to athletic development and training adaptations. For example, a common variable assessed in strength and conditioning–related research is muscle protein synthesis (MPS), which is the process of creating new proteins from bioavailable amino acids within the body on a molecular level. Testing this sort of phenomenon requires invasive biochemical procedures performed in biological laboratories called *wet labs*. Following long-term training, basic research has shown that MPS is the driver behind muscle hypertrophy (i.e., an increase in muscle mass and cross-sectional area). In this sense, the findings of basic research have a potential application that can aid strength and conditioning professionals and other members of the sport performance team. However, basic research findings must be tested with an applied, in vivo (in living animal) research approach within humans to further assess the impact of training adaptations and performance outcomes among athletes. The strength and conditioning professional should be familiar with the best training methods and nutritional practices to maximize MPS and subsequent muscular hypertrophy within specific athletes.

The basic research conducted in laboratories has produced the key findings that serve as a foundation for sport performance services and can be found in peer-reviewed journals or presentations at national and international conferences. Theoretical knowledge is then tested in applied or translational research that documents if interventions work in the more complex world of sport. This "lab to bench to game" evolution of sport performance has continued to expand, and today the field is coming to the forefront as an important area of exercise science research and application in professional careers.

Hot Topic

Assessing the Causes of Delayed Onset Muscle Soreness

Basic and applied exercise physiology research can provide interesting and exciting findings that could explain the mechanisms of certain responses in the body after exercise is performed. Athletes (and other physical training populations) tend to experience soreness, pain, and tenderness following eccentric (muscle resisting or braking motion from an external force) bouts of exercise in the active muscles, and the onset begins approximately 8 hours after the bout. The associated symptoms can continue peaking 1 to 2 days after the exercise and can last up to 7 days. Since the early 1900s delayed onset muscle soreness (DOMS) has been thought to be a result of microruptures in the muscle, leading to decreased range of motion and increased pain. A riveting article by Sonkodi and colleagues (2020) offers a new potential mechanism behind the cause of DOMS. Researchers proposed repetitive eccentric loading can incur damage upon axons innervating the muscle spindles along with microdamage to the surrounding tissue, which can be exacerbated by an immune-mediated pro-inflammatory response. Damage to the muscle spindles has been proposed as a new mechanistic contributor to DOMS. More research on this specific mechanism is warranted to help provide possible further explanations behind this phenomenon.

KEY POINT

Strength and conditioning research is designed to make decisions about training methods to elicit specific responses. Basic research focuses on finding the best mechanistic or theoretical explanations, while applied research focuses on specific, pragmatic questions of use of knowledge in specific strength training situations.

Sport Nutrition Research

Since the 1990s research on sports nutrition interventions has increased, with a goal of finding the best nutritional strategies to help athletes and individuals who exercise regularly gain a competitive edge. It is important to consider the different types of nutritional research methods used to convey the efficacy of certain nutritional practices.

Preeminent exercise and sport nutrition researchers and laboratories from across the world conduct research with methods of the highest quality to establish and continually shape the most up-to-date, valid, and reliable nutritional recommendations for athletes and the general population. High-quality exercise and sport nutrition research is essential, especially with the rapid progression of new nutritional supplements and diets being touted. Few legal limits are in place on what supplement manufacturers can claim about their products or what fitness trainers can claim about their performance or weight loss diets.

It takes arduous work on the part of the exercise and sport nutrition scientist to identify the best dietary strategies for certain populations. Various styles of research design can be used to show the effects of nutrition on health, well-being, exercise, performance, and recovery. Nutrition research can be conducted through **placebo-controlled**, **randomized controlled (RCTs)**, crossover design studies, reviews, and meta-analyses. In the nutrition field, RCTs are considered the gold standard to determine causality between exposure to foods, nutrients, or various nutritional practices and physiological or biological outcome measures such as body composition, performance metrics, and biomarkers (Lichtenstein et al., 2021). It is common practice to include two or more groups in a study examining the effects of a certain dietary behavior or supplement. One group of participants usually serves as a placebo (control) group for comparison. Doing so allows observations of possible effects the dietary intervention might, or might not, yield. Studies that incorporate this method are usually referred to as single-arm, double-arm, or three-arm studies depending on the number of groups. A single-arm study uses one group of individuals, who receive the same treatment. In a double-arm study, participants are allocated into either a control or intervention group. A three-arm study involves a control group, an intervention group, and an active control group. Examples of variables observed during a nutrition intervention in RCT research might include physiological and biomarker changes, sport performance effects, and exercise capacity effects. Nutrition RCTs can be carried out in a double-blind manner, meaning (in the context of supplementation studies) neither the participants nor the researchers are aware of the treatment being administered. Depending on the blinding protocol, a third-party sponsor, or an individual not part of the actual study process, typically blinds the treatments with either a code or method to ensure double blinding is executed properly.

A **crossover design** study uses one cohort or group of participants with whom the experimental and control intervention are carried out in a randomized fashion. A crossover study details that participants receive any treatment given and serve as their own control. Once the participants finish the experimental or control treatment (whichever comes first) a washout period of 1 to 4 weeks is universally observed. Afterward, the participants then follow the opposite treatment (crossover). Performing such research methods allows the participants to act as their own controls, which can negate any variation between participants that might exist in non-crossover studies.

Different reviews of previous original research strive to integrate these results, so they are useful to the reader. A **systematic review** of nutritional research constitutes the synthesis of existing evidence found in RCTs and other research designs (see chapter 9) that can exist on certain dietary practices or supplements. Secondly, a **meta-analysis** uses statistics to combine the results from existing studies conducted on different diets and supplements either in conjunction with exercises or for overall well-being and health. Systematic reviews and meta-analyses are said to exhibit the highest-quality evidence (Ahn & Kang, 2018). What differentiates a meta-analysis from a systematic review is that a meta-analysis combines and analyzes the results from several studies, and a systematic review assimilates the findings from studies.

Case-control studies are studies conducted with usually one participant throughout the course of a cycle or phase, leading to the development of a certain outcome. A common case-control study in sport nutrition entails researchers following and recording the strict dietary, exercise, sleep, and recovery plans (or cycles) of a competitive athlete. Common case controls one might see in sport nutrition are conducted in bodybuilders, weight class athletes, or athletes requiring strict body weight regulation (e.g., combat sports, gymnastics, Olympic weightlifters, and powerlifters). An athlete's practices or behaviors can be observed and recorded by the researchers as the athlete prepares for a competitive event. Each research method can provide some of the strongest evidence available in the sport nutrition literature. These research methods should be considered when designing or planning a study involving nutrition and health.

KEY POINT

Research methods vary when assessing the effects of different nutritional strategies. Double-blinded, placebo-controlled RCTs are considered the gold standard when conducting research on the effectiveness of supplements or diets.

Sport Analytics Research

Sport analytics encompasses a field of advanced statistics, data visualization, and data management, which sport scientists use in training athletes to gain a competitive advantage (Alamar, 2013). Early exercise science and kinesiology research related to sport primarily focused on movements, muscles, and physiology, and less on the tactical and strategic data specific to the athlete and sport. The recent expansion of sport analytics comes from improvements in data collection, computing, storage, and analysis capabil-

EXERCISE SCIENCE COLLEAGUES

Dr. Melvin "Mel" H. Williams

Courtesy of Mel Williams.

Before the 1970s, research on nutrition and its effects on the body and performance was rather scant. Mel Williams pioneered research on nutritional and pharmacological ergogenic aids used in sport. As both a scholar and an athlete who competed in football, wrestling, and distance running, Williams shaped the state of sport nutrition research as it is today. In the late 1960s Dr. Williams founded the Human Performance Laboratory at Old Dominion University, where he would eventually become Eminent Scholar Emeritus. His work on performance-enhancing substances influenced the International Olympic Committee's decision to ban blood doping in the 1980s. In fact, Dr. Williams helped usher in the current World Anti-Doping Agency (WADA). He published more than 20 textbooks on exercise physiology and sport nutrition and was the founding editor of the *International Journal of Sport Nutrition*. Mel passed away in 2016 at the age of 78; however, Mel's contributions to the field of sport performance and nutrition live on.

ity. Use of advanced statistical analysis and computing power to handle big data sets have contributed to the advancement of knowledge and application of sport analytics. Databases provide functionality and organization that can house large amounts of data for data analysis and even for entertainment purposes for professional, collegiate, and high school athletes.

KEY POINT

Sport science research and practice includes sport analytics, which is the collection, storage, and advanced statistical analysis of performance data that can be used to inform athletes and coaches about improving performance and decreasing risk of injury.

Sport coaches traditionally have made practice and strategic moves based on experience and qualitative judgments, not rigorous collection and analysis of performance data. All sports had traditions of collecting and reporting of game statistics for recognition and records; however, the actual scientific study of sport performance data that constitute sports analytics is a fairly recent development. This is not to denigrate traditions and the work of early sport statistics lovers but to emphasize the fact that a science is a formalized body of theoretical knowledge based on established research noted in chapter 1. The acknowledgment of the utility of sport analytics and sport science to assist high-level athletes with improving performance and decreasing risk of injury has been more accepted by collegiate and professional sport in the United States since 2010. Now next-generation sport statistics and analytics are part of broadcasting and advertising.

Sport analytics provide a direct method of studying athletes during competition. It has been proposed that teams or athletes supported by a strong, real-time analytics program gain a competitive edge over opponents that do not (Alamar, 2013). The use of such programs gives instantaneous, real-time results of performance. Not only is this important for performance monitoring during training and competition, but also for research and development. **Wearable sensor technology** epitomizes the recent

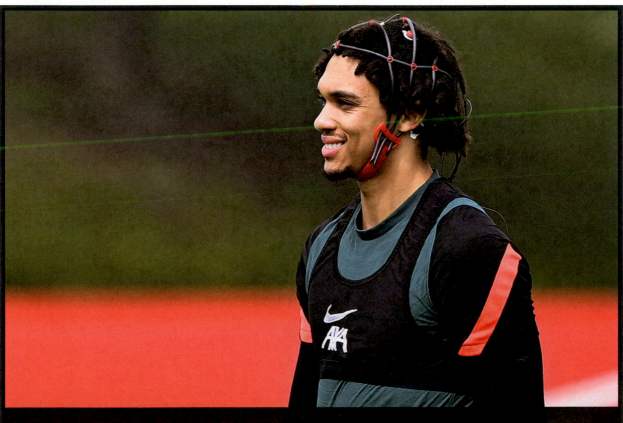

New and innovative technologies such as analytical data headgear are being developed and used to augment the responsiveness to training in athletes.

PAUL ELLIS/AFP via Getty Images

growth of real-time sports analytics. Advances in wearable monitoring technology provide immediate feedback to athletes and coaches during training sessions and data for sport scientists to analyze training and performance. Small sensors can be attached to the athlete's body or equipment or integrated into smart apparel. Advances in wearable sensors, miniature computer chips, and wireless technology and software (including virtual reality; VR) are leading to the rapid expansion of the sport performance field (Faure et al., 2020). A review of the use of VR in sport training and rehab was outlined by Faure and colleagues (2020) and shows how VR can be used to study and train athletes on specific skills.

This technology also has significantly advanced strength and conditioning for sport (Gilgien et al., 2018; Macadam et al., 2017; Orange et al., 2019). Most teams use these types of devices to track performance gains in addition to assessing intensity, load volume, and rest and recovery time needed to optimize player readiness prior to important competitions and to prevent overtraining. This includes training in power platform systems with force plate technology (e.g., VALD Performance and Tekscan) and cameras that provide instant visual feedback to athletes and their coaches. Feedback includes the power flow to the floor and vertical force generated during lifts, monitoring and displaying heart rates of those training and ongoing training volume of each exercise, and even the linear path of a lifted barbell.

Electromyography (EMG), the use of electrodes, amplification, processing, and presentation software, provides a measure and visual representation of the electrical activity of muscle. This specific tool gives real-time feedback on the activation of motor units near the electrodes in muscles during a particular movement. In addition, EMG gives a general representation of the neuromuscular control an athlete uses in movements, with the onset and relative amplitude of the signal being meaningful. Changes in body geometry and muscle mechanical properties do not allow the size of the EMG signal to be considered proportional to muscle tension or force. Any comparison of the size of EMG signals between muscles or athletes requires normalization (processed signal expressed as a percentage of another condition). Even with normalized EMG one cannot obtain a 0% to 100% scale of muscle activation. Advanced frequency domain processing of EMG can be used to assess rate coding (the frequency of motor unit activation) and amount of motor units recruited for a task, which is important to several muscular performance variables (Brooks et al., 2005).

Some wearable technologies give athletes real-time feedback on training intensity and volume. This can be a major benefit in monitoring training overload and balancing this against risk of injury. One variable of interest is the **acute-chronic workload ratio (ACWR)**, which is a ratio between acute volume or workload (usually 1 week average) and chronic volume or workload (usually average over 4 weeks) of training that could indicate the status of recovery and might be associated with the risk of injury (Gabbett, 2016). Besides ACWR, some professionals monitor percent increases of training volume in weekly training programs to help minimize injury and maximize progression and overload. Gabbett (2016) also outlines the importance of monitoring intensity progression, which can be assessed through the use of some of the previously discussed wearable technology, on a weekly basis to predict the risk of injury. Using wearable technology during training seems to be beneficial to sport medicine professionals for tracking recovery, competition readiness, and injury possibility. Monitoring workload or volume of training and competition is an important practice to maximize preparedness for competition and to better manage injury risk. However, appropriate training volumes and ACWRs relative to recovery or injury risk are not clearly defined (Maupin et al., 2020; Wang et al., 2020).

Additionally, the use of GPS athletic software has been shown to be valuable when monitoring athletic performance. Systems such as IsoLynx (Haverhill, Massachusetts), Catapult (Melbourne, Australia), STATSports (Newry, Northern Ireland), and Sonda Sports (Wrocław, Poland) provide sport scientists real-time data on athlete velocity and distance traveled (IsoLynx, n.d.) in addition to cardiovascular response and even risk of injury calculation, among other metrics (Sonda Sports, n.d.). Media industry employs these systems for entertainment purposes for viewers to marvel at the athletic feats that can be accomplished by elite athletes.

In terms of a more biomechanics-related focus, single-plane (two-dimensional) video, with careful setup, can be used for actual measurement and calculations of many biomechanical parameters of movement. Training in biomechanics is usually required for accurate two- or three-dimensional measurement of kinematics (displacements, trajectories, angles, velocity, acceleration) and kinetic (forces, moments or torques, impulse) parameters to analyze the technique. A variety of force and pressure platforms allow for the measurement of three-dimensional forces and moments created by athletes in movements.

The devices mentioned earlier are integral to growth of exercise and sport science knowledge and application of these data with sports analytics. The vast amounts of performance data from biomechanics, nutrition, physiology, psychology, and training or competition pose a problem for sport coaches and scientists. The problem of understanding and integrating all these sources of knowledge about athletes requires interdisciplinary collaboration. Some research groups and sport organizations with large analytics teams have begun to use big data computing and artificial intelligence (AI) to analyze and integrate the data (Claudino et al., 2019). Advanced statistics and AI can discover invisible constructs that underlie and influence performance. For example, early exercise science research used factor analysis to discover the domains of physical fitness from hundreds of performance tests. A review article by Claudino and colleagues (2019) included over 50 articles that used a form of AI to monitor sport performance and injury risk. For example, Qilin and associates (2016) determined the magnitude of various factors and how they contributed to knee injury risk. Artificial intelligence is seemingly going to be a staple modality to monitor performance in the coming years (Fernández-Echeverría et al., 2021; Milbrath et al., 2016; Nicholls et al., 2018; O'Donoghue, 2006).

Analytics also has been integrated within the field of sport nutrition. **Nutrition analytics** has seen a major development over recent years, making it more convenient to track caloric and nutrient intake and potentially to house nutritional information on ingredients, supplements, and meals. Different applications, programs, and functions in the digital age make nutritional information more easily accessible. For example, the United States Department of Agriculture (USDA) offers FoodData Central, an integrated nutrient database system that offers expanded information on micro- and macronutrient distribution of various foods. FoodData Central also gives information regarding experimental research. Additionally, Elizabeth Stewart Hands and Associates (ESHA research) offers an analytical nutrient and ingredient database software called Food Processor, which houses an abundance of nutrient distribution on countless ingredients and meals. Both FoodData Central and Food Processor can be used in research to analyze the nutrient distribution of participants in the analysis of food log data. Such programs also are useful for RDs to provide recommendations to athletes or the general population when making food choices.

Overview of Knowledge in Sport Performance

Sport performance professionals must obtain foundational knowledge of exercise training, nutrition, and analytics to train athletes appropriately, gauge nutrition, and analyze and apply large amounts of data efficiently. Practitioners use the theories that arise from sport performance research and application to prepare and train athletes successfully.

Theories

Each subtype of sport performance relies on various theories for proper application in practice. In the following sections, foundational theories that are integral for proper strength and conditioning, nutrition, and data analytics application are discussed in detail. Strength and conditioning entails various training ideals and components that pertain to different exercise training modalities and health. Sport nutrition theory contains foundational information on the basis of food as energy and how various eating behaviors can affect health in a multitude of ways. Furthermore, one must understand the basics of macronutrients and micronutrients to apply sports nutrition theory optimally. Sport analytics theory encompasses proper data management, accurate data interpretation, and presentation of data in a comprehensible manner, all to ensure success of a team, organization, or individual athlete.

Strength and Conditioning Theories

Theories and principles of strength and conditioning exist to explain adaptations to training, and they can be applied when designing training programs for athletes with specific adaptations in mind (e.g., muscular hypertrophy, speed, strength). Various modalities of training also can yield different physiological and anthropometric changes following exercise training. While reading this chapter, it is important to realize strength and conditioning encapsulates the morphological changes and the physiological adaptations that occur in response to strength, endurance, and concurrent kinds of exercise training. These various modes of training are based on theories that predict training adaptations specific to the mode of training. Training for a specific goal (e.g., muscular hypertrophy) requires thoughtful planning of a specific phase or cycle of training targeted to maximize outcomes. It is typical to see strength and conditioning practitioners plan periods of different training within a year to induce certain responses. Some expected outcomes of common strength and conditioning programs are improved **health-related fitness** components (cardiovascular endurance, muscular strength, muscular endurance, flexibility, and body composition) and **skill-related fitness** components (agility, balance, coordination, reaction time, speed, and power) (Bounds, 2012). The major theories and principles considered by strength and conditioning professionals to elicit specific physiological outcomes are outlined in this section.

The principles of training—specificity, overload, progression, and variation—are followed in resistance training programs to confer optimal training adaptations as one progresses through a prescribed evidence-based training program. **Specificity** of training was coined by Thomas Delorme in 1945. The term indicates that an athlete trains in a certain manner to elicit specific training adaptations favorable for the sport in which they participate (Haff & Triplett, 2016). The strength and conditioning professional must consider exercises that target specific muscle groups commonly used during the respective sport. Training specificity also considers how the main motions of the sport are conducted, thus allowing the members of the sport performance team to work together to improve performance and reduce the risk of injury. It is prudent to mimic the movements and skills conducted in training exercises, which likely will result in better success during competition. However, it is important to note mimicry does not have to be taken literally in following the principles of specificity. For example, an athlete who conducts a lot of vertical jumps in competition (e.g., a high jumper or basketball athlete) still should perform squats and lunges even though they are not the same movement as a countermovement jump or static vertical jump. With regards to applications of specificity of training, not one athlete of one sport will conduct the exact training program of another athlete in a different sport. For example, an Olympic or collegiate sprinter looks to decrease their race time to become more successful during competition. During training, the sprinter should perform movements specific to the actions performed during a race. Therefore, squats, lunges, plyometrics, and power-based training are imperative modes of training to focus on in a prescribed exercise program. Conversely, an American football quarterback looking to add distance to his deep pass should perform more unilateral training along with core training and lower-body explosiveness drills.

Another principle of training is **overload**, which is the prescription of an exercise or stimulus that is greater than what the athlete is used to (Haff & Triplett, 2016). The athlete must be subjected to overload in training because this is how training adaptation to stimuli is reached. Overload is used to increase the demands placed on the muscle to cause optimal adaptation and subsequent performance in a given competition. Regarding resistance training and muscle characteristics, overload can cause **myoplasticity**, the adaptability of physical characteristics of skeletal muscle in response to a stressor.

A third principle of exercise training is progression. **Progression** defines the gradual increase in stimulus intensity to ensure appropriate long-term enhancements in performance. This training principle is used to enable long-term success in exercise and sport. Progression not only applies to intensity, but also can apply to frequency, duration, and volume of exercise.

- *Frequency* is the number of exercise bouts conducted in a specific time frame (usually seen as per week). One can increase the frequency of exercise sessions per week to ensure optimal progression and adaptation to training.
- *Duration* is the length of a single exercise bout. One can increase the duration of exercise sessions (in case of exercise intolerance at the beginning of a training plan) to allow for optimal adaptation to training.
- *Volume* is the total amount of training stimulus performed in an exercise bout. Volume will be discussed more shortly, but note that in resistance training volume is usually expressed as the product of the repetitions performed, sets, and resistance lifted in exercise bouts.

All can be progressively increased in training, which can lead to gradual improvements in fitness or sports performance.

Variation of training is an important component of an exercise program to help push through and prevent plateaus and to avoid overtraining. Variation entails incorporating varying modes and intensities of exercise to negate the repetitive or stale nature of a long-term exercise program. In other words, it is important to prescribe exercises that allow slightly different movements. This could mean prescribing different variations of a certain exercise or performing movements with increased or decreased volume. It is important to remember that people have genetic predispositions in response to exercise, so some planned variations in exercise prescription might fit a person's particular psychological and physiological responses more effectively.

Another important principle to consider is **reversibility**, which occurs when the exercise stimulus or training is removed completely. This can result in the loss of gains in fitness that were seen in the exercise program and can result in decreased performance in competition or training sessions. Training-induced adaptations can be temporary, which shows the importance of overload and progression, because muscles without a stimulus placed on them can return to a pretrained state. *Detraining* is a term sometimes used when describing the principle of reversibility. **Detraining** refers to the decrease in performance and loss in performance adaptation when ceasing a training program, and encompasses a less severe, planned method of detraining, called *deloading*. **Deloading** is used by athletes to decrease the amount of training stimulus within a given time frame to allow the body to recover for an upcoming training cycle.

Each of these principles details the general programming behind strength and conditioning programs. A strength and conditioning professional should take each principle into consideration when constructing a plan of periodized training. The principles also do not strictly apply to one type of athlete or one type of training. Each principle is used with all athletes and all individuals looking to start an exercise plan. Endurance, speed and power, intermittent sport, and combat athletes all train with the principles in mind. General population individuals seeking to become healthier by starting an exercise program also should consider the principles because they provide the greatest opportunity for positive health and fitness attainment.

Strength and conditioning professionals seek to prescribe exercise programs strategically over a long-term time frame such as a year or a competitive season through a method called *periodization*. **Periodization** is a systematic and sequential approach to programming exercise and training programs in different periods of the athlete's season to induce appropriate training adaptations (Haff & Triplett, 2016). Leonid Matveyev, considered the father of sports periodization, brought the strategic plan of training into fruition (Haff & Triplett, 2016). Essentially, periodization delineates an annual training plan that, when appropriately used, can offer proper exercise intensity and volume, which, along with rest and recovery during certain periods of a given year, can translate into peak performance at competitions.

KEY POINT

The key strength and conditioning principles are specificity, overload, progression, and reversibility.

Sport Nutrition Theory

Exercise and nutrition have a common problem: the public's susceptibility to misinformation. In the current age of the Internet and social media, an abundance of inaccurate nutritional advice is spread, usually without the proper scientific context. It is important to know that sound, evidence-based nutritional recommendations must be based on the consensus of several decades' worth of research, not miraculous discoveries from the latest study. Over many years of research, dietetics and sport nutrition scholars have compiled theoretical knowledge of effective use of foods and supplements to support health and high-level performance. Many diets and supplements have been studied to examine how nutrition plays a role in exercise, health, and performance. Results from countless studies have given us promising recommendations to suggest to athletes and the general population alike with specific health and fitness goals in mind.

The energy-yielding nutrients are the macronutrients. **Macronutrients** consist of carbohydrates, proteins, and fats and are the nutrients in food that contain energy. The kilocalorie (kcal), which is colloquially referred to as the *calorie*, is the energy consumed through food. A calorie refers to the amount of heat needed to raise 1 kilogram (2.2 lbs) of water by 1 degree Celsius (34 °F) (Kreider, 2019). The macronutrients contain different amounts of calories per gram of nutrient. For example, carbohydrates and protein contain about 4 kilocalories per gram and fat contains about 9 kilocalories per gram. When summed together in certain food products, the amount of energy contained in each macronutrient gives us an estimate of the amount of energy required to metabolize the food. The energetic basis provided allows individuals to follow certain dietary plans that yield the optimal number of calories respective to their fitness goals. Another nutrient that is essential for survival is water, and it is sometimes considered a macronutrient due to its presence in the body and the wide-ranging functions it plays in cellular survival. Water makes up about 80% of a living cell and is a transportation medium in the body to allow passage of drugs, hormones, and nutrients. Additionally, certain nutrients do not yield energy for muscle contraction or other metabolic processes but serve to assist in numerous energy and biological pathways (Kreider, 2019). The **micronutrients** are organic compounds and consist of vitamins, minerals, and trace elements. It is imperative to meet the dietary guidelines of vitamin and mineral intake predominantly through nutrient-dense, whole foods but also through supplementation if need be. A wide variety of nutrient-dense foods provides an expansive food matrix that can optimize health.

In order to elicit performance-enhancing gains or to maintain healthy fitness goals, it is essential to follow dietary guidelines that promote either success in competition or a healthy lifestyle. Each macronutrient and micronutrient provides health and performance benefits in its own regard. **Protein** is an important nutrient for growth and tissue repair as well as an essential structural component for every cell in a living organism. Interestingly, the body is about 18% protein, and approximately 45% of adult body weight is skeletal muscle, which highlights the need for protein consumption. Proteins are complex organic molecules that are composed of amino acids. **Amino acids** are the building blocks for proteins and are classified as essential, nonessential, or conditionally essential.

- *Essential* amino acids are not synthesized in the body and must be obtained through the diet.

- *Nonessential* amino acids are synthesized in the body from other amino acids.

- *Conditionally essential* amino acids can be synthesized by the body, but the metabolic pathways to do so are not efficient; therefore, the diet is the main source of these amino acids.

Carbohydrates are a main source of energetic substrate during high-intensity exercise and are classified as either simple or complex. A simple carbohydrate (e.g., glucose) is more refined and is digested more quickly than complex carbohydrates (e.g., grains and starchy vegetables). Consumption of carbohydrates is essential to the levels of muscle and liver glycogen levels present during exercise, which is a determining factor in fatigue onset. Also, carbohydrates are the brain's preferred fuel source.

Fats are an integral component of the diet and are essential for cellular structure and function. Dietary fats are classified into two categories, saturated and unsaturated fats, with unsaturated fats containing two types, polyunsaturated and monounsaturated. Saturated, unsaturated, and trans fats are dietary fats found in the diet. High consumption of saturated fats, which are present in high amounts in refined foods, sweets, and meat, has been linked to the onset of various cardiometabolic diseases. Unsaturated fats contain fats important for health and are mostly present in nuts, seeds, various cooking oils, and avocados. Dietary fats are important for maintaining cellular membrane integrity and health throughout the body. Trans fats are another form of dietary fat that is considered a less healthy form of fat. Trans fats are the product of manufacturers adding hydrogen to low-cost unsaturated fats to make a firmer form of fat for foods and spreads. Chronically high consumption of trans fats has been linked to several chronic diseases (Harris, 2004).

It has been theorized that certain dietary behaviors that manipulate the macronutrient distribution in total daily caloric intake can incur benefits specific to the macronutrient being prioritized. For example, a high-protein diet, one containing 1.8 to 2.2 grams per kilogram of body weight (0.002-0.04 oz/lb) per day, has been proposed to be the most optimal daily protein dosage to promote muscle mass gain if the individual is in a caloric surplus, or it can help maintain muscle mass when an individual is in a caloric deficit. Also, carbohydrate loading—a proposed dietary practice that prioritizes high carbohydrate intake, 8 to 12 grams per kilogram of body weight (1.3-1.9 oz/lb) per day, with a concomitant decrease in training volume in days leading up to the exercise event—has been shown to yield performance-enhancing benefits due to the buildup of muscle and liver glycogen stores (Jeukendrup, 2004). Carbohydrate loading is often undertaken in the days leading up to a prolonged endurance event such as a marathon or triathlon. High-fat, low-carb diets (ketogenic diets) might have beneficial effects on some populations with diseases such as epilepsy (D'Andrea Meira et al., 2019) and on the body's ability to use fat as a fuel during exercise, but have yielded equivocal findings on exercise and sport performance enhancement (Burke et al., 2017; Cox et al., 2016). However, long-term adherence to a ketogenic diet for weight loss purposes might not be adherable and can be dangerous when considering the alterations in carbohydrate and fat metabolism and substrate availability. While the ketogenic diet might be another tool in the toolbox, more research is needed (Ludwig, 2019).

KEY POINT

Professionals working as dietitians or in sport nutrition should be familiar with the biochemistry of metabolism, macronutrients (carbohydrates, proteins, fats, and water), and micronutrients (minerals, vitamins, and trace elements) and how manipulating them in the diet can support health and performance.

Sport Analytics Theory

The integration of analytics in competitive sport has shown to be an important tactic for performance enhancement in individual athletes and teams. It has been proposed that teams using sport analytics as part of their strategy for competition planning will have a competitive advantage over opponents that might not invest in sport analytics programs. While sport analytics is a relatively new field with an evolving definition, several major textbooks have been published (Alamar, 2013; Jayal et al., 2018).

Alamar (2013) provides a detailed sport analytics framework that underpins the importance of the three main characteristics of a sound data analytics program: data management, analytic models, and information systems for individuals in decision-making roles in sport. Data management incorporates the method of how data is stored for future interpretation of results. It is the job of the sport analyst to gather data and store it in a safe way. This is also true for primary investigators in research studies. Typically, data is stored in computerized programs to allow for easier access when one wants to acquire information on a study or a particular athlete. Analytic models involve the process of data interpretation to help answer research questions of the decision makers (Jayal et al., 2018). Various analytic models are used in the context of showing the connections between the results and the performance of the athlete performing an exercise being measured. In the context of professional sport, visual analytic models are insightful models for drafting collegiate players. However, analytic models can be useful in the realm of exercise and health research. Wearable devices are used for this reason. These devices show data in real time and allow practitioners and researchers to make inferences about the modality or intensity of exercise being performed and how performance can be affected. Information systems encapsulate the idea of presenting data in an understandable and coherent manner that allows sport analysts, sport scientists, and investigators to make inferences about the data (Alamar, 2013). Real-time data-sharing software displays data in an understandable manner to researchers and sport scientists, which is also important for communicating recommendations to athletes, coaches, and administrators.

Applications

The following section discusses the theories of sport performance and how these theories are applied in practice. Sport performance professionals commonly use these practices to train athletes properly, gauge injury risk, optimize dietary intake, and properly analyze and use data to improve athletic performance.

Strength and Conditioning Applications

Certain combinations of repetitions, sets, intensity, duration, volume, and modality of training are set to target specific adaptations an athlete or team is looking to improve. It is important to note the differences in volume and volume-load between hypertrophy, strength, power, and muscular endurance resistance training. Schoenfeld and colleagues (2017) indicated that a spectrum of low to high loads can induce significant increases in muscle hypertrophy, while greater increases in strength are seen with high loading protocols (heavier weights). Volume refers to the total amount of weight lifted during an exercise session, and volume-load is the total amount of sets multiplied by the number of repetitions per set, which is then multiplied by the weight lifted per repetition (Haff & Triplett, 2016).

Various training styles can be undertaken depending on the specific adaptations an individual is seeking. Typically, it is recommended to perform high-volume resistance training to elicit muscular hypertrophy. Three to six sets of 6 to 12 repetitions at 67% to 85% 1RM on compound (multijoint movements) or core exercises is a standard volume to train with when seeking gains in muscle mass. This involves a great amount of training volume, but it is imperative to stimulate muscle growth in trained athletes. Strength training regimens are typically prescribed with moderate to low amounts of volume, with two to six sets of less than six repetitions at high (85% 1RM) loads. In contrast, resistance training for **muscular endurance** uses high volume with light resistance. A typical recommendation is to train with two to three sets, more than 12 repetitions per set, and at less than or equal to 67% 1RM. A third kind of training, blending both muscular strength and movement speed, is called **power training**, with a training volume typically set around three to five sets of one to five repetitions at 75% to 90% 1RM. Power training also sometimes uses smaller volume, which results from fewer repetitions and lighter loads to maximize the quality of the exercise (Haff & Triplett, 2016).

Rest intervals are also an important consideration when creating exercise programs. It is best to rest 2 minutes between sets and exercises for hypertrophy training, 2 to 5 minutes for strength and power training, and less than or equal to 30 seconds for muscular endurance training. Some research indicates resting for longer periods of time between sets and exercises (8 minutes) can help maintain volume and intensity during sets, which leads to greater increases in strength and hypertrophy over time (Hernandez et al., 2021). It is up to the strength and conditioning professional to determine the most effective and practical rest period within a given workout, because athletes are usually on a schedule. Sport conditioning programs typically also involve the following:

- Speed training (e.g., linear, sprinting technique, reactive, active acceleration, frequency, complex speed drills)
- Agility training (e.g., perceptual and decision-making drills, deceleration, technical agility drills)
- Plyometrics (e.g., exercises with dynamic, bouncing countermovement)
- Jump training (e.g., jumping in place, multiple hops, distance jumps)
- Interval training (e.g., repeated sprints of varying lengths with recovery periods)
- Classical repeated sprint training with progressively decreasing work-to-rest ratios (e.g., 1:10, 1:8, 1:6) as training adaptations occur

Different variations of training also can be used to enhance cardiovascular endurance performance, typically long-slow distance, pace or tempo, interval, or Fartlek training.

- Long-slow distance training refers to prolonged training at exercise intensities at or near 70% maximal oxygen consumption ($\dot{V}O_2$max). Long-slow distance exercise is best used with paces that are slower than race or competition pace.

- Pace or tempo training is a type of endurance training in which the participant or athlete trains at or near race pace, a high intensity pace.
- Interval training can be used as a method for endurance training with the appropriate intensities, as can high-intensity interval training. Interval training might involve performing a prolonged bout of exercise that is broken up into higher-intensity sprint bursts interspersed with a less intense recovery burst. For example, an athlete might perform a 20-minute run that consists of 15-second sprints with 45 seconds of slower jogs.
- Fartlek training consists of a combination of long-slow distance training and high-intensity training and can be used with a combination of different types of endurance exercise such as running, swimming, and biking (Haff & Triplett, 2016).

KEY POINT

Strength and conditioning specialists apply a variety of training theories and styles to elicit responses specific to athletes' goals.

Sport Nutrition Application

Assessment of the dietary needs of athletes or any exercising population involves obtaining measurements of body composition and energy expenditure with physical activity levels factored in. The results of this assessment are integrated with results from sport nutrition research and dietary recommendations to inform individuals about important dietary needs and practices essential for health and success

Research and Evidence-Based Practice in Exercise Science

Plyometrics Training and Hypertrophy

Maximizing muscle growth through training is often a primary goal of exercise. Different modes of resistance training have been studied extensively over the years, detailing mechanisms of how hypertrophy occurs during a resistance training program. A review by Grgic and collaborators (2021) indicates potential beneficial effects of plyometric training on muscular hypertrophy. Not only were the effects of plyometric training on hypertrophy shown, but plyometrics also were shown to be equally effective as resistance training on muscle mass growth. The authors synthesized eight studies directly assessing the effects of plyometrics versus resistance training or plyometric–resistance combined training versus isolated resistance training on hypertrophy and found similar effects on lower-extremity muscle hypertrophy. It should be noted that studies directly assessing plyometrics versus resistance training on hypertrophy are sparse, and more research is warranted on the area to further explore the possibility of plyometrics inducing prominent changes in muscle morphology. However, studies like this one allow strength and conditioning practitioners to implement certain training methods that not only lead to gains in external power output but gains in muscle mass as well.

in competition. Researchers in dietetics and sport nutrition provide empirical evidence, extending our knowledge and informing dietary recommendations by scientific panels.

Sport dietitians can assist the team or client in making diet decisions; are involved in the catering process of food during and after competitive events; and teach about nutrient timing, nutritional supplements, hydration, and appropriate nutritional recovery methods. Dietitians are particularly important when athletes have comorbidities and medical treatment that should be coordinated with medical professionals. Sport dietitians are essential for the in-competition success of teams or individual athletes; however, it is important for the public to maintain proper dietary behaviors when looking to seek a healthier lifestyle.

The dietary needs of individuals and athletes participating in different sports or exercise programs vary because the use of metabolic systems vary in different exercises and sports. Strength and power athletes typically expend around 3,000 to 7,300 kilocalories per day while typically needing to ingest 5 to 7 grams per kilogram (0.08-0.11 oz/lb) of body weight per day of carbohydrate along with 1.7 to 2.2 grams per kilogram (0.03-0.04 oz/lb) of body weight per day of protein during periods of heavy training. This subpopulation of athletes includes powerlifters, Olympic weightlifters, short-distance sprint athletes, and any other athlete whose events are not long enough to rely on aerobic metabolism (longer than 3 minutes) for energetic needs. Endurance athletes can expend anywhere from 2,500 to 6,400 kilocalories per day and should ingest 7 to 13 grams per kilogram (0.1-0.2 oz/lb) of body weight per day of carbohydrate and 1.5 to 2.0 grams per kilogram (0.02-0.03 oz/lb) of body weight per day of protein. The high carbohydrate recommendation supports the need for sustained muscular contractions seen in long endurance training and competition. This allows for maintenance of glycogen stores to support the need for long-duration, high-intensity races or exercise.

For elite athletes, energy expenditure during heavy training or competition might be greater (Barrero et al., 2014) than other athletes. Exercise science defines *elite athletes* as qualifying for the highest level of performance, typically operationalized as international, Olympic, or professional competition. Energy expenditure for professional, competitive cyclists has been estimated as high as 12,000 kilocalories per day (Brouns et al., 1989a, 1989b). Moreover, larger elite athletes (200-331 lb [100-150 kg]) typically require increased caloric needs, ranging between 6,000 and 12,000 kilocalories

per day depending on the intensity and volume of training during given training phases (Heydenreich et al., 2017). It is imperative for athletes engaged in heavy training to eat enough nutrient-dense foods to supply calories to offset energy expenditure, which keeps the athlete in an energetic surplus. They must consume the proper amount of carbohydrates (5-8 g/kg/day [0.08-0.13 oz/lb/day] during normal training and 8-10 g/kg/day [0.13-0.16 oz/lb/day] or more during heavy training) to ensure optimal amount of muscle and liver glycogen stores. Protein intake should be high (1.5-2.2 g/kg/day [0.02-0.04 oz/lb]) to support muscle growth, recovery, and repair while maintaining a relatively low-fat diet (1.0-1.5 g/kg/day [0.016-0.024 oz/lb/day]). Meals and snacks should be ingested at appropriate time intervals before, during, and following exercise or competition to provide energy and promote recovery. Athletes training at high intensities and frequencies also should ensure they are properly hydrated before exercise and competition (and during exercise and competition in longer events). Rest and nutritional strategies should be incorporated to optimize recovery. Also, athletes involved in heavy training should only consider using nutritional supplements that have been found to be an effective and safe means for improving performance capacity and enhancing recovery (Kerksick et al., 2018; Kreider, 2019).

No matter the type of athlete, proper nutrition habits are important in supporting rigorous exercise training regimens and competition. Figure 10.2 illustrates the interaction between training and diet and their complex association with performance (Kreider, 2019). The strength and conditioning specialist is responsible for implementing a training program that contains the optimal amount of exercise volume and intensity that will lead to gains in practice that translate to gains in performance. Too little training

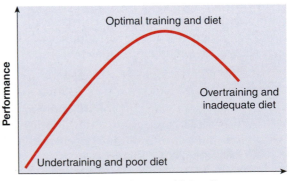

Figure 10.2 The training and diet paradigm.
Reprinted from R.B. Kreider, *Essentials of Exercise and Sport Nutrition: Science to Practice* (Morrisville, NC: Lulu Publishers, 2019).

(intensity and volume) will underprepare athletes for competitive seasons. Too much volume and intensity can increase the chances of overtraining and injury, which is why periodization is a popular method for seasonal training programs. A sport dietitian is responsible for ensuring athletes meet the nutrient and energy demands of their training and competition through dietary habits. Athletes must not neglect their diet due to its direct relationship to fueling the body with foods that allow for energy production, supporting training volume and intensity, body adaptation, and eventually performance. If athletes neglect their diet over a prolonged period while training, it will increase the likelihood of exercise intolerance, injury, early-onset fatigue, eating disorders, **amenorrhea** (lack of a menstrual cycle, in the case of females), and many other medical conditions. A diet too high in calories can cause undesirable outcomes for athletes competing in weight-class sports such as powerlifting or combat sports or athletes needing to maintain certain body fat percentages such as swimmers and gymnasts. Therefore, the collaborative work of the strength and conditioning specialist and the sport nutritionist is imperative in making sure the athlete is operating at a high level with the most optimal training and nutritional strategies.

Many athletes rely on nutritional ergogenic aids to help improve performance. A **nutritional ergogenic aid** is any nutritional supplement, food, or diet that has the potential to enhance performance. Various types of ergogenic aids exist such as mechanical (e.g., weight belts or wrist wraps), psychological (i.e., using different models of psychology to enhance responsiveness to training or even imagined training), or pharmacological methods (e.g., anabolic steroids or selective androgen receptor modulators), which can enhance exercise and sport performance. Sport nutrition professionals should work with athletes interested in nutritional ergogenic aids such as supplements. Sport dietetics and nutrition research and recommendations are essential for effective and safe use of many supplements. Several supplements have been shown to enhance muscular performance in various ways.

- Evidence strongly shows that creatine monohydrate, supplemental protein, and essential amino acids can enhance development of muscle mass during training (Jäger et al., 2017; Kerksick et al., 2018; Kreider et al., 2017; Kreider & Stout, 2021; Wax et al., 2021).
- There is convincing evidence that beta alanine (a nonproteinogenic amino acid produced through the breakdown of uracil) can enhance exercise performance in vigorous activities lasting 1 to 4 minutes in duration (Trexler et al., 2015). It is important to note that beta alanine supplementation is associated with paresthesia (tingly sensation) in higher dosages (Trexler et al., 2015) but has been shown to be a safe and effective supplement for anaerobic exercise performance outcomes (Kerksick et al., 2018).
- Caffeine is arguably one of the most popular and effective ergogenic aids, as well as one of the most highly consumed substances in the world (Guest et al., 2021). Caffeine supplementation has been shown to reduce ratings of perceived exertion (how hard an individual perceives they are working or exercising) as well as enhance endurance exercise performance outcomes (Guest et al., 2021). Additionally, data support some positive effects on resistance training (Goldstein et al., 2010; Guest et al., 2021) as well as enhancement to cognitive function, especially during periods of chronic sleep loss and sleep-deprived states (Goldstein et al., 2010; Gonzalez et al., 2022; Guest et al., 2021). The most common methods for caffeine consumption are in the forms of capsules, energy drinks, and coffee and other caffeinated beverages.
- Sodium bicarbonate supplementation has been shown to delay the onset of fatigue and enhance exercise lasting 1 to 10 minutes in duration (Christensen et al., 2017; Grgic, Pedisic, Saunders, et al., 2021).
- Sodium phosphate has been shown to play a significant role in one's buffering capacity during exercise (making the cellular environment less acidic), which in turn enhances aerobic capacity and the anaerobic threshold (Buck et al., 2013; Kerksick et al., 2018).
- Preworkout supplements and energy drinks have been shown to enhance exercise performance primarily due to their caffeine content along with beta alanine, B vitamins, and citrulline content with possible synergism among the nutrients (Campbell et al., 2013; Kerksick et al., 2018; Smith et al., 2010).

It is important to note that marketing claims about the conferred benefits of dietary supplements are sometimes exaggerated. Therefore, results from sport nutrition research should be reviewed critically to ensure that marketing claims are based on the results of studies cited. Additionally, it should be noted that comments about safety and efficacy are based

Sport dietetics and nutrition research and recommendations are essential for effective and safe use of supplements.

Milan_Jovic/E+/Getty Images

on available scientific evidence and not anecdotal claims or experience. We recommend sport scientists, coaches, and athletes refer to the consensus of research on the supplement being considered relative to the goal as opposed to relying on testimonials from others.

KEY POINT

It is generally more beneficial for people wanting to improve their general health and physical performance to adopt a food-first approach to their diet. Consuming whole foods is a more effective practice than relying on supplements due to the dense and diverse nutrient matrix whole foods offer. Supplements should be considered in consultation with a sport dietitian and nutrition professional because some ingredients are banned by various governing bodies.

Sport Analytics Applications

The convenience of analytics and miniature computer technology provides nearly real-time monitoring, processing, storage, and presentation of extensive training data that assist coaches and sport scientist specialists in making decisions regarding strength and conditioning. Incorporating analytics of performance data in individual or team sports can be advantageous in examining the effects of training and are of tactical and sport-specific value to the coach. Researchers are starting to integrate analytics into their research processes to make the data collection process easier. Sport analytics is a rapidly evolving field; more and more analytic tools and software are being developed. Sporting markets also have a heavy interest in the entertainment aspect that analytics provides by enabling viewers to see real-time performance metrics on athletes during paid television events. Metrics such as peak and average power and velocity and distance traveled are typical variables seen in analytic

Professional Issues in Exercise Science

Myths of Creatine Monohydrate Supplementation

Creatine monohydrate is one of the most popular ergogenic aids on the market and is a topic of prolific research in the scientific community. Many individuals taking creatine monohydrate might discuss common myths surrounding the supplement like increased water retention, increased risk of renal damage, or even questioning if creatine monohydrate is an anabolic steroid. A review article (Antonio et al., 2021) addresses the myths and common concerns about creatine monohydrate supplementation. Some of the most common myths and misconceptions of creatine monohydrate addressed in this review article include the following: (1) Is creatine monohydrate an anabolic steroid? (2) Does taking creatine monohydrate lead to kidney damage or renal dysfunction? (3) Does taking creatine monohydrate result in hair loss and baldness? (4) Does taking creatine monohydrate cause dehydration and muscle cramping? The conclusion from available scientific evidence is that creatine monohydrate supplementation does not cause any of these side effects and is well tolerated even at high doses for years in some clinical populations and that it acts as a pleiotropic (multifunctioning) molecule that can benefit serious athletes, recreational exercise undertakers, the elderly, tactical or occupational athletes, and even certain clinical populations (those who are combatting neurological disorders). Thus, creatine monohydrate is a very applicable supplement that provides a multitude of health and performance benefits (Gonzalez et al., 2022; Kerksick et al., 2018; Kreider et al., 2017; Kreider & Stout, 2021). The reader is directed to these review articles to familiarize themselves with more up-to-date research regarding creatine monohydrate.

Citation style: APA

Antonio, J., Candow, D.G., Forbes, S.C., Gualano, B., Jagim, A.R., Kreider, R.B., Rawson, E.S., Smith-Ryan, A.E., VanDusseldorp, T.A., Willoughby, D.S., & Ziegenfuss, T.N. (2021). Common questions and misconceptions about creatine supplementation: What does the scientific evidence really show? *Journal of the International Society of Sports Nutrition, 18*(1), 13. doi.org/10.1186/s12970-021-00412-w

Gonzalez, D.E., McAllister, M.J., Waldman, H.S., Ferrando, A.A., Joyce, J., Barringer, N.D., Dawes, J.J., Kieffer, A.J., Harvey, T., Kerksick, C.M., Stout, J.R., Ziegenfuss, T.N., Zapp, A., Tartar, J.L., Heileson, J.L., VanDusseldorp, T.A., Kalman, D.S., Campbell, B.I., Antonio, J., & Kreider, R.B. (2022). International society of sports nutrition position stand: Tactical athlete nutrition. *Journal of the International Society of Sports Nutrition, 19*(1), 267-315. doi.org/10.1080/15502783.2022.2086017

Kerksick, C.M., Wilborn, C.D., Roberts, M.D., Smith-Ryan, A., Kleiner, S.M., Jäger, R., Collins, R., Cook, M., Davis, J.N., Galvan, E., Greenwood, M., Lowery, L.M., Wildman, R., Antonio, J., & Kreider, R.B. (2018). ISSN exercise & sports nutrition review update: Research & recommendations. *Journal of the International Society of Sports Nutrition, 15*(1), 38. doi.org/10.1186/s12970-018-0242-y

Kreider, R.B., Kalman, D.S., Antonio, J., Ziegenfuss, T.N., Wildman, R., Collins, R., Candow, D.G., Kleiner, S.M., Almada, A.L., & Lopez, H.L. (2017). International Society of Sports Nutrition position stand: Safety and efficacy of creatine supplementation in exercise, sport, and medicine. *Journal of the International Society of Sports Nutrition, 14*, 18. doi.org/10.1186/s12970-017-0173-z

Kreider, R.B., & Stout, J.R. (2021). Creatine in health and disease. *Nutrients, 13*(2), 447.

models used in intermittent activity sports such as basketball, soccer, hockey, and rugby. These metrics also are assessed in track and field events through GPS-provided analytics.

The evolution of the sports entertainment market (going from print to digital) was an important component of the progression of data analytics software. Much of the analyses used during televised sporting events are prime examples of sport analytics. For example, ESPN broadcasts athletic events (team and individual sports) along with programs on which sport scientists and analysts analyze sport statistics. For example, during a Major League Baseball (MLB) telecast, a viewer might see launch angle, exit velocity, and distance displayed when a batter is successful. The earned run average (ERA) data of pitchers are formulated and presented to the viewers by the work of sport analysts, analytic software, and broadcasting production staff. This type of information is used for entertainment purposes in addition to aiding athletes and coaches to gauge performance and modify game strategy. Sometimes entertainment drives technological developments that are later used in sport analytics. Dartfish video software was originally developed in Sweden for sport television broadcasts but is now used by many sport performance professionals.

Exercise and sport analytics are used in high-level competition and research as well as in the development of young athletes. Secondary schools across the globe implement video streaming services to monitor practice sessions and competition events to study the opposition and potentially enhance performance. Hudl (located in Lincoln, Nebraska) provides coaches and athletes video of other teams to monitor technique and strategy, giving athletes and coaches knowledge about how to formulate a game plan. Sport clubs also use it to monitor practice to assist in breaking down skill fundamentals. This tool is widely used in various team and individual sports. Video feedback technology such as Hudl is an influential tool in determining success in sport (Fernández-Echeverría et al., 2021; Milbrath et al., 2016; Nicholls et al., 2018; O'Donoghue, 2006). Additionally, the integration of team and individual athlete statistics into one online format has been shown to be an influential method of logging and monitoring game statistics. For example, MaxPreps (El Dorado Hills, California) is an online platform that includes the United States' high school game statistics from past years to the present. Statistics from a wide variety of team and individual sports are included on MaxPreps. The integration of this amount of data into one large platform results in an interconnectedness between media personnel, sports analysts, coaches, and athletes.

Wearable analytic devices with wireless communication allow for easier data attainment of real-time physiological responses occurring in the body during an exercise assessment. COSMED (Rome, Italy), a worldwide supplier of metabolic and cardiopulmonary research instruments, constructed the K5, a portable metabolic system used to examine exercise and resting metabolic activity along with heart rate and electrocardiography (the electrical activity of the heart). The K5 can be helpful in monitoring the metabolic cost associated with certain types of exercise or sporting events. Additionally, more devices and software such as the GymAware (Canberra, Australia) can give immediate feedback to athletes performing velocity-based weight training. GymAware is used to monitor the vertical velocity and power flow to the ball during velocity-focused movements such as a clean or snatch. Video analysis programs such as Dartfish (Fribourg, Switzerland) and Siliconcoach (Dunedin, New Zealand) are used to improve the coach's qualitative diagnosis of athletes' technique with video capture, storage, and replay. These two web-based programs have numerous features and remote storage of large video files. Several of these video systems now exist as apps, both through subscription and for free with basic features.

Exercise and sport performance will be affected positively by the integration of analytics software with noninvasive sensors. As these systems are done in real time and scientifically validated, they will help athletes reach closer to their true potential. Nutrition analyses also can be made more convenient, with governing bodies incorporating widely used nutrition analysis programs that make dietary monitoring possible. Apps and software available to the public also garner more interest in public health awareness. People find it more convenient to track calories in an app compared to writing out ingredients and meals, which makes dietary choices easier to make.

> **KEY POINT**
>
> Using sport analytics to track athletic performance is a useful tool for sport scientists when implementing training programs.

Collaboration With Medicine and Rehabilitation Professionals

As mentioned earlier, a collaborative effort is needed from several key personnel of a sport team or organization (e.g., sports medicine doctor and strength and conditioning coaches). Strength and conditioning, sports medicine, and rehabilitation often are thought of as separate entities, but all professionals in the sport performance team work together when athletes are injured. Athletes participating in training, practice, and game play are inherently at risk of serious injury and sometimes even death. Therefore, it is critical that all professionals of the sport organization work together to ensure the safety of the athletes. In fact, sports medicine professions (e.g., athletic trainers, dietitians, physicians, physical therapists) emphasize and train for this collaborative, interprofessional practice (Breitbach & Richardson, 2015).

One of the more popular areas of collaboration has been the efforts made to translate research results on nutritional supplements in strength and conditioning into use in medicine and rehabilitation. One of the most researched supplements, creatine, has been shown to have several health and therapeutic benefits including positive effects on heart health, cognitive function, diabetes and glucose management, neurodegenerative diseases and muscular dystrophy, brain and spinal cord neuroprotection, enhanced rehabilitation outcome, cancer, and pregnancy (Kreider & Stout, 2021).

Other sport nutrition supplements and dietary protocols are being researched for their benefits in both the health and performance of athletic populations. Additionally, sport performance analytics are being used to improve the recovery and sleep of athletes such as the NormaTec PULSE Leg Recovery System, which is part of a physical therapy approach to improving recovery. Other devices such as the OURA rings and WHOOP are useful in tracking recovery, sleep, and other metrics. Moreover, the OURA rings are capable of detecting COVID-19 symptoms, even when only subtle (Smarr et al., 2020). Lastly, with wearable performance devices (e.g., Apple Watch or Garmin), individuals can track heart rate, blood oxygen levels, running performance metrics, and even take an electrocardiogram simply by pushing a button at the wrist. These wearables have become significant in monitoring an athlete with respect to sport performance and sports medicine (Li et al., 2016).

Sport scientists and analysts should consult with physicians, coaches, and RDs by confidential sharing and real-time physiological and movement parameters during training and competition. Diet, sleep, and other physiological data can be advantageous for screening (e.g., ACWR or fatigue, concussion) potential risk of injury (Li et al., 2016). The advancements made within the sport performance field are far reaching and will continue to expand in the future. The benefit of the sport performance team is maximized with interprofessional collaboration with sports medicine and rehabilitation professionals. It is important to note that in the collaboration between sport scientists and physicians, keeping the shared data confidential is paramount. The Health Insurance Portability and Accountability Act of 1996 (HIPAA) was enacted to protect the patient's health information and requires the information to not be disclosed without the consent of the patient.

More Information on Sport Performance

Organizations

Academy of Nutrition and Dietetics (AND)

British Association of Sport and Exercise Sciences (BASES)

Collegiate Strength and Conditioning Coaches Association (CSCCa)

European College of Sport Science (ECSS)

International Federation of Sports Medicine (FIMS)

International Society of Performance Analysis of Sport (ISPAS)

International Society of Sports Nutrition (ISSN)

National Strength and Conditioning Association (NSCA)

Journals

European Journal of Sport Science

International Journal of Performance Analysis in Sport

International Journal of Sports Nutrition and Exercise Metabolism

International Journal of Sports Physiology and Performance

International Journal of Sports Science & Coaching

KEY POINT

The sport performance team is a model of professional collaboration or interprofessional practice. Research results and knowledge also is shared between sport analytics, nutrition, strength and conditioning, and sports medicine.

Wrap-Up

Strength and conditioning coaches, researchers, and other sport performance personnel are continuing to advance exercise science knowledge. Today, nearly all competitive university and professional sport organizations have employed various sport science professionals that make up the sport performance team. These key personnel help to deliver the best evidence-based recommendations to athletes, coaches, and sport staff to improve athletic performance and minimize the risk of injury. Recent advancements in sport nutrition and strength and conditioning, all enhanced by sport analytics, will continue to help sport performance teams help athletes reach their full potential and reduce risk of injury. Students working toward a sport performance career can expect an educational focus on many subdisciplines of exercise science such as biomechanics, exercise physiology, exercise and sport psychology, sport analytics, sport nutrition, and statistics. As the sport performance field develops, sport performance personnel will only grow to assist coaches and athletes in achieving greatness.

International Journal of Sports Medicine

Journal of the International Society of Sports Nutrition

Journal of Science and Medicine in Sport

Journal of the Society of NeuroSports

Journal of Sports Sciences

Journal of Strength and Conditioning Research

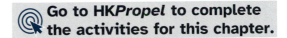

Go to HK*Propel* to complete the activities for this chapter.

Review Questions

1. List and discuss the types of research that are conducted in strength and conditioning, sport nutrition, and analytics.

2. What are the principles of training a strength and conditioning specialist should implement when designing a training program for athletes?

3. How would a sport nutritionist be helpful to an individual athlete or sport organization?

4. What are the three main characteristics of a sound data analytics program? Give an example for each characteristic.

5. Which organizations are focused on strength and conditioning research? Name their offered certifications. Which organizations focus on sport supplement research, and which certifications do they offer?

6. How do sport scientists use data analytics to increase performance or recovery from training?

Suggested Readings

Brooks, G., Fahey, T., & Baldwin, K. (2005). *Exercise physiology: Human bioenergetics and its applications* (4th ed.). McGraw Hill.

French, D.N. & Torres-Ronda, L. (2021). *Essentials of sport science.* Human Kinetics.

Gabbett, T.J. (2016). The training—injury prevention paradox: Should athletes be training smarter and harder? *British Journal of Sports Medicine, 50*(5), 273-280.

Haff, G.G. & Triplett, N.T. (Eds). (2016). *Essentials of strength training and conditioning* (4th ed.). Human Kinetics.

Jayal, A., McRobert, A., Oatley, G., & O'Donoghue, P. (2018). *Sports analytics: Analysis, visualization and decision making in sports.* Routledge.

Plisk, S.S., & Jeffreys, I. (2021). Effective needs analysis and functional training principles. In. I. Jeffreys & J. Moody (Eds.), *Strength and conditioning for sports performance* (2nd ed., pp. 193-207). Routledge.

Schoenfeld, B.J. (2010). The mechanisms of muscle hypertrophy and their application to resistance training. *Journal of Strength and Conditioning Research, 24*(10), 2857-2872.

Williams, M.H. (1976). *Nutritional aspects of human physical and athletic performance.* Ross Laboratories.

References

Acs, P. (2015). Research methodology in sport sciences. *University of PECS Faculty of Health Sciences.*

Ahn, E., & Kang, H. (2018). Introduction to systematic review and meta-analysis. *Korean Journal of Anesthesiology, 71*(2), 103-112.

Alamar, B. (2013). *Sports analytics: A guide for coaches, managers, and other decision makers.* Columbia University Press.

Antonio, J., Candow, D.G., Forbes, S.C., Gualano, B., Jagim, A.R., Kreider, R.B., Rawson, E.S., Smith-Ryan, A.E., VanDusseldorp, T.A., Willoughby, D.S., & Ziegenfuss, T.N. (2021). Common questions and misconceptions about creatine supplementation: What does the scientific evidence really show? *Journal of the International Society of Sports Nutrition, 18*(1), 13.

Applegate, E.A., & Grivetti, L.E. (1997). Search for the competitive edge: A history of dietary fads and supplements. *The Journal of Nutrition, 127*(5 Suppl), 869S-873S.

Aragon, A.A., Schoenfeld, B.J., Wildman, R., Kleiner, S., VanDusseldorp, T., Taylor, L., Earnest, C.P., Arciero, P.J., Wilborn, C., Kalman, D.S., Stout, J.R., Willoughby, D.S., Campbell, B., Arent, S.M., Bannock, L., Smith-Ryan, A.E., & Antonio, J. (2017). International Society of Sports Nutrition position stand: Diets and body composition. *Journal of the International Society of Sports Nutrition, 14*(1), 16.

Barrero, A., Erola, P., & Bescós, R. (2014). Energy balance of triathletes during an ultra-endurance event. *Nutrients, 7*(1), 209-222.

Bergström, J., & Hultman, E. (1972). Nutrition for maximal sports performance. *JAMA, 221*(9), 999-1006.

Borms. (2008). *Directory of sport science: A journey through time: the changing face of ICSSPE.* Human Kinetics.

Bounds, L., Shea, K.B., Agnor, D., & Darnell, G. (2012). *Health and fitness: A guide to a health lifestyle.* Kendall Hunt Publishing.

Breitbach, A.P., & Richardson, R. (2015). Interprofessional education and practice in athletic training. *Athletic Training Education Journal, 10*(2), 170-182.

Brooks, G.A., Fahey, T.D., & Baldwin, K.M. (2005). *Exercise physiology: Human bioenergetics and its application.* McGraw Hill.

Brouns, F., Saris, W.H., Stroecken, J., Beckers, E., Thijssen, R., Rehrer, N.J., & ten Hoor, F. (1989a). Eating, drinking, and cycling. A controlled Tour de France simulation study, part I. *International Journal of Sports Medicine, 10*(Suppl 1), S32-40.

Brouns, F., Saris, W.H., Stroecken, J., Beckers, E., Thijssen, R., Rehrer, N.J., & ten Hoor, F. (1989b). Eating, drinking, and cycling. A controlled Tour de France simulation study, part II. Effect of diet manipulation. *International Journal of Sports Medicine, 10*(Suppl 1), S41-48.

Brown, T.G., & Cathcart, E.P. (1909). The effect of work on the creatine content of muscle. *Biochemical Journal, 4*(9), 420-426.

Buck, C.L., Wallman, K.E., Dawson, B., & Guelfi, K.J. (2013). Sodium phosphate as an ergogenic aid. *Sports Medicine, 43*(6), 425-435.

Buford, T.W., Kreider, R.B., Stout, J.R., Greenwood, M., Campbell, B., Spano, M., Ziegenfuss, T., Lopez, H., Landis, J., & Antonio, J. (2007). International Society of Sports Nutrition position stand: Creatine supplementation and exercise. *Journal of the International Society of Sports Nutrition, 4*(1), 6.

Burke, L.M., Ross, M.L., Garvican-Lewis, L.A., Welvaert, M., Heikura, I.A., Forbes, S.G., Mirtschin, J.G., Cato, L.E., Strobel, N., Sharma, A.P., & Hawley, J.A. (2017). Low carbohydrate, high fat diet impairs exercise economy and negates the performance benefit from intensified training in elite race walkers. *Journal of Physiology, 595*(9), 2785-2807.

Büttner, J. (2000). Justus Von Liebig and his influence on clinical chemistry. *Ambix, 47*(2), 96-117.

Butts, J., Jacobs, B., & Silvis, M. (2018). Creatine use in sports. *Sports Health, 10*(1), 31-34.

Cade, R., Spooner, G., Schlein, E., Pickering, M., & Dean, R. (1972). Effect of fluid, electrolyte, and glucose replacement during exercise on performance, body temperature, rate of sweat loss, and compositional changes of extracellular fluid. *Journal of Sports Medicine and Physical Fitness, 12*(3), 150-156.

Campbell, B., Kreider, R.B., Ziegenfuss, T., La Bounty, P., Roberts, M., Burke, D., Landis, J., Lopez, H., & Antonio, J. (2007). International Society of Sports Nutrition position stand: Protein and exercise. *Journal of the International Society of Sports Nutrition, 4*(1), 8.

Campbell, B., Wilborn, C., La Bounty, P., Taylor, L., Nelson, M.T., Greenwood, M., Ziegenfuss, T.N., Lopez, H.L., Hoffman, J.R., Stout, J.R., Schmitz, S., Collins, R., Kalman, D.S., Antonio, J., & Kreider, R.B. (2013). International Society of Sports Nutrition position stand: Energy drinks. *Journal of the International Society of Sports Nutrition, 10*(1), 1.

Christensen, E.H., & Hansen, O. (1939). V. Respiratorischer quotient und O2-aufnahme. *Acta Physiologica, 81*(1), 180-189.

Christensen, P.M., Shirai, Y., Ritz, C., & Nordsborg, N.B. (2017). Caffeine and bicarbonate for speed. A meta-analysis of legal supplements potential for improving intense endurance exercise performance [Review]. *Frontiers in Physiology, 8*(240). https://doi.org/10.3389/fphys.2017.00240

Claudino, J.G., Capanema, D.d.O., de Souza, T.V., Serrão, J.C., Machado Pereira, A.C., & Nassis, G.P. (2019). Current approaches to the use of artificial intelligence for injury risk assessment and performance prediction in team sports: A systematic review. *Sports Medicine - Open, 5*(1), 28.

Costill, D.L., Kammer, W.F., & Fisher, A. (1970). Fluid ingestion during distance running. *Archives of Environmental Health: An International Journal, 21*(4), 520-525.

Cox, P.J., Kirk, T., Ashmore, T., Willerton, K., Evans, R., Smith, A., Murray, A.J., Stubbs, B., West, J., McLure, S.W., King, M.T., Dodd, M.S., Holloway, C., Neubauer, S., Drawer, S., Veech, R.L., Griffin, J.L., & Clarke, K. (2016). Nutritional ketosis alters fuel preference and thereby endurance performance in athletes. *Cell Metabolism, 24*(2), 256-268.

Coyle, E.F., Costill, D.L., Fink, W.J., & Hoopes, D.G. (1978). Gastric emptying rates for selected athletic drinks. *Research Quarterly. American Alliance for Health, Physical Education and Recreation, 49*(2), 119-124.

D'Andrea Meira, I., Romão, T.T., Pires do Prado, H.J., Krüger, L.T., Pires, M.E.P., & da Conceição, P.O. (2019). Ketogenic diet and epilepsy: What we know so far. *Frontiers in neuroscience, 13*, 5-5.

Dijkstra, H.P., Pollock, N., Chakraverty, R., & Alonso, J.M. (2014). Managing the health of the elite athlete: A new integrated performance health management and coaching model. *British Journal of Sports Medicine, 48*(7), 523-531.

Faure, C., Limballe, A., Bideau, B., & Kulpa, R. (2020). Virtual reality to assess and train team ball sports performance: A scoping review. *Journal of Sports Sciences, 38*(2), 192-205.

Fernández-Echeverría, C., Mesquita, I., González-Silva, J., & Moreno, M.P. (2021). Towards a more efficient training process in high-level female volleyball from a match analysis intervention program based on the constraint-led approach: The voice of the players [Original Research]. *Frontiers in Psychology, 12*(563). https://doi.org/10.3389/fpsyg.2021.645536

Fry, A.C., & Newton, R.U. (2000). A brief history of strength training and basic principles and concepts. In W.J. Kraemer & H.K. (Eds.), *Handbook of sports medicine and science: Strength training for sport* (pp. 1-19). Wiley & Sons.

Gabbett, T.J. (2016). The training—injury prevention paradox: Should athletes be training smarter *and* harder? *British Journal of Sports Medicine, 50*(5), 273-280.

Gilgien, M., Kröll, J., Spörri, J., Crivelli, P., & Müller, E. (2018). Application of dGNSS in alpine ski racing: Basis for evaluating physical demands and safety. *Frontiers in Physiology, 9*(145). https://doi.org/10.3389/fphys.2018.00145

Goldstein, E.R., Ziegenfuss, T., Kalman, D., Kreider, R., Campbell, B., Wilborn, C., Taylor, L., Willoughby, D., Stout, J., Graves, B.S., Wildman, R., Ivy, J.L., Spano, M., Smith, A.E., & Antonio, J. (2010). International Society of Sports Nutrition position stand: Caffeine and performance. *Journal of the International Society of Sports Nutrition, 7*(1), 5.

Gonzalez, D.E., McAllister, M.J., Waldman, H.S., Ferrando, A.A., Joyce, J., Barringer, N.D., Dawes, J.J., Kieffer, A.J., Harvey, T., Kerksick, C.M., Stout, J.R., Ziegenfuss, T.N., Zapp, A., Tartar, J.L., Heileson, J.L., VanDusseldorp, T.A., Kalman, D.S., Campbell, B.I., Antonio, J., & Kreider, R.B. (2022). International Society of Sports Nutrition position stand: Tactical athlete nutrition. *Journal of the International Society of Sports Nutrition, 19*(1), 267-315.

Grgic, J., Schoenfeld, B.J., & Mikulic, P. (2021). Effects of plyometric vs. resistance training on skeletal muscle hypertrophy: A review. *Journal of Sport and Health Science, 10*(5), 530-536.

Grgic, J., Pedisic, Z., Saunders, B., Artioli, G.G., Schoenfeld, B.J., McKenna, M.J., Bishop, D.J., Kreider, R.B., Stout, J.R., Kalman, D.S., Arent, S.M., VanDusseldorp, T.A., Lopez, H.L., Ziegenfuss, T.N., Burke, L.M., Antonio, J., & Campbell, B.I. (2021). International Society of Sports Nutrition position stand: Sodium bicarbonate and exercise performance. *Journal of the International Society of Sports Nutrition, 18*(1), 61.

Guest, N.S., VanDusseldorp, T.A., Nelson, M.T., Grgic, J., Schoenfeld, B.J., Jenkins, N.D.M., Arent, S.M., Antonio, J., Stout, J.R., Trexler, E.T., Smith-Ryan, A.E., Goldstein, E.R., Kalman, D.S., & Campbell, B.I. (2021). International Society of Sports Nutrition position stand: Caffeine and exercise performance. *Journal of the International Society of Sports Nutrition, 18*(1), 1.

Haff, G.G., & Triplett, N.T. (Eds.). (2016). *Essentials of strength training and conditioning* (4th ed.). Human Kinetics.

Harris, H.A. (1966). Nutrition and physical performance. The diet of Greek athletes. *Proceedings of the Nutrition Society, 25*(2), 87-90.

Harris, W.S. (2004). Fish oil supplementation: Evidence for health benefits. *Cleveland Clinic Journal of Medicine, 71*(3), 208-210, 212, 215-208 passim.

Heffernan, C. (2015). *Creatine: A short history.* Physical Culture Study. https://physicalculturestudy.com/2015/02/26/creatine-a-short-history/

Hermansen, L., Hultman, E., & Saltin, B. (1967). Muscle glycogen during prolonged severe exercise. *Acta Physiologica Scandinavica, 71*(2), 129-139.

Hernandez, D.J., Healy, S., Giacomini, M.L., & Kwon, Y.S. (2021). Effect of rest interval duration on the volume completed during a high-intensity bench press exercise. *Journal of Strength and Conditioning Research, 35*(11), 2981-2987.

Heydenreich, J., Kayser, B., Schutz, Y., & Melzer, K. (2017). Total energy expenditure, energy intake, and body composition in endurance athletes across the training season: A systematic review. *Sports Medicine - Open*, *3*(1), 8.

IsoLynx. (n.d.). IsoLynx. Retrieved September 8, 2022, from www.finishlynx.com/isolynx

Jäger, R., Kerksick, C.M., Campbell, B.I., Cribb, P.J., Wells, S.D., Skwiat, T.M., Purpura, M., Ziegenfuss, T.N., Ferrando, A.A., Arent, S.M., Smith-Ryan, A.E., Stout, J.R., Arciero, P.J., Ormsbee, M.J., Taylor, L.W., Wilborn, C.D., Kalman, D.S., Kreider, R.B., Willoughby, D.S., Hoffman, J.R., Krzykowski, J.L., & Antonio, J. (2017). International Society of Sports Nutrition Position Stand: Protein and exercise. *Journal of the International Society of Sports Nutrition*, *14*(1), 20.

Jäger, R., Mohr, A.E., Carpenter, K.C., Kerksick, C.M., Purpura, M., Moussa, A., Townsend, J.R., Lamprecht, M., West, N.P., Black, K., Gleeson, M., Pyne, D.B., Wells, S.D., Arent, S.M., Smith-Ryan, A.E., Kreider, R.B., Campbell, B.I., Bannock, L., Scheiman, J., . . . Antonio, J. (2019). International Society of Sports Nutrition position stand: Probiotics. *Journal of the International Society of Sports Nutrition*, *16*(1), 62-62.

Jayal, A., McRobert, A., Outley, G., & O'Donoghue, P. (2018). *Sports analytics: Analysis, visualisation and decision making in sports performance*. Routledge.

Jeukendrup, A.E. (2004). Carbohydrate intake during exercise and performance. *Nutrition*, *20*(7-8), 669-677.

Juzwiak, C.R. (2016). Reflection on sports nutrition: Where we come from, where we are, and where we are headed. *Revista de Nutrição*, *29*(3), 435-444.

Karlsson, J., & Saltin, B. (1971). Diet, muscle glycogen, and endurance performance. *Journal of Applied Physiology*, *31*(2), 203-206.

Kerksick, C.M., Harvey, T., Stout, J., Campbell, B., Wilborn, C., Kreider, R., Kalman, D.S., Ziegenfuss, T.N., Lopez, H., Landis, J., Ivy, J.L., & Antonio, J. (2008). International Society of Sports Nutrition position stand: Nutrient timing. *Journal of the International Society of Sports Nutrition*, *5*(1). https://doi.org/10.1186/1550-2783-5-17

Kerksick, C.M., Arent, S., Schoenfeld, B.J., Stout, J.R., Campbell, B., Wilborn, C.D., Taylor, L., Kalman, D., Smith-Ryan, A.E., Kreider, R.B., Willoughby, D., Arciero, P.J., VanDusseldorp, T.A., Ormsbee, M.J., Wildman, R., Greenwood, M., Ziegenfuss, T.N., Aragon, A.A., & Antonio, J. (2017). International Society of Sports Nutrition position stand: Nutrient timing. *Journal of the International Society of Sports Nutrition*, *14*(1), 33.

Kerksick, C.M., Wilborn, C.D., Roberts, M.D., Smith-Ryan, A., Kleiner, S.M., Jäger, R., Collins, R., Cooke, M., Davis, J.N., Galvan, E., Greenwood, M., Lowery, L.M., Wildman, R., Antonio, J., & Kreider, R.B. (2018). ISSN exercise & sports nutrition review update: Research & recommendations. *Journal of the International Society of Sports Nutrition*, *15*(1), 38.

Kreider, R.B. (2019). *Essentials of exercise and sport nutrition: Science to practice*. Lulu Publishers.

Kreider, R.B., & Stout, J.R. (2021). Creatine in health and disease. *Nutrients*, *13*(2), 447.

Kreider, R.B., Kalman, D.S., Antonio, J., Ziegenfuss, T.N., Wildman, R., Collins, R., Candow, D.G., Kleiner, S.M., Almada, A.L., & Lopez, H.L. (2017). International Society of Sports Nutrition position stand: Safety and efficacy of creatine supplementation in exercise, sport, and medicine. *Journal of the International Society of Sports Nutrition*, *14*(1), 18.

Kreider, R.B., Wilborn, C.D., Taylor, L., Campbell, B., Almada, A.L., Collins, R., Cooke, M., Earnest, C.P., Greenwood, M., Kalman, D.S., Kerksick, C.M., Kleiner, S.M., Leutholtz, B., Lopez, H., Lowery, L.M., Mendel, R., Smith, A., Spano, M., Wildman, R., . . . Antonio, J. (2010). ISSN exercise & sport nutrition review: Research & recommendations. *Journal of the International Society of Sports Nutrition*, *7*(1). https://doi.org/10.1186/1550-2783-7-7

La Bounty, P.M., Campbell, B.I., Wilson, J., Galvan, E., Berardi, J., Kleiner, S.M., Kreider, R.B., Stout, J.R., Ziegenfuss, T., Spano, M., Smith, A., & Antonio, J. (2011). International Society of Sports Nutrition position stand: Meal frequency. *Journal of the International Society of Sports Nutrition*, *8*, 4.

Li, R.T., Kling, S.R., Salata, M.J., Cupp, S.A., Sheehan, J., & Voos, J.E. (2016). Wearable performance devices in sports medicine. *Sports Health*, *8*(1), 74-78.

Lichtenstein, A.H., Petersen, K., Barger, K., Hansen, K.E., Anderson, C.A.M., Baer, D.J., Lampe, J.W., Rasmussen, H., & Matthan, N.R. (2021). Perspective: Design and conduct of human nutrition randomized controlled trials. *Advances in Nutrition*, *12*(1), 4-20.

Ludwig, D.S. (2019). The ketogenic diet: Evidence for optimism but high-quality research needed. *The Journal of Nutrition*, *150*(6), 1354-1359.

Macadam, P., Simperingham, K.D., Cronin, J.B., Couture, G., & Evison, C. (2017). Acute kinematic and kinetic adaptations to wearable resistance during vertical jumping. *European Journal of Sport Science*, *17*(5), 555-562.

Massengale & Swanson. (1997). Current and future directions in exercise and sport science. *The History of Exercise and Sport Science*, 439-450.

Maupin, D., Schram, B., Canetti, E., & Orr, R. (2020). The relationship between acute: chronic workload ratios and injury risk in sports: a systemic review. *Open Access Journal of Sports Medicine, 11*, 51.

McLean, S., Kerherve, H.A., Stevens, N., & Salmon, P.M. (2021). A systems analysis critique of sport-science research. *International Journal of Sports Physiology and Performance, 16*(10), 1385-1392.

Milbrath, M., Stoepker, P., & Krause, J. (2016). Video analysis tools for the assessment of running efficiency. *Track and Cross Country Journal, 2*, 279-283.

Nicholls, S.B., James, N., Bryant, E., & Wells, J. (2018). The implementation of performance analysis and feedback within Olympic sport: The performance analyst's perspective. *International Journal of Sports Science & Coaching, 14*(1), 63-71.

O'Donoghue, P. (2006). The use of feedback videos in sport. *International Journal of Performance Analysis in Sport, 6*(2), 1-14.

Orange, S.T., Metcalfe, J.W., Liefeith, A., Marshall, P., Madden, L.A., Fewster, C.R., & Vince, R.V. (2019). Validity and reliability of a wearable inertial sensor to measure velocity and power in the back squat and bench press. *Journal of Strength and Conditioning Research, 33*(9), 2398-2408.

Paton, D.N., & Mackie, W.C. (1912). The liver in relation to creatine metabolism in the bird. *Journal of Physiology, 45*(1-2), 115-118.

Qilin, S., Xiaomei, W., Xiaoling, F., Yuanping, C., & Shaoyong, W. (2016). Study on knee joint injury in college football training based on artificial neural network. *RISTI - Revista Iberica de Sistemas e Tecnologias de Informacao, 2016*, 197-211.

Schoenfeld, B.J., Grgic, J., Ogborn, D., & Krieger, J.W. (2017). Strength and hypertrophy adaptations between low- vs. high-load resistance training: A systematic review and meta-analysis. *Journal of Strength and Conditioning Research, 31*(12), 3508-3523.

Shurley, J.P., & Todd, J.S. (2012). "The strength of Nebraska": Boyd Epley, Husker power, and the formation of the strength coaching profession. *Journal of Strength and Conditioning Research, 26*(12), 3177-3188.

Smarr, B.L., Aschbacher, K., Fisher, S.M., Chowdhary, A., Dilchert, S., Puldon, K., Rao, A., Hecht, F.M., & Mason, A.E. (2020). Feasibility of continuous fever monitoring using wearable devices. *Scientific Reports, 10*(1), 21640.

Smith, A.E., Fukuda, D.H., Kendall, K.L., & Stout, J.R. (2010). The effects of a pre-workout supplement containing caffeine, creatine, and amino acids during three weeks of high-intensity exercise on aerobic and anaerobic performance. *Journal of the International Society of Sports Nutrition, 7*(1), 10.

Smith, J., & Smolianov, P. (2016). The high performance management model: From Olympic and professional to university sport in the United States. *The Sports Journal, 21*(February), 1-12.

Sonda Sports. (n.d.). https://sportstechworldseries.com/

Sonkodi, B., Berkes, I., & Koltai, E. (2020). Have we looked in the wrong direction for more than 100 years? Delayed onset muscle soreness is, in fact, neural microdamage rather than muscle damage. *Antioxidants (Basel), 9*(3). doi.org/10.3390/antiox9030212

Thomas, D.T., Erdman, K.A., & Burke, L.M. (2016). Position of the Academy of Nutrition and Dietetics, Dietitians of Canada, and the American College of Sports Medicine: Nutrition and athletic performance. *Journal of the Academy of Nutrition and Dietetics, 116*(3), 501-528.

Tiller, N.B., Roberts, J.D., Beasley, L., Chapman, S., Pinto, J.M., Smith, L., Wiffin, M., Russell, M., Sparks, S.A., Duckworth, L., O'Hara, J., Sutton, L., Antonio, J., Willoughby, D.S., Tarpey, M.D., Smith-Ryan, A.E., Ormsbee, M.J., Astorino, T.A., Kreider, R.B., . . . Bannock, L. (2019). International Society of Sports Nutrition position stand: Nutritional considerations for single-stage ultra-marathon training and racing. *Journal of the International Society of Sports Nutrition, 16*(1), 50.

Trexler, E.T., Smith-Ryan, A.E., Stout, J.R., Hoffman, J.R., Wilborn, C.D., Sale, C., Kreider, R.B., Jäger, R., Earnest, C.P., Bannock, L., Campbell, B., Kalman, D., Ziegenfuss, T.N., & Antonio, J. (2015). International Society of Sports Nutrition position stand: Beta-alanine. *Journal of the International Society of Sports Nutrition, 12*(1), 30.

Wang, A., Healy, J., Hyett, N., Berthelot, G., & Okholm Kryger, K. (2021). A systemic review on methodological variation in acute: chronic workload research in elite male football players. *Science and Medicine in Football, 5*(1), 18-34.

Wax, B., Kerksick, C.M., Jagim, A.R., Mayo, J.J., Lyons, B.C., & Kreider, R.B. (2021). Creatine for exercise and sports performance, with recovery considerations for healthy populations. *Nutrients, 13*(6), 1915.

Williams, M.H. (1976). *Nutritional aspects of human physical and athletic performance.* Thomas.

Wilson, J.M., Fitschen, P.J., Campbell, B., Wilson, G.J., Zanchi, N., Taylor, L., Wilborn, C., Kalman, D.S., Stout, J.R., Hoffman, J.R., Ziegenfuss, T.N., Lopez, H.L., Kreider, R.B., Smith-Ryan, A.E., & Antonio, J. (2013). International Society of Sports Nutrition position stand: Beta-hydroxy-beta-methylbutyrate (HMB). *Journal of the International Society of Sports Nutrition, 10*(1), 6.

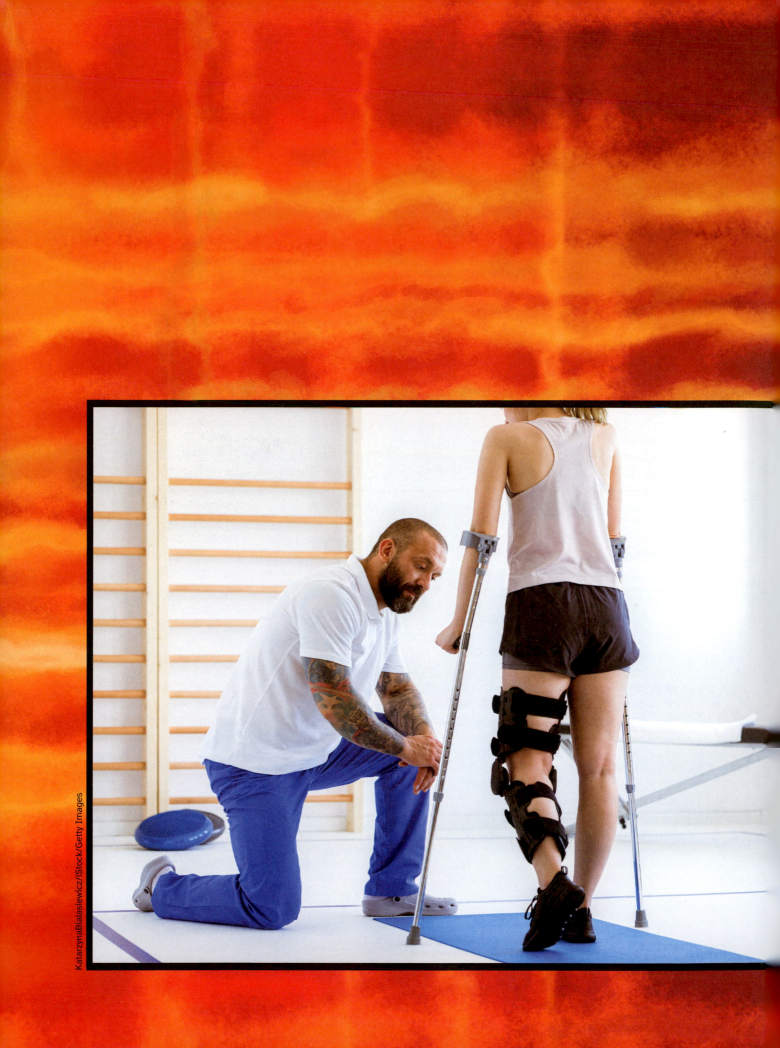

CHAPTER 11

Medicine and Allied Health

Chad Starkey and Julie Cavallario

CHAPTER OBJECTIVES

In this chapter, we will

- acquaint you with the wide range of professional opportunities and research in medicine and allied health;

- familiarize you with the purpose and types of work done by professionals in diagnosis management of a wide array of conditions; and

- review a variety of therapeutic interventions, including exercise, and present how contemporary research influences patient care to obtain optimal results.

While going in for a layup, a basketball player lands awkwardly, injuring her knee. The athletic trainer (AT) examines the player's knee and suspects that she has torn her anterior cruciate ligament (ACL). From here, the AT will guide the immediate care of this patient and refer her to a physician, physician assistant, or nurse practitioner for diagnostic imaging and, probably, surgery. After the injury, the AT will work with the patient to develop as much strength and function as possible prior to the patient's surgery. Following surgery, the AT and a physical therapist (PT) will develop an intervention program to enable the player to return safely to sport. Each step in this patient's care—clinical diagnosis, immediate care, medical diagnosis, prehabilitation, surgery, and rehabilitation—should incorporate evidence-based practice (EBP), integrating research, clinical experience, and patient values.

This scenario depicts situations faced by medical and allied health professionals who are employed in settings such as hospitals, laboratories, physician offices, clinics, and sports medicine facilities. Although their patients and clients might differ, these professionals—including ATs, clinical exercise physiologists, chiropractors, dentists, occupational therapists, physicians, physician assistants, nurse practitioners, and PTs—all incorporate some form of diagnostic evaluation, and many of these professions apply exercise and movement interventions to improve a person's physical functioning. All these professionals rely on exercise science and specialized clinical research to identify the most effective patient care (outcomes).

KEY POINT

Medical and allied health professionals have an ethical responsibility to evaluate the quality of their care, making sure to assess client or patient perspectives.

Many allied health professionals help people acquire skills and functions considered normal or expected levels needed for specialized fitness (**habilitative therapeutic exercise**) or to restore lost function (**rehabilitative therapeutic exercise**). Most medical and allied health professionals receive advanced graduate training beyond an undergraduate major such as exercise science that builds on their knowledge of human anatomy, human physiology, exercise physiology, biomechanics, and neurology. The scientific study of programmed exercise makes exercise science and kinesiology a preferred major for students seeking admission to graduate programs in medical and allied health professions. Exercise science also plays a key role in applied research on habilitative and rehabilitative therapeutic exercise in these professions.

Almost all medical and allied health professions have a diagnostic component and an exercise science–based intervention component to their practice, although the proportion of each can vary. Diagnostics identify the physiological, anatomical, or biomechanical limitations that must be addressed. Both clinical and applied research are important sources of EBP knowledge for diagnosing injuries and illnesses and prescribing interventions (treatment). Interventions strive to correct the problem, restore function, or proactively promote wellness.

Diagnostics in Medicine and Allied Health

In the most basic form, **diagnosis** is the process of collecting and analyzing information and then comparing the findings to norms to identify if a problem exists and, if so, the nature and potential causes of that problem. In the context of medicine and allied health, diagnosis is identifying a person's current health and well-being and matching the current level of function to the patient's needs and goals.

To determine a diagnosis, the clinician first conducts a patient interview (history taking) to identify the onset of illness or injury, any current symptoms and functional limitations and to obtain a personal and family medical history. The clinician then conducts a physical examination to assess the current level of physical function and orders laboratory, imaging, or clinical testing if needed. As the clini-

Research and Evidence-Based Practice in Exercise Science

Better Than a Coin Flip?

Many clinical diagnostic tests have a binary result: positive or negative, yes or no, heads or tails. In clinical diagnostics, we need to ensure that the conclusion provides us with a better answer than a flip of a coin.

The determination of diagnostic accuracy of these clinical tests is an example of EBP in the medical and allied health professions. Here, the result of a clinical test is compared to a **gold standard**, a procedure that best identifies the presence of a condition, using a 2 × 2 contingency table.

The best tests have a high proportion of TP and TN results. Those that have a high percentage of FN findings should be avoided.

Two of the most common statistics are sensitivity, which represents the proportion of TP results, and specificity, which represents the proportion of TN results. To be clinically useful, a test's sensitivity or specificity should be 70% or better—in other words, 20% better than a flip of a coin.

	CONDITION	
Clinical test	**Present**	**Absent**
Positive	True positive (TP)	False positive (FP)
Negative	False negative (FN)	True negative (TN)

cian works through the diagnostic process, a series of conditions or problems, known as the *differential diagnoses*, are ruled in or ruled out to derive the final diagnosis. The careful consideration of this large set of patient-specific information often is supplemented by specific tests that sequentially test various hypotheses of causes of disease or disability to make a diagnosis. Research provides the knowledge that helps professionals evaluate evidence gathered during the diagnostic process.

KEY POINT

Medical and allied health professionals use diagnostic techniques that are supported by research to make evidence-based recommendations to clients.

After the clinician derives a diagnosis, a plan of care is developed that identifies one or more interventions to correct the problem. On the medical side, such as with physicians or physician assistants, the intervention might be entirely pharmaceutical or might include a referral to an allied health care provider for additional interventions. Many professions described in this chapter might incorporate therapeutic agents (e.g., heat, cold, electrical stimulation),

manual therapy, and exercises into the patient's plan of care. The clinician then should assess the effectiveness of these interventions.

Patient or client input is a key part of EBP. Clinicians should use the results of their care to determine the effectiveness of interventions, and one way to determine intervention effectiveness is through patient outcome measures. **Patient outcome measures (POMs)** or **patient-reported outcome measures (PROMs)** are forms of EBP that is used to evaluate the patient's perceived response to the interventions used to treat their condition. Two types of POMs exist: disease or condition-specific and general quality of life measures.

- *Disease or condition-specific* outcome measures include the signs and symptoms associated with a given body part, injury, illness, or condition and how the condition affects the patient's function. An example of this type of POM is the Foot and Ankle Ability Measure (FAAM), which evaluates changes in self-reported physical function of patients with foot, ankle, or leg musculoskeletal injuries or disorders.

- *General* POMs evaluate a patient's quality of life, generalized experience with pain, and overarching functional limitations or disability.

Core Professional Competencies

The following are core professional competencies developed by the Health and Medicine Division (formerly the Institute of Medicine) of the National Academies of Sciences, Engineering, and Medicine for those who provide patient care (Greiner & Knebel, 2003):

- *Provide patient-centered care.* Efforts should be taken to educate and communicate with patients in a compassionate manner that includes advocating for the patient's needs and best interests. Caregivers should respect patients' differences, their values, and their need to improve function and overall quality of life.

- *Work in interdisciplinary teams.* Patients are treated by a variety of medical and health care personnel, and this team should communicate and work in unison in order to provide continuous and reliable care that ultimately improves patient outcomes.

- *Employ evidence-based practice.* The integration of clinician expertise, the best available research, and patient values in the delivery of care form the basis of EBP. Caregivers should incorporate patient care techniques that have been scientifically validated.

- *Apply quality improvement.* Caregivers should continually evaluate the health care delivery process and systems of care to decrease errors and improve the level and efficiency of care.

- *Use informatics.* Caregivers should gather, process, and analyze a range of data (e.g., electronic patient records, patient outcomes, satisfaction surveys) to support individual and operational decision making. Caregivers also should optimize the use of technology to communicate effectively and privately with other caregivers and the patient.

You probably will be introduced to some of these **core competencies** during your undergraduate education. You will become more fully immersed in them when you begin professional training and experience to enter a medical or allied health profession.

Hot Topic

Quality Improvement

While not new to health care, **quality improvement** (QI) is a current emphasis in medical and allied health professions. QI is a systematic approach to the analysis of professional practice performance, identifying areas that need improvement, and implementing the findings to improve. QI is not designed to answer formal research questions. It involves clinicians collecting and analyzing data to address patients' care concerns or errors that affect patients' well-being.

As an example, because of the patient volume in athletic training facilities, some clinics were reporting a high rate of bacterial infections among their patients. The resulting QI plan involved gaining baseline data by swabbing surfaces in four facilities and measuring the number of bacteria present. Three interventions were then implemented: (1) using surface disinfectants, (2) using posters to remind clinicians and patients of proper hygiene and checklists for facility cleaning, and (3) staff education and training sessions. Surfaces were then reswabbed to see if the prevalence of bacteria decreased, and patient records were reviewed to identify if any new facility-acquired infections occurred during the QI period.

Following the single-semester, three-phase QI effort, overall bacterial load in the facility was reduced by nearly 95%, and no cases of facility-acquired infections were found among patients during the period. This is an example of QI of processes performed at a facility to reduce patient risk while in the facility (LaBelle et al., 2019).

An example of a general POM is the Health-Related Quality of Life (HRQOL) questionnaire, a 14-item scale developed by the Centers for Disease Control and Prevention that provides an overview of health and well-being (Martin et al., 2005).

All POMs incorporate patient-rated scores that indicate the level of function considering the condition of interest. All POMS should have an identified **minimal clinically important difference** (MCID) score. The MCID score allows the clinician to distinguish how much of an increase in the score on the POM is needed to indicate true clinical improvement in the patient's condition (McGlothlin & Lewis, 2014). Clinicians then can use the POM to evaluate change over time in a patient under their care and determine if the plan of care being used has effectively improved the health and well-being of the patient.

Types of Interventive Exercise in Medicine and Allied Health

Many medical and allied health professionals design and implement exercise prescriptions to restore or improve motor function to a level that enables people to reach personal or occupational goals unencumbered by physical limitations. Medical and allied health professionals have additional training beyond undergraduate exercise science to prescribe specific exercises that address pathomechanical

deficits people have resulting from disease, injury, or disability. To develop therapeutic goals, clinicians must call on their knowledge of the effects of exercise on the muscular, nervous, skeletal, and cardiorespiratory systems and relate those effects to the patient's needs and expectations. Depending on the patient, the workplace, and the conditions being treated, therapeutic goals should be both short and long term and might include restoring muscular function and strength, joint range of motion, proprioception, cardiovascular and pulmonary function, or metabolic function so that the patient can participate in activities they deem important.

Reassessing "Disability"

Historically, health care and medicine tended to focus on what patients and clients were unable to do as the result of injury, illness, or inactivity. This focus on limitations, reflected in the term *disability*, carries a stigma that emphasizes the negative aspects of the person's state of well-being. In 2002 the World Health Organization (WHO) developed a system that places the emphasis on what the patient *can* do—the **International Classification of Functioning, Disability, and Health (ICF)** (WHO, 2002). The ICF system incorporates both the medical approach of resolving pathology and the social model of reducing a condition's negative effect on the person's life (figure 11.1). This model illustrates the relationship between function and disability as the result of the interactions between health conditions (injury and disease) and contextual factors. Examples of contextual factors include

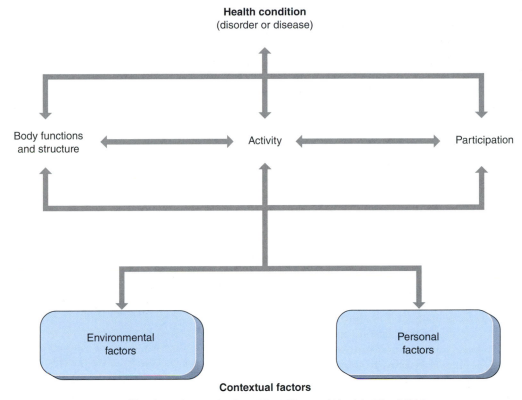

Health condition
(disorder or disease)

Body functions
and structure

Activity

Participation

Environmental
factors

Personal
factors

Contextual factors

Figure 11.1   International Classification of Functioning, Disability, and Health. The ICF focuses on a person's level of function and identifies interventions that maximize function and lead to a more individualized course of care. Thus, activity (function) is the core of the model.

Reprinted by permission from "Towards a Common Language for Functioning, Disability and Health: The International Classification of Functioning, Disability and Health," World Health Organization, accessed November 11, 2022, https://cdn.who.int/media/docs/default-source/classification/icf/icfbeginnersguide.pdf

- external environmental factors (e.g., social attitudes),
- physical characteristics of the places where the person lives and works, and
- personal factors (e.g., gender, profession, life experiences).

In the years to come, exercise science and related health professions will place more emphasis on documenting physical ability, identifying physical barriers to activity (e.g., issues of strength, range of motion, and cardiovascular limitation), and biological and biomechanical factors that are related to the risk of injury.

KEY POINT

Medical and allied health professions primarily focus on restoring lost function or disability; however, current best practice is to focus on a client's ability rather than disability.

Habilitative Exercise

Habilitation involves developing the processes and treatments that lead to the acquisition of skills and abilities that are normal and expected for someone based on their age and status (Olivares et al., 2011). The history of exercise science, kinesiology, and physical education is rooted in the concept of habilitative exercise. The research generated in these professions forms the framework for contemporary habilitative exercise theories and techniques.

The standards or expectations that signal a need for habilitation can differ vastly for people of the same age. For example, we would expect physical performance standards for lawyers to differ from those of professional athletes. A lawyer who is physically fit probably is not in need of habilitation; their state of fitness is desirable for good health, and special physical abilities are not necessary for any rigorous occupational needs. In contrast, a competitive athlete who is fit but lacks the advanced cardiorespiratory endurance or strength to perform the tasks needed in their sport is a candidate for habilitation that involves intensive conditioning, training, and other skill-specific exercise.

Medical and allied health professionals working in this area must understand the muscular and cardiovascular capabilities needed by people in these special groups. Often, the health care provider has a professional background in the specialty area. For example, a former baseball player with a degree in physical therapy might specialize in organizing habilitative exercise regimens for spring training camps. Such experience can help keep the exercise program in line with physical expectations for the stakeholders involved.

Habilitation can be applied to a variety of targeted populations. Although not recognized as such, specialized habilitation is the most common form of habilitative exercise. Specialized habilitation involves bringing specific groups of people in line with standards that exceed rather than merely meet those of the general population. Specialized habilitative exercise takes place in settings such as preseason sport training camps, military boot camps, and police and firefighter academies. Habilitative exercise also is used by overweight and obese populations. Exercise that is used to manage weight and body mass is considered habilitative because of its role in bringing people in line with established standards.

Children with physical developmental delays also are candidates for habilitative treatment because of the benefits of exercise. The goal of this type of habilitative exercise is to help the child adapt to, or compensate for, functional anatomical and physiological deficits. In certain cases, the underlying condition might have been surgically corrected. Other cases might involve teaching the child how to use a prosthetic device or how to perform basic skills such as rolling over, walking, or eating.

The general population also benefits from habilitative exercise. Allied health professionals play an important role in helping people achieve healthy standards of physical fitness by introducing them to exercise through hospital-based fitness centers, employer-based wellness centers, and community-based fitness and wellness centers. The use of habilitative exercise in this manner also can serve to benefit the mental and behavioral well-being of the participant. People who engage in regular physical activity tend to sleep better and suffer fewer emotional disorders than those who do not engage in such activity.

KEY POINT

"An ounce of prevention is worth a pound of cure." In this spirit, habilitative exercise focuses on developing many of the body's systems so that injury and disease are less likely. Exercise science is the field that leads in creating knowledge in habilitative exercise for medical and allied health professionals.

A variety of allied health professionals, with a range of education levels, engage in the delivery of habilitative exercise programming. ATs, PTs, occupational therapists, and physicians work cooperatively to develop specialized habilitative activities. For example, an occupational therapist might work in a secondary school setting developing habilitative

Hot Topic

Clinical Application of Habilitative Exercise and Exercise Science Research

One area in which exercise science specialists and researchers have excelled is in the realm of applying habilitative exercise to injury-prevention programming to reach optimal performance and mitigate injury risks associated with physically demanding occupations. The Warrior Model for Human Performance and Injury Prevention: Eagle Tactical Athlete Program (ETAP) (Sell et al., 2010a, 2010b) provides an example of researchers identifying the risks associated with active-duty military personnel performing daily job tasks and then developing and implementing habilitative programming to prevent these injuries from occurring. The first step of this process was to examine injury surveillance data (epidemiology) to identify prevalent injuries in the military population. The next phase involved biomechanical analysis to determine musculoskeletal and physiological demands of the respective roles within the target population, termed a *task and demand analysis*. Researchers then developed and validated habilitative interventions to address modifiable injury and performance factors identified during the task and demand analysis. Lastly, researchers monitored and determined the efficacy of the interventions. The goals of programs like this are to reduce the number of days missed due to injury and to reduce health care costs associated with physically demanding jobs. This is just one example of how exercise science researchers and clinicians can collaborate to apply habilitative exercise to benefit populations.

programs for students with developmental disabilities. This clinician would work to develop skills and adaptations for their students to participate in expected activities such as learning to grip a pencil.

Rehabilitative Exercise

Rehabilitation describes processes and treatments (interventions) that restore skills or function that were previously acquired but have been lost because of injury, disease, or behavior such as voluntary inactivity. Clinicians involved in rehabilitation must address muscular strength, endurance, or flexibility; neuromuscular coordination; cardiovascular efficiency; or other health and performance factors. In practice, exercise is aimed at improving or restoring the quality of life.

Physical activity such as sport, work, or activities of daily living (ADL) carries varying risk of injury. Likewise, inactivity, either as the result of injury, illness, or lifestyle, can cause detrimental effects to the musculoskeletal, cardiorespiratory, and nervous systems. Consider a patient who fractures a bone in the lower leg, and as a part of the plan of care the

leg is immobilized non-weight-bearing with a brace or cast for 4 to 6 weeks. In this scenario muscles of the leg will atrophy due to disuse. The lack of ability to walk, run, or do other cardiovascular activity will cause decreased cardiorespiratory efficiency. Therefore, the goal of orthopedic rehabilitation programs is twofold: to restore symptom-free movement and to restore the function of the cardiopulmonary system. Restoring the function of a limb consists of increasing joint range of motion, increasing muscular strength and endurance, reeducating neuromuscular pathways, and building cardiovascular endurance. Clinicians involved in this type of rehabilitation use both passive and active exercise to restore limb function. The next step involves muscular strength and endurance training, which protects the limb. These exercises can be augmented by various forms of heat, cold, electrical stimulation, therapeutic ultrasound, and manual therapy. Throughout this process, the clinician tries to maintain the patient's level of cardiovascular endurance. Most types of neuromuscular rehabilitation occur in outpatient physical therapy clinics, hospitals, and athletic training clinics.

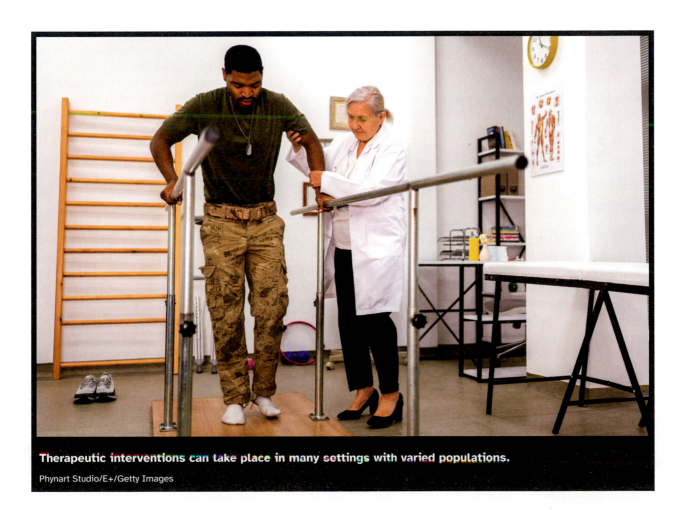

Therapeutic interventions can take place in many settings with varied populations.

Phynart Studio/E+/Getty Images

Like habilitation, rehabilitation is used with a variety of patients. Perhaps most commonly, rehabilitation is used to address musculoskeletal injuries. Many musculoskeletal injuries occur in the participation of physical activity or athletics, while others occur in older populations due to falls. Rehabilitation also is used after surgery. Although surgery is performed to restore a person's health and function, the process of surgery itself has detrimental effects on the body. The incision affects the involved muscles, soft tissues, and possibly bones and nerves. The decreased activity associated with recovery leads to decreased muscle mass and function and decreased cardiovascular efficiency. Although this effect is most evident when surgery involves the limbs, the heart and lungs also are affected. Just as skeletal muscle wastes away when it is not used, cardiac muscle responds similarly. Therefore, in most cases, people recovering from long-term disability must participate in neuromuscular rehabilitation to restore normal cardiopulmonary function.

A more specialized area in which rehabilitation is used is during cardiopulmonary rehabilitation. Diseases of the cardiopulmonary system include, but are not limited to, coronary artery disease, arrhythmia, hypertension, heart attack, and emphysema. Collectively, these diseases constitute the leading cause of death and long-term disability among adults in the United States. Undiagnosed heart disease also is a leading cause of death among young athletes (Maron et al., 2016). Working closely with a physician, a professional involved in exercise therapy participates in planning, implementing, and supervising physical activity programs that are designed to help restore individuals to normal function (Chang et al., 2004). Exercise specialists must be aware of certain conditions that can make rehabilitative exercise a potentially deadly activity.

Rehabilitation specialists need a thorough knowledge of the pathological aspects of injury and disease, of the limitations they impose on human performance, and of the types of treatments required to meet the patient's functional needs. Because people are more than just muscles and bones, rehabilitation specialists also must consider the psychological and social effects of the injury, the patient's personal goals and expectations, and the course of care leading to recovery. One example of rehabilitative exercise can be found in the regimen used by an athletic trainer to restore a football running back's injured knee so that he can return to competition. ATs, PTs, occupational therapists, and physicians are all graduate-degreed professionals who regularly deliver rehabilitative exercise programs. PT assistants, occupational therapy assistants, and exercise physiologists receive their professional education at the undergraduate or certificate level. These professionals deliver care under the supervision of other providers.

Providers of behavioral health care often prescribe exercise regimens as part of their patients' therapy

Professional Issues in Exercise Science

ACL Prehabilitation Programs

One of the most publicized sports injuries is tearing of the knee's ACL. While nonoperative care is an appropriate option for some patients, those anticipating return to high levels of physical activity often choose to undergo ACL reconstruction, a surgical procedure in which the ACL is rebuilt using tissue from either the patient's own body (autograft) or a cadaver-donated tissue (allograft). Several studies have examined the timing of surgical ACL reconstruction, and most recommend that surgery be performed at least 3 weeks after the initial injury, because earlier reconstruction can result in a complication known as *arthrofibrosis*.[1] The 3-week window between injury and surgery is the ideal time to implement a prehabilitation program. ACL prehabilitation programs focus on the development and maintenance of quadriceps and hamstring strength, the attainment of full knee range of motion, and reducing swelling in and around the joint. Studies indicate that 2 years after surgery, patients who participated in prehabilitation prior to surgery have higher return-to-sport rates, have increased self-reported knee function, and experience fewer long-term physical limitations including the need for additional surgeries.[2]

Citation style: AMA

1. Evans S, Shaginaw J, Bartolozzi A. ACL reconstruction—it's all about timing. *Int J Sports Phys Ther*. 2014;9(2):268-273.

2. Giesche F, Niederer D, Banzer W, Vogt L. Evidence for the effects of prehabilitation before ACL-reconstruction on return to sport-related and self-reported knee function: A systematic review. *PLoS One*. 2020;15(10):e0240192.

Dustin Grooms

Dustin Grooms had the opportunity to serve as an intern AT for a professional football team. During this experience, he assisted with the rehabilitation of a quarterback who had sustained a torn ACL. He noticed that although the intervention focused on the knee, subtle differences appeared in how the patient moved his leg. He began to wonder if injury to the ACL affects how the brain processes information to and from the knee.

While earning his doctorate in neuroscience, Grooms worked with colleagues to study how the brain processes information from the knee in people who have intact ACLs versus those who have torn an ACL. Using a special form of magnetic resonance imaging (MRI), the team identified definite changes in how the brain processes information following an ACL tear. The results of this research have led to improvements in rehabilitation (and subsequent function) for people who suffer ACL tears by reeducating and remolding their brains to function as they did prior to injury.

© Lauren Bowers

Dr. Grooms' research in neuroplasticity has helped shape both ACL injury-prevention programs and how physicians, ATs, and PTs form their ACL rehabilitation programs.

programs. The older population referred to in the preceding section also can gain psychological benefits from exercise. Elderly people who engage in physical activity exhibit fewer signs of depression, enhanced self-image, better physical health, and improved morale, all of which lead to a better quality of life (Knapen et al., 2014). Exercise also is included in many smoking and alcohol cessation programs, although scientific data have yet to show that this is effective (Ussher et al., 2019).

> **KEY POINT**
>
> Exercise is as important for the mind as it is for the body. Improving one's physical health through exercise also improves one's mental state.

Prehabilitative Exercises

No longer unique to the sports medicine setting, a specialized form of therapeutic exercise has emerged: prehabilitation. Whereas rehabilitation occurs following an injury or surgery, **prehabilitation** occurs following the injury but before surgery. When a physician deems that surgery is necessary for an orthopedic injury such as an ACL tear, part of the rehabilitation process must focus on restoring lost function. Prehabilitation involves developing as much strength, range of motion, and cardiovascular health as possible before surgery. As a result, the patient emerges from surgery at a higher baseline than if no prehabilitation was conducted. This

approach also is used in some cases to strengthen the heart and improve the circulation of people who will undergo cardiac surgery.

Roles and Scope of Practice for Medical and Allied Health Professions

Because exercise science majors develop a background in the anatomical and physical principles of human movement, they are well prepared to enter many medical and allied health care professions that use exercise in habilation or rehabilitation treatment regimens. These professions include athletic training, cardiac rehabilitation, dentistry, occupational therapy, medicine, physician assistantships, and physical therapy. This section briefly summarizes these professions, their scope of professional practice, and the educational requirements and credentials needed to practice them. Many of these professionals assist scholars with clinical research studies that contribute to the body of scientific evidence for practice and will be summarized in the last section of the chapter.

The professional roles or scope of practice discussed here often overlap. Indeed, many people pursue multiple credentials, such as athletic training and physical therapy (table 11.1). The need for multiskilled and multicredentialed individuals will increase as competition in the health care industry increases and the health care job market tightens. Many of the professions described in this section participate in both interprofessional practice and research.

Table 11.1 **Skill Matrix**

	Athletic trainer	Clinical exercise physiologist	Occupational therapist	Physical therapist	Therapeutic recreation specialist	Orthotists and prosthetists
Fitness assessment		X				
Analysis of function	X	X	X	X	X	X
Neuromuscular conditioning	X	X	X	X		
Diagnostic skill	X		X	X		
Rehabilitation						
Developing intervention programs	X	X	X	X	X	
Restoring range of motion	X		X	X		
Restoring strength	X	X	X	X		
Improving cardiovascular function	X	X	X	X		
Restoring activities of daily living	X	X	X	X	X	X
Restoring work or sport function	X	X	X	X	X	X
Technical design and fabrication						X

KEY POINT

Interprofessional practice involves multiple health care professions—physicians, PTs, and occupational therapists, for example—working together to improve the quality of patient care and support.

All medical and allied health professions require a postbaccalaureate education to enter the profession. PTs in the United States, for instance, must graduate from a clinical doctorate in physical therapy (DPT) degree program. In contrast, the entry-level education requirement for PT assistants and occupational therapy assistants is currently an associate's degree.

An undergraduate education in exercise science should provide you with comprehensive knowledge of human movement, the effects of exercise, and normal anatomical and physiological functions.

An undergraduate exercise science degree provides an ideal basis for advanced studies in many graduate-level medicine and allied health programs. You should always check the current educational requirements for any profession that you intend to pursue and use elective courses to fulfill any specific program requirements.

Medical and allied health professions are characterized by differing levels of regulation and required preparation. Therefore, health care professionals must pay close attention to state licensure requirements, especially when changing locations.

KEY POINT

The strong anatomical and scientific prerequisites in exercise science majors make them good preparatory degrees for graduate training in medicine and allied health professions.

Primary Medical Professions

Primary medical care is provided by licensed clinicians with professional doctorate degrees. Physicians receive medical doctor (MD) or doctor of osteopathy (DO) degrees, while dentists typically receive doctor of dental surgery (DDS) degrees. Completion of these degrees must be followed up by passing national board exams before these clinicians can practice. Outstanding kinesiology graduates who have taken additional courses that meet admissions requirements for these professional programs have become successful medical professionals. Kinesiology graduates also have been successful in admission, completion, and licensure as physician assistants (PA) and chiropractors (DC).

Physician

The American Medical Association recognizes MDs and DOs as physicians. Both types of physicians can diagnose diseases and injuries, prescribe medications, and perform surgeries. Although this is now more generalization than fact, the primary difference between MDs and DOs is philosophical—MDs diagnose and treat conditions and symptoms, whereas DOs use a more holistic approach by recognizing the interrelatedness of the body's systems. DOs also incorporate osteopathic manual medicine (OMM) into their plan of care. Over time, the original distinctions between these two forms of medicine have blurred with the development of patient-centered care. Physicians work collaboratively with, and might supervise or direct, many of the allied health professionals described in this chapter. Many times, physicians and allied health professionals perform additional training to specialize in a particular area of health care such as emergency medicine, geriatrics, internal medicine, pain management, or pediatrics.

Dentist

Physicians who specialize in the treatment of diseases and injuries to the mouth earn a doctor of dental medicine (sometimes referred to as *doctor of medical dentistry*) (DMD) or doctor of dental surgery (DDS) degree. These two degrees represent a historical nuance rather than a difference in practice. Although dentistry often conjures images of having a cavity filled, dentists are responsible for a wide range of medical and surgical procedures involving diseases or injuries to the mouth, gums, tongue, and salivary glands. X-rays or other imaging techniques are used to identify pathologies involving the teeth, jaw, and roof of the mouth.

The educational plan for dentists is similar to that of physicians. Dentists complete a 4-year dental degree program that includes both classroom and clinical education. Undergraduate courses in anatomy, physiology, mathematics, and psychology are common prerequisites. Prospective dental students should sit for the Dental Admission Test, which is a prerequisite for admission into dental school. Before they can practice, dentists must pass the written National Board Dental Examination. A clinical (practice-based) examination and state licensure also are required.

Hot Topic

Telemedicine

The COVID-19 pandemic illustrated the potential for **telemedicine** in all aspects of health care. To counter the risks inherent to in-person consultations, health care providers turned to a technology with which most students are familiar: videoconferencing. The use of videoconferencing to replace in-person meetings is especially meaningful in medically underserved communities. Clinicians can remotely assess the patient's current state of health and well-being, monitor the progress of interventions, and provide other forms of care without the patient being in physical proximity. Telehealth programs also allow clinicians to prescribe, monitor, and evaluate exercise programs by means of video and voice connections, web-based forms, and electronic monitoring devices.

Telehealth has demonstrated that patient outcomes are at least equally as good—sometimes even better—as in-person health care because of increased patient accessibility. For example, a meta-analysis compared patient outcomes for those patients who had in-person physician visits for diabetes and those who used telehealth visits. Those patients who were seen via telehealth had greater improvement in the management of their condition than those who had in-person visits only (Wu et al., 2018). The improvement was attributed to the increased frequency and duration of telehealth visits.

Physician Assistant

The physician assistant (PA) profession was founded in the mid-1960s to address a shortage of physicians in the United States. The first PAs were predominantly military medical corpsmen—medics—who were retrained to enter mainstream medicine in the United States (Carter, 2001). Initially, the official title of PAs was physician's assistants, with an emphasis on the apostrophe and *s*, because they literally assisted the physician. As their education and role evolved, their title changed to the current nomenclature. The removal of the apostrophe and *s* was more than a grammatical change; it signaled the ability to practice relatively free of physician supervision. Through this transition, the PA was reclassified from an allied health profession to a medical profession.

PA professional practice resembles that of physicians and is largely diagnostic in nature. The PA relies on history taking, physical examination, imaging (e.g., X-rays, CT scans), and laboratory work to arrive at a diagnosis. From this diagnosis the appropriate interventions can be identified.

Chiropractor

Chiropractic care is based on the premise that improperly functioning spinal joints and other spinal structures negatively affect other body systems, resulting in poor mechanics, pain, and decreased health. The base of care involves spinal adjustments to decrease pain and improve function, especially in the back and neck.

Radiographs (X-rays) are ordered routinely to obtain precise information on spinal alignment prior to care. Appropriate adjustments are then made to alleviate the patient's symptoms. Most chiropractors will follow up spinal adjustments with interventions such as heat, cold, electrical stimulation, ultrasound, massage, and flexibility and strengthening exercises to improve posture. Many employ a PT or an AT to implement the exercise portion of the patient's plan of care. Braces, orthotics, or shoe inserts might be prescribed to improve posture and function. Beware of chiropractic clinicians who try to stretch their scope of practice beyond the immediate musculoskeletal alignment and health of the spine.

Chiropractic education involves a 4-year professional doctoral degree. A bachelor's degree is required for admission into a doctor of chiropractic degree program, but some institutions might blend the undergraduate with the doctoral program to allow the degree to be completed in 6 years. Once in the chiropractic program, in addition to classroom and laboratory education, students engage in clinical rotations. Graduates receive the doctor of chiropractic (DC) degree and then must pass the four-part exam administered by the National Board of Chiropractic Examiners. These requirements and the continuing education needed to maintain licensure vary by state.

Allied Health Professions

The allied health professions described in this chapter might work under the direction of or in cooperation with medical professionals. This interprofessional cooperation forms a health care team that is focused on restoring the patient to the highest level of function.

Athletic Trainer

High school, college, and professional athletes become injured at staggering rates; for example, as many as one-third of the one million high school football players in the United States suffer an injury each year (King et al., 2015). Combining the excitement of athletics with the demands of health care, athletic training is the only profession that falls entirely within the realm of sports medicine. Athletic training is a health care profession that addresses prevention, risk management, clinical diagnosis, immediate care, treatment, and rehabilitation of inju-

---

Practice Guidelines

One type of clinical research that is commonly seen in medical and allied health professions is **practice guidelines**. Practice guidelines are systematically developed statements, based on assimilated high-quality evidence, that guide the decision making and actions to be taken by a clinician in a given set of patient circumstances. Practice guidelines aim to reduce error in the treatment of conditions or can be used to increase the quality of care provided. For example, the *Journal of the American Academy of Physician Assistant*s published a practice guideline relative to the screening of adolescent athletes for a condition called *hypertrophic cardiomyopathy* (enlarged heart) (Pydah et al., 2021). This practice guideline recommends that any adolescent athlete with previous cardiorespiratory symptoms or with a family history of cardiac conditions should be screened using a 14-point questionnaire, an electrocardiogram, and an echocardiogram prior to engaging in physical activity. Clinicians often look to their overseeing professional organization to release practice guidelines.

ries and other conditions experienced by athletes and other physically active people. An AT works in collaboration with or under the direction of a licensed MD or DO physician. Besides attending to the direct health care of athletes and others engaged in physical activity, an AT coordinates referrals to appropriate medical and health care specialists.

Traditionally, ATs have been employed by high schools, colleges and universities, and professional sport teams. Roles also exist for ATs in hospitals, sports medicine clinics, military clinics, industrial rehabilitation clinics, police or fire academies, performing arts medicine, and other allied medical environments. Thus, an AT's roles and responsibilities vary depending on the work setting and population being treated.

ATs seeking national certification are required to graduate from a master's degree program accredited by the Commission on Accreditation of Athletic Training Education (CAATE). Academic coursework is supplemented by clinical education that permits students to affiliate with athletic training clinics at high schools, colleges, and professional sport teams and gives them the opportunity to learn in private clinics and hospitals. Many ATs also go on to pursue postprofessional education in the field through additional graduate programs, residencies, or fellowships and might specialize in areas of athletic training practice such as orthopedics or pediatrics. National certification testing of ATs is conducted by the Board of Certification (BOC). BOC certification is required to work as an AT at a major college or university, with a professional sport team, or with the U.S. Olympic Committee. It is also required by 49 U.S. states as a prerequisite for practicing athletic training (Board of Certification, 2021).

Clinical Exercise Physiologist

Exercise supervised by a Certified Exercise Physiologist (ACSM EP-C) is an effective treatment for people with chronic cardiovascular, pulmonary, and metabolic diseases. ACSM EP-Cs administer and interpret fitness assessments and from those data develop appropriate exercise programs. ACSM EP-Cs administer cardiopulmonary exercise tests in hospital laboratories, administer fitness testing, and implement and deliver cardiovascular conditioning programs both for individuals who are apparently disease free and for those with medically controlled diseases such as diabetes.

The next level of exercise physiologist is the Certified Clinical Exercise Physiologist (ACSM CEP). Exercise testing such as graded exercise tests and exercise prescription and fitness counseling for those individuals with cardiovascular, respiratory, or metabolic disorders all fall in the ACSM CEP's domain.

The Registered Clinical Exercise Physiologist (ACSM RCEP) is the highest level of certification for exercise physiologists. A hospital might offer a cardiac rehabilitation exercise program for people recovering from heart surgery. The ACSM RCEP and the physician, working as a team, collect the client's medical history and evaluate their cardiopulmonary function. Using these data, the two clinicians develop a structured exercise program, whereupon the client works with an exercise specialist to progress through the program. The patient's progress is evaluated at regular intervals by the program director, physician, and exercise specialist. This type of exercise physiologist is also responsible for administering the rehabilitation center and educating the cardiac rehabilitation staff and is often engaged in research.

Epidemiological Data

Epidemiology encompasses the study of trends and patterns in the occurrence of injuries, illnesses, or conditions in a given population. By identifying these trends, clinicians can better predict who in a specific population is more susceptible to a condition occurring or has a better prognosis after a condition occurs. ATs often have relied on the epidemiological data collected through the National Collegiate Athletic Association (NCAA) injury surveillance system. ATs who practice in an NCAA institution's athletics program participate in the injury surveillance system by tracking data about injuries that occur within their patient populations as well as reporting the number of practices and games in which all patients potentially could have been injured. The injury surveillance system has allowed for a deeper understanding of the risks associated with specific sports, and the findings are published in the *Journal of Athletic Training*. For example, between 2014 and 2019 in men's football, the risk of injury was 9.31 per 1,000 exposures (i.e., situations in which a person participates in the activity), the most common injury was a concussion, and injuries were significantly more likely to occur during games as compared to practice (Chandran et al., 2021). Epidemiological data in all health professions allow clinicians to learn from past trends to make informed decisions about the risk, diagnosis, and management of conditions.

Occupational Therapist

An occupational therapist (OT) helps people with physical, emotional, or mental disability to restore or develop as much independence as possible in daily living and work throughout their lives. (The term *occupational* has roots in the word *activity*.) Some of the physical care rendered by OTs closely mirrors some of the rehabilitative exercises used by PTs. Indeed, in many rehabilitation centers, a PT and an OT work together on a single patient's case.

As with many health professions, the roots of occupational therapy can be traced to postwar rehabilitation of veterans; specifically, pioneers in the profession taught craft skills to soldiers with disability (Christiansen & Haertl, 2014). Today, OTs specialize in functional bracing and the modification of everyday items for the special needs of their patients. Functional bracing is the use of a supportive or assistive device that allows a joint to function despite anatomical or biomechanical limitations. Examples include knee braces and splints of various kinds, such as wrist and hand splints that allow people to eat with a fork. Occupational specialists also work with people to improve their concentration, motor skills, and problem-solving ability.

OTs can specialize in task-specific, work-related rehabilitation to help clients reacquire the motor and cognitive skills they need to return to work. This type of activity might range from teaching basic skills such as coordinated movement to specific skills such as hammering, typing, or driving a car. To assist people with disability, an OT might be called on to evaluate the layout of schools, homes, and workplaces and to suggest methods of eliminating functional barriers. An OT often employs a Certified Occupational Therapy Assistant (COTA) as a clinical assistant in carrying out rehabilitation plans.

OTs and occupational therapy assistants (OTAs) work in hospitals, rehabilitation centers, nursing homes, and orthopedic clinics and provide outpatient service in secondary schools and colleges. They instruct people with disabilities (e.g., muscular dystrophy, cerebral palsy, spinal cord injury) in the use

OTs assist those individuals with physical or mental challenges to maximize their ability to function independently or with moderate assistance.

Aldomurillo/E+/Getty Images

of adaptive equipment such as wheelchairs, walkers, and aids for eating and dressing. They also might design special equipment for home or work use and cooperate with employers or supervisors to modify the work environment for a patient.

OTs generally attain their education at the master's degree level, and doctoral-level programs also are an option. Certification or licensure as an OTA requires the successful completion of a 2-year program. Both OTs and OTAs must be licensed in the states where they practice. The basic requirement for licensure is to complete an accredited program and pass the National Board for Certification in Occupational Therapy examination. Those who pass the examination earn the title of Occupational Therapist Registered (OTR) or Certified Occupational Therapist Assistant (COTA).

Health Care Data Analytics and Visualization

Through the course of their normal operations, medical and allied health professionals collect a large amount of information. When stored separately, this information is useful only to the patient and the provider. When this information is pooled, it creates valuable data.

Data analytics is the science and process of extracting meaningful and useful knowledge from the database. When extracted in a purposeful manner, these data can identify health care trends, determine the accuracy of diagnostic techniques, ascertain the efficacy of various interventions and their outcomes, and provide predictive insights. Trends can be used for budget management and personnel needs. When data analytics is combined with data visualization (presentation of data in a graphical format) or infographics, findings can be communicated easily to the target audience.

To provide context to the findings, many health care data scientists have professional training in a medical or allied health field. Additional education in statistics, computer programming, and data mining techniques is needed to become a competent data analyst. Doctoral or specialized certificates are two routes to become a data scientist. An artistic or creative mindset is useful in developing data visualizations.

Physical Therapist

The PT is educated to provide rehabilitative care to a diverse patient population with a wide range of injuries, illnesses, and diseases. Patients include accident victims; people with low-back pain, arthritis, or head injury; and those with a congenital or acquired disease state such as cancer. A PT might treat a range of patients from infants to seniors with conditions such as joint injury, burns, cardiovascular disease, and neurological deficits. In addition, PTs often are in daily contact with other health professionals such as nurses and physicians.

> **KEY POINT**
>
> PTs administer the patient's program and perform the required functional evaluations. Physical therapy assistants function under a PT's supervision and assume much of the hands-on patient care.

PTs combine diagnostic tests with passive interventions, active interventions, and manual techniques such as joint mobilization to rehabilitate or habilitate their clients. The therapy used depends on the condition being treated. Therapy for orthopedic injuries relies heavily on resistance training and proprioception activities (i.e., activities that improve knowledge of the position, weight, and resistance of objects in relation to the body). In contrast, therapy designed to treat people with disease states and neurological conditions tends to emphasize cardiovascular aspects and neuromuscular control.

Entry-level education prepares students to enter the workforce as generalists, but once a person begins to practice professionally, the tendency is to begin to specialize. The employment settings for PTs and physical therapy assistants are as diverse as their patient base. Because they tend to specialize in their practice, PTs might work in specialized private clinics or serve specialized (e.g., pediatric or neurological) roles in hospitals, although some professionals still fill the generalist role.

About two-thirds of PTs work in a hospital, skilled nursing facility, health and wellness center, or clinic. PTs also might be in private practice or provide home care services. In addition, they can be employed by school systems, colleges, or universities to provide on-site services to students. Some PTs hold more than one job—one in private practice and one in a health care facility.

Entry-level physical therapy education takes place at the clinical doctorate level. Students can enter these doctoral programs after completing a bachelor's degree. Physical therapy programs are accredited by the Commission on Accreditation in Physical Therapy Education (CAPTE). Becoming a physical therapy assistant involves completing a 2-year associate's degree accredited by CAPTE. In many states, though not all, physical therapy assistants must be licensed in order to practice. Regardless of whether they are licensed, physical therapy assistants work under the supervision of a PT to carry out the prescribed protocol. Patients who are under the care of a physical therapy assistant must be reevaluated regularly by the PT. The national professional organization for PTs is the American Physical Therapy Association (APTA).

> **KEY POINT**
>
> PTs are educated as generalists but tend to develop specialties while in the workforce.

Clinical Prediction Rules

Clinical prediction rules (CPRs) are a type of research that combines clinical findings that statistically demonstrate predictability of an outcome. The outcome of a CPR could be diagnostic, therapeutic, or prognostic. Diagnostic prediction rules outline the patient presentation that most likely indicates the presence of a given condition or warrants specific diagnostic testing to determine the presence of a condition. Therapeutic CPRs outline the types of patients who most likely will benefit from certain treatment interventions. Prognostic CPRs describe the types of patient presentations who are most or least likely to recover from a condition within a specific time period. For example, a therapeutic clinical prediction rule published in the *Annals of Internal Medicine* indicates that a patient with low back pain that presents with four of five listed symptoms will have greater benefit from spinal manipulation and exercise (Childs et al., 2004). CPRs have varying levels of validation, and it is important that clinicians understand that some CPRs only work with very specific patient populations.

PTs are licensed to practice on a state-by-state basis, and the APTA offers voluntary certification for each specialty area. Licensed professionals also can seek advanced education in physical therapy via residency or PhD programs.

Orthotist and Prosthetist

Orthotist and prosthetist (OP) careers are a blend of health care, engineering, and fabrication. These individuals develop, build, and fit braces and artificial limbs to enable patients to improve function. Generally speaking, orthotists fabricate braces and splints, whereas prosthetists create and fit artificial limbs. An emerging area, pedorthics, specializes in foot care. The devices OPs create are referred to collectively as *appliances*. OPs tend to specialize or focus on a specific body area or limitation.

Often working in conjunction with a medical or allied health professional, the OP assesses the patient's limitations and desired goals. The appliance is then designed, typically using computer software, to fit the patient with a functionality to meet the patient's goal. Often, multiple appliances are needed to meet multiple goals. For example, one type of leg prosthetic might be required for walking and another for running.

OP education occurs at the master's degree level in an accredited program, with the typical academic program being 2 years long. The typical OP program includes courses on the materials used to develop orthotics and prosthetics. Other courses might include neurology, orthopedics, and biomechanics. Following graduation, the OP must complete a 1-year residency focused on orthotics or prosthetics. After completing the residency, the OP should become certified by the American Board for Certification in Orthotics, Prosthetics & Pedorthics. The examination

consists of multiple parts focusing on knowledge and skill. State licensure for OP is varied.

Applied and Translational Exercise Science Research in Medicine and Allied Health

In chapter 9 you learned about basic, applied, and translational research. Each medical and allied health profession described in this chapter is well positioned to engage in applied clinical research. Not all clinicians are clinical scholars, but all clinicians should incorporate applied contemporary research into their practice. Here we provide some examples of how research has influenced practice in medical and allied health professions.

ACL Injury Prevention

Epidemiologists found a relatively high rate of ACL injuries and identified that females have a much greater risk of ACL rupture than males, with the risk ranging from two to nine times more likely to occur in females. Other factors that have been studied include leg dominance, increased quadriceps angle, decreased intraarticular notch width, decreased ACL cross-sectional fiber area, increased valgus angulation during landing, and hormonal factors. Regardless of the cause, experts agree that females are at increased risk of ACL rupture, and in some cases, it has been demonstrated that females have worse outcomes following reconstruction (Tan et al., 2016).

These studies were conducted primarily outside of the clinical setting, and clinicians have applied the research findings to benefit their respective patient populations. Preventive programs have been developed by exercise science specialists to address landing mechanics and to strengthen knee stabilizers. These preventive programs have gone through dozens of iterations because it was found that laboratory research findings did not directly translate to a diverse population of athletes.

As physicians began to develop new surgical approaches, clinicians refined their rehabilitative techniques, which emphasized function in all planes of motion and addressed imbalances between dominant and nondominant leg strength, with special attention given to identifying and modifying landing mechanics. Advances in the clinical prevention and treatment of ACL rupture now are commonly accepted as best practices as a result of the research that has been done in this area.

Fall Prevention

While many exercise scientists specialize in athletic populations, many more are focused on other populations that can benefit from exercise science research. Older adults are at increased risk of falling due to decreased muscle strength, balance, and proprioception associated with the aging process. The increased incidence of falls is particularly concerning in populations over the age of 65 because of the corresponding decrease in bone density that also occurs in this population, increasing their risk of fracture and decreasing fracture healing rate. As a result, falls in those over the age of 65 can result in significant loss of mobility and independence, decreasing patients' longevity and quality of life. Studies have identified that, similar to our prior discussion of ACL ruptures, falls are caused by many factors, including drug interactions of multiple medications (polypharmacy), decreased visual acuity, environmental hazards (e.g., uneven walking surfaces), vestibular disturbances, and the prevalence of orthostatic hypotension (i.e., a drop in blood pressure while standing from sitting) in this population (Chang et al., 2004). As a result of this research, clinicians have developed multifactorial screenings and fall prevention programs to address the patient's unique impairments.

Biometric Data

The use of biometric data to address performance enhancement and injury or illness prevention is another example of applied research. **Biometrics** are measurements of an individual's physical or behavioral characteristics that are collected via wearable (smart watches) or consumable technology (ingestible pills) and then analyzed. Heart rate variability (HRV) is a common example of data collected by smart watches. HRV is the normal modulation of heart rhythms controlled by both the sympathetic and parasympathetic nervous systems. HRV has long been studied relative to its influence and diagnostic link to cardiovascular disease and mortality (Rajendra Acharya, 2006). More recently, research has sought to determine the relationship between HRV and athletic performance or, more specifically, training thresholds and training recovery. Exercise physiologists and exercise science experts apply this biometric research to adapt training and recovery schedules for athletes at all levels to improve performance (Dong, 2016).

Wrap-Up

Exercise science knowledge contributes to many medical and allied health professions, so exercise science graduates often successfully compete for graduate education in these fields. Although each medical and allied health profession has discrete scopes of practice, sufficient overlap encourages interprofessional practice. Almost all these professions have a diagnostic component, where the clinician collects information and compares it to normative or historic data, that you have been (or will be) taught in your exercise science program. Interventions then are used to restore or maximize the person's physical function via habilitative or rehabilitative exercise that might or might not be accompanied by other interventions. Many of these interventions are based on applied and translational research conducted by exercise scientists.

While they are directed by clinicians, interventions should incorporate the patient's needs and have clear goals that factor in all aspects of the patient's life and occupational circumstances. Just as you, as a student, can evaluate your instructors, patients must be provided the opportunity to evaluate their providers and the improvements made while receiving care.

Evidence-based practice should drive all facets of the patient's care from the initial episode to return to the highest possible function. Regardless of the medical or allied health specialty, research should inform the patient's care. Throughout this chapter we have presented examples of how research informs patient care and how individual medical and allied health specialties use said research in practice. Exercise scientists contribute the research that informs clinical practice and can be involved in the collaborative delivery of high-quality patient care.

More Information on Medicine and Allied Health

Organizations

American Academy of Orthotists and Prosthetists

American Academy of Physician Assistants

American Association of Cardiovascular and Pulmonary Rehabilitation

American Board of Physician Specialties

American Chiropractic Association

American College of Sports Medicine

American Dental Association

American Kinesiotherapy Association

American Medical Association

American Occupational Therapy Association

American Osteopathic Association

American Physical Therapy Association

Canadian Association of Occupational Therapists

Canadian Athletic Therapists Association

National Athletic Trainers' Association

National Strength and Conditioning Association

Journals

American Journal of Occupational Therapy

Annual in Therapeutic Recreation

Canadian Journal of Occupational Therapy

Clinical Kinesiology

International Journal of Athletic Therapy and Training

Journal of Athletic Training

Journal of Cardiopulmonary Rehabilitation and Prevention

Journal of Chiropractic Medicine

Journal of Clinical Exercise Physiology

Journal of Orthopaedic & Sports Physical Therapy

Journal of Prosthetics and Orthotics

Journal of Sport Rehabilitation

Journal of Strength and Conditioning Research

Journal of the American Academy of Physician Assistants

Physical Therapy

Physical Therapy in Sport

Physiotherapy

Physiotherapy Canada

Sports Health

Sports Medicine

Strength and Conditioning Journal

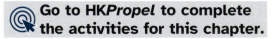

Go to HK*Propel* to complete the activities for this chapter.

Review Questions

1. Describe methods where exercise science research has influenced clinical patient care in medical and allied health professions.

2. In what ways is quality improvement in health care similar to research? How is it different?

3. Explain the difference between sensitivity and specificity of diagnostic tests.

4. Compare and contrast habilitation, prehabilitation, and rehabilitation and which populations each is used for.

5. Describe how exercise scientists and clinicians can use epidemiological data in their respective fields.

Suggested Readings

Bridges, D.R., Davidson, R.A., Odegard, P.S., Maki, I.V., & Tomkowiak, J. (2011). Interprofessional collaboration: Three best practice models of interprofessional education. *Medical Education Online, 16*(1). https://doi.org/10.3402/meo.v16i0.6035

Freeman, B.S. (2019). *The ultimate guide to choosing a medical specialty.* McGraw Hill Medical.

Maiorana, A., Levinger, I., Davison, K., Smart, N., & Coombes, J. (2018). Exercise prescription is not just for medical doctors: The benefits of shared care by physicians and exercise professionals. *British Journal of Sports Medicine, 52*(13), 879-880.

Topol, E.J. (2019). *Deep medicine: How artificial intelligence can make healthcare human again.* Basic Books.

References

Board of Certification (BOC). (2021). Map of state regulatory agencies. Board of Certification for the Athletic Trainer. Retrieved September 8, 2022, from https://bocatc.org/state-regulation/state-regulation

Carter, C. (2001). Physician assistant history. *Perspective on Physician Assistant History, 12*(2), 130-132.

Chandran, A., Morris, S.N., Powell, J.R., Boltz, A.J., Robison, H.J., & Collins, C.L. (2021). Epidemiology of injuries in National Collegiate Athletic Association men's football: 2014-2015 through 2018-2019. *Journal of Athletic Training, 56*(7), 643-650.

Chang, J.T., Morton, S.C., Rubenstein, L.Z., Mojica, W.A., Maglione, M., Suttorp, M.J., Roth, E.A., & Shekelle, P.G. (2004). Interventions for the prevention of falls in older adults: Systematic review and meta-analysis of randomised clinical trials. *BMJ, 328*(7441), 680.

Childs, J.D., Fritz, J.M., Flynn, T.W., Irrgang, J.J., Johnson, K.K., Majkowski, G.R., & Delitto, A. (2004). A clinical prediction rule to identify patients with low back pain most likely to benefit from spinal manipulation: A validation study. *Annals of Internal Medicine, 141*(12), 920-928.

Christiansen, C.H., & Haertl, K. (2014). A contextual history of occupational therapy. In B. Schell & G. Gillen (Eds.), *Willard & Spackman's Occupational Therapy* (pp. 9-34). Lippincott Williams & Wilkins.

Dong, J.G. (2016). The role of heart rate variability in sports physiology. *Experimental and Therapeutic Medicine, 11*(5), 1531-1536.

Greiner, A.C., & Knebel, E. (Eds.). (2003). The core competencies needed for health care professions. In *Health professions education: A bridge to quality.* National Academies Press. www.ncbi.nlm.nih.gov/books/NBK221519/

King, H., Campbell, S., Herzog, M., Popoli, D., Reisner, A., & Polikandriotis, J. (2015). Epidemiology of injuries in high school football: Does school size matter? *Journal of Physical Activity & Health, 12*(8), 1162-1167.

Knapen, J., Vancampfort, D., Moriën, Y., & Marchal, Y. (2014). Exercise therapy improves both mental and physical health in patients with major depression. *Disability and Rehabilitation, 37*(16), 1490-1495.

LaBelle, M.W., Knapik, D.M., & Voos, J.E. (2019). Impact of a quality improvement infection risk reduction program on pathogen presence in high school and collegiate athletic training room facilities. *Orthopaedic Journal of Sports Medicine.* https://doi.org/10.1177/2325967119S00410

Maron, B.J., Haas, T.S., Ahluwalia, A., Murphy, C.J., & Garberich, R.F. (2016). Demographics and epidemiology of sudden deaths in young competitive athletes: From the United States national registry. *American Journal of Medicine, 129*(11), 1170-1177.

Martin, R.L., Irrgang, J.J., Burdett, R.G., Conti, S.F., & Van Swearingen, J.M. (2005). Evidence of validity for the Foot and Ankle Ability Measure (FAAM). *Foot & Ankle International, 26*(11), 968-983.

McGlothlin, A.E., & Lewis, R.J. (2014). Minimal clinically important difference: Defining what really matters to patients. *JAMA, 312*(13), 1342-1343.

Olivares, P.R., Gusi, N., Parraca, J.A., Adsuar, J.C., & Del Pozo-Cruz, B. (2011). Tilting Whole Body Vibration improves quality of life in women with fibromyalgia: a randomized controlled trial. *Journal of Alternative and Complementary Medicine, 17*(8), 723-728.

Pydah, S.C., Mauck, K., Shultis, C., Rolfs, J., Schmidt, E., & Nicholas, J. (2021). Screening for hypertrophic cardiomyopathy. *JAAPA: Official Journal of the American Academy of Physician Assistants, 34*(10), 23-27.

Rajendra Acharya, U., Paul Joseph, K., Kannathal, N., Lim, C.M., & Suri, J.S. (2006). Heart rate variability: A review. *Medical & Biological Engineering & Computing, 44*(12), 1031-1051.

Sell, T.C., Abt, J.P., Crawford, K., Lovalekar, M., Nagai, T., Deluzio, J.B., Smalley, B.W., McGrail, M.A., Rowe, R.S., Cardin, S., & Lephart, S.M. (2010a). Warrior Model for Human Performance and Injury Prevention: Eagle Tactical Athlete Program (ETAP) part I. *Journal of Special Operations Medicine, 10*(4), 2-21.

Sell, T.C., Abt, J.P., Crawford, K., Lovalekar, M., Nagai, T., Deluzio, J.B., Smalley, B.W., McGrail, M.A., Rowe, R.S., Cardin, S., & Lephart, S.M. (2010b). Warrior Model for Human Performance and Injury Prevention: Eagle Tactical Athlete Program (ETAP) part II. *Journal of Special Operations Medicine, 10*(4), 22-33.

Tan, S.H., Lau, B.P., Khin, L.W., & Lingaraj, K. (2016). The importance of patient sex in the outcomes of anterior cruciate ligament reconstructions: A systematic review and meta-analysis. *The American Journal of Sports Medicine, 44*(1), 242-254.

Ussher, M.H., Faulkner, G.E.J., Angus, K., Hartmann-Boyce, J., & Taylor, A.H. (2019). Exercise interventions for smoking cessation. *Cochrane Database of Systematic Reviews*. https://doi.org/10.1002/14651858.CD002295.pub6

World Health Organization (WHO). (2002). *Towards a common language for functioning, disability, and health: The International Classification of Functioning, Disability, and Health*. WHO. Retrieved September 8, 2022, from https://cdn.who.int/media/docs/default-source/classification/icf/icfbeginnersguide.pdf

Wu, C., Wu, Z., Yang, L., Zhu, W., Zhang, M., Zhu, Q., Chen, X., & Pan, Y. (2018). Evaluation of the clinical outcomes of telehealth for managing diabetes: A PRISMA-compliant meta-analysis. *Medicine, 97*(43), e12962.

GLOSSARY

abduction—Joint rotation away from the midline of the body in the frontal plane.

action (muscle)—Neuromuscular activation of muscle resulting in tension and joint moments of force to influence body motion.

active tension—The tensile force resulting from use of chemical energy in the contractile proteins of muscle.

acute-chronic workload ratio (ACWR)—A ratio between acute and chronic volume or workload of training that could indicate the status of recovery and might be associated with risk of injury.

adduction—Joint rotation toward the midline of the body in the frontal plane.

adenosine triphosphate—Also called ATP, this molecule stores and transfers energy in the cells of the body.

aerobic—Metabolism in the presence of oxygen.

agonist—Anatomical classification of a muscle's action that tends to create a specific joint rotation.

amenorrhea—The absence of a menstrual cycle.

amino acids—Biochemical subcomponents of protein.

anaerobic—Metabolism in the absence of oxygen.

analysis of variance (ANOVA)—A statistical test used to determine the simultaneous equality of two or more sample means.

anatomical position—A standardized beginning body position used in anatomy to allow qualitative description of positions and motions.

anatomy—The study of the structure of the human body.

anemia—Condition marked by hemoglobin concentration below 12 grams per deciliter in women and below 13 grams per deciliter in men.

antagonist—Anatomical classification of a muscle's action that opposes a specific joint rotation.

anterior—Anatomical directional positions toward the front of the body.

anthropometrics—The science measuring the physical properties of the body and their relationship to movement.

applied research—Scientific research that focuses on questions about real-world application of knowledge.

applied science—Scientific research that focuses on questions about real-world application of knowledge.

ATP-PC system—An anaerobic energy system that serves as an immediate source of energy for the body.

attention—The capability to process specific information actively in the motor performance environment while tuning out other details.

autonomy—A sense of choice and control in an activity.

autonomy-supportive interpersonal style—Communication style in which a coach or instructor provides opportunities for choice within limits and explains the rationale behind activities.

basic science—Scientific research that focuses on theory and the ultimate mechanisms of reality or function.

biochemistry—Chemistry of living organisms.

biomechanical principle—A law, principle, or theory from physics or engineering that explains how forces produce an effect on an organism's movements or structure.

biomechanics—Subdiscipline of exercise science and kinesiology that applies mechanical principles to understand how forces affect the structures and function of living things.

biometrics—The collection and analysis of a person's physical and behavioral characteristics obtained via wearable or consumable technology.

Boolean operator—Simple logic words (e.g., *AND*, *OR*, *NOT*, or *AND NOT*) used to combine or exclude key words in an Internet search.

carbohydrates—The primary energy macronutrient for high-intensity exercise.

cardiac output—Amount of blood pumped out of one side of the heart per minute.

case-control studies—A type of observational study used to help determine what factors are associated with disease or outcomes that looks at two groups of individuals (one with the disease and another similar group without the disease).

case report—A type of study that provides an in-depth or intensive investigation of a single individual or specific group. Also referred to as a *case study*.

causality—The relationship between cause and effect.

coactivation—Neuromuscular activation of muscles on opposite sides of the joint.

coefficient of determination—The square of the correlation coefficient (r^2) documents the strength of association between two variables (percentage of variance accounted for in one variable by the other variable).

cohesion—Tendency for groups to stick together and remain united in pursuing goals.

cohort—A type of longitudinal study that follows a specific group of individuals over a period of time to evaluate a particular outcome.

concentric—A muscle action where activated muscles create a moment or torque greater than resistance creating segment rotation.

controlling interpersonal style—Communication style in which a coach or instructor uses rewards to control behavior and reduces support if exercisers make mistakes.

core competencies (health care)—Behaviors performed by medical and allied health care providers associated with evidence-based practice, patient-centered care, interprofessional education and collaborative practice, health information technology, and quality improvement.

cosmetic exercise—Evidence-based prescription of movements intended to reshape a person's body for aesthetic reasons.

critically appraised topic—A short summary of evidence on a topic of interest, usually focused around a specific clinical question.

crossover design—A research design in which participants receive both control and experimental conditions with time in between conditions.

cross-sectional—A type of study that looks at data for a population at a single time point.

degrees of freedom—The number of independent anatomical motions that must be controlled simultaneously during the execution of a motor skill.

deloading—Part of a long-term training strategy to decrease training stimulus for a time to allow recovery for an upcoming increase.

dependent variable—The outcome variable in an experiment.

detraining—The decrease in fitness or performance when training is stopped.

diagnosis—The process of collecting and analyzing information and then comparing the findings to norms to identify if a disease or injury exists and, if so, the nature and potential causes of that problem.

discipline—A formal body of human knowledge organized around a certain theme or focus.

dynamometer—An instrument used to measure muscular force, moment, or torque.

eccentric—A muscle action where activated muscles create a moment or torque that resists segment motion in the opposite direction.

ectomorph—An anthropometric body type describing a tall and thin person.

electromyography (EMG)—Instruments and methodology for amplifying, processing, and recording electrical signals in muscles to study their activation.

endomysium—Connective tissue that covers each muscle fiber.

epimysium—Connective tissue around each individual muscle.

evidence-based practice—A problem-solving approach that integrates the best available research, professional experience, and client values into practice.

exercise—A form of physical activity consisting of evidence-based prescription of structured and repetitive movements that sustain or improve specific health or fitness objectives. Integrated knowledge from kinesiology and exercise science is essential to exercise prescription.

exercise physiology—Subdiscipline of exercise science or kinesiology that focuses on the physiology of physical activity.

exercise science—Discipline or body of knowledge focusing on evidence-based exercise prescribed by physical activity professionals.

extension—Joint rotation in the sagittal plane that brings anterior surfaces farther away from each other.

external rotation—Joint rotation in the transverse plane whereby the anterior portion of a segment moves away from the midline of the body.

external validity—A form of validity related to the extent to which research results can be generalized to other contexts, situations, or groups.

fascicle—Distinct bundles of muscle fibers that combined make up an individual muscle.

fats—The energy macronutrient with the highest energy-per-unit mass.

feedback (intrinsic and extrinsic)—Information about movement that is provided to a learner either during or after a movement, which might derive from either an external source (e.g., instructor or video footage) or an internal source (e.g., muscles, joints, nervous system).

flexion—Joint rotation in the sagittal plane that brings anterior surfaces closer together.

focus group interview—An interview conducted with a small group of participants who share similar experiences or other commonalities.

force—A linear push or pull effect that acts between two objects and tends to create linear acceleration to modify motion or deform an object.

frontal plane—The anatomical plane of motion using both medial and lateral and superior and inferior directions.

function—In biomechanics, refers to a specific physical purpose or goal for an organism or one of its component parts to achieve.

glycogen—Storage form of glucose in the liver and skeletal muscles. Used to produce energy.

gold standard—The "best" method to determine the presence or absence of an injury or illness. In orthopedics the gold standard is often diagnostic imaging or surgery.

gray literature—Research and materials that are produced by organizations outside of the traditional or commercial publishing. Common types include reports, working papers, government documents, white papers, and evaluations.

ground reaction force—An equal and opposite force applied to a person by the ground, in response to the person pushing against the ground.

groupness—The extent to which group members perceive they are a unit and the degree to which they interact to form group roles and norms.

habilitation—Processes and treatments leading to acquisition of skills and functions that are considered normal and expected for an individual of a particular age and status.

habilitative therapeutic exercise—Exercise treatments leading to acquisition of skills and abilities that are considered normal and expected for an individual of a certain age, status, and occupation.

health-related exercise—Evidence-based prescription of movements to develop or maintain a sound working body and reduce the risk of disease for the purpose of healthy longevity.

health-related fitness—Fitness developed through physical activity experience and characterized by capacities and traits associated with low risk of hypokinetic disease. Common health-related fitness components are cardiovascular endurance, muscular strength, muscular endurance, flexibility, and body composition.

health-related quality of life—An evaluation of the goodness of the factors that can be affected by health and by health interventions, such as physical, social, and emotional functioning.

hemoglobin—Iron-containing protein found in red blood cells that carries most of the oxygen in the blood.

holism—Philosophical belief of the interdependence of mind, emotion, body, and spirit. A human is a synergistic whole, not a separate body and mind.

human movement biomechanics—The study of the effects of forces on the structures and functions of the human body and its components.

hyperextension—Joint extension in the sagittal plane beyond the anatomical position or some typical end range of motion.

hypertrophy—Increase in muscle mass due to enlargement of muscle fibers.

hypothermia—Condition in which body temperature falls below 95 °F (35 °C), which causes heart rate and metabolism to slow and can be life threatening.

incidence—The number of people in a given population who develop a specific disease (new case) during a specified period such as a month or year, usually expressed as person-years.

independent variable—The predictor variable. In an experiment, this is the variable that is of interest to, controlled by, or manipulated by the researcher.

inferior—Anatomical directional positions toward the feet.

internal rotation—Joint rotation in the transverse plane whereby the anterior portion of a segment moves toward the midline of the body.

internal validity—A form of validity related to the level of confidence that results of research are not biased or influenced by other factors or variables.

International Classification of Functioning, Disability, and Health (ICF)—Devised by the World Health Organization, this model frames a person's condition in the contextual factors such as external environment, physical barriers, societal influences, and personal factors.

intraclass correlation coefficient—A correlation coefficient that assesses the consistency between multiple measurements of the same variable.

isokinetic—Muscle action at a joint involving nearly constant joint angular velocity on an isokinetic dynamometer.

isometric—A muscle action where activated muscles create a moment or torque equal to a resistance moment and therefore does not result in change in joint angle.

isotonic—Involving muscle action to move a constant external (usually gravitational) resistance.

kappa (K) coefficient—A statistic used to estimate the degree of agreement between categorical variables, indicating the consistency of classification.

kinematics—Branch of mechanics that describes or measures motion.

kinesiology—Discipline or body of knowledge that studies all human physical activity.

kinetics—The branch of mechanics that explains how forces and moments cause motion.

lateral—Anatomical directional positions away from the midline of the body.

learning—Permanent alteration in nervous system functioning that enables performers to achieve a predetermined goal consistently. Three major domains of learning are knowledge, skills, and attitudes or values.

life span motor development—Examines the changes observed in movement during childhood and adolescence and throughout the aging process.

ligament—Connective tissues that connect bones at joints.

linked segment model—A biomechanical model of a series of rigid body segments linked by frictionless joints.

literature review—A comprehensive summary of the previous research on a topic. Also called *narrative review.*

macronutrients—The energy-producing nutrients of carbohydrates, fats, and proteins in food.

maximal oxygen uptake ($\dot{V}O_2$max)—Highest rate of oxygen uptake during heavy dynamic exercise.

mean—A measure of central tendency of a distribution of scores; the arithmetic average obtained by dividing the sum of scores by the number of scores.

medial—Anatomical directional positions toward the midline of the body.

median—A measure of central tendency of a distribution of scores; the middle score of a set of ordered scores.

memory drum theory—Theory developed by Franklin Henry (1960) proposing that rapid and well-learned movements are not consciously controlled but are run off automatically (as an older computer uses a memory drum to store and retrieve data).

mental skills training—The consistent practice of various psychological skills to enhance sport performance and enjoyment.

mesomorph—An anthropometric body type describing a muscular and moderate-sized person.

meta-analysis—A kind of systematic review of previously published research that mathematically combines their results into a net effect and its variability.

metabolic equivalent of task (MET)—Units of energy expenditure based on resting metabolic rate.

metabolic rate—Rate at which the body uses chemical energy.

micronutrients—Organic compounds in food needed in small amounts for metabolism, including vitamins, minerals, and trace elements.

minimal clinically important difference (MCID)—The smallest change in a patient-reported outcome measure that signifies improvement or worsening of the patient's function.

mode—A measure of central tendency; the most frequently occurring score in a distribution.

moment of force—The tendency of an off-center force to create angular acceleration to modify rotation.

motivation—The direction and intensity of one's effort.

motivational orientation—A person's reasons for doing an activity, with regulations ranging from less self-determined to more self-determined.

motor behavior—A broad subdiscipline of exercise science and kinesiology focused on motor learning, motor control, and life span motor development.

motor equivalence—The capability of the motor control system to enable a person to perform a motor skill by adapting another skill, with similar characteristics, to the specific conditions (see *transfer*).

motor performance—The degree to which someone can demonstrate or perform a motor skill at any given time.

movement technique—The sequence of limb and body motions used to accomplish a particular goal.

muscle action—The mechanical effect of activated muscle to contribute to movement by acting to either stabilize (isometric, or act as a strut), shorten (concentric, or act as a motor creating motion), or lengthen (eccentric, or act as a brake on other forces). Many biomechanics scholars recommend this term rather than "contraction."

muscle glycogen—Stored glucose.

muscle group—A set of muscles classified to tend to create the same joint rotation based on anatomy (e.g., quadriceps and hamstrings).

muscular endurance—Ability of a muscle to exert force repeatedly over a prolonged period.

myofibril—The part of a muscle fiber that contains contractile elements.

myoglobin—A protein found in muscle tissue that assists storage and transport of oxygen.

myoplasticity—The adaptability of physical characteristics of skeletal muscle to physical activity and exercise stressors.

normal curve—A probability distribution that is perfectly symmetrical and known to have certain properties.

norms—Shared beliefs and behaviors of a team or group.

novel learning task—Movement task with which the participant does not have prior experience and that usually involves simple movement (e.g., linear positioning or tracking).

nutrition analytics—Collection and advanced statistical analysis of dietary data.

nutritional ergogenic aid—A nutritional supplement, food, or diet that has the potential to enhance performance.

odds ratio—A measure of the association of the exposure–disease relationship typically employed in cross-sectional and case-control studies, commonly calculated as a ratio of the odds of a disease occurring in the exposed group (e.g., inactive people) versus the odds of the disease in the nonexposed group (e.g., active people).

outcomes research—A broad umbrella term related to the study of results of treatments and interventions on health care practices, patients, and populations.

overload—A training principle that exercise prescribed must be greater than the client's current fitness level to elicit improvement.

oxygen debt—Oxygen consumed above resting levels following the cessation of exercise.

parallel—Muscle structure in which fibers and fascicles are in line with the tendon.

passive dynamics—The transfer of forces across joints in linked-segment biomechanical systems.

passive tension—The tensile force of muscle resulting from the stretch of connective tissue elements.

patient outcome measure (POM) or patient-reported outcome measure (PROM)—A method of data collection and evaluation of the patient's experiences and efficacy of the intervention program.

pennate—Muscle structure in which fibers and fascicles are angled into the aponeurosis or tendon on the long axis of the muscle.

perceived competence—One's belief about their ability in a life domain (e.g., social or physical) or subdomain (e.g., friendships or yoga).

percentile rank—A value that indicates the percentage of scores below a given score.

perceptual-motor integration—The use of sensory information during a movement.

perimysium—A layer of connective tissue that surrounds each fascicle.

periodization—A long-term plan to adjust training stimulus to maximize performance during athletes' competitive season.

physical activity—Movement that is voluntary, intentional, and directed toward an identifiable goal.

physical activity epidemiology—The study of the distribution and determinants of physical activity, its associations with health-related outcomes, and the application of this study to disease prevention and health promotion.

placebo-controlled trials—A research design, typically in medicine or nutrition studies, that includes two or more groups, with one group blindly receiving a nonexperimental treatment (placebo) and the other group(s) blindly receiving the experimental treatment.

plasma volume—Volume of extracellular fluid in the blood.

post hoc tests—A set of comparisons between group means; usually used after a significant finding in an analysis of variance.

posterior—Anatomical directional positions toward the back of the body.

power—The rate of doing mechanical work.

power training—A strength and conditioning strategy referring to exercises designed to increase both movement force and speed.

practice guidelines—Systematically developed statements that guide the decision making and actions to be taken by a clinician in a given set of patient circumstances.

prehabilitation—Therapy that occurs following an injury but before surgery in order to develop as much strength and range of motion as possible.

prevalence—The number of people in a given population who have a disease at a particular point in time, usually expressed as a percentage of the population.

primary performance outcomes—Any type of sport performance metric that directly translates to competitive events.

principle of specificity—Training adaptations or changes in fitness are specific to the type of training performed.

progression—The gradual increase of training stimulus (overload) to ensure long-term improvement in fitness or performance. Three components of training stimulus are duration, frequency, and volume.

progressive overload—Gradual increases in weight, repetitions, or frequency to maintain a stress on the muscular system to stimulate training adaptations.

protein—An essential macronutrient needed for growth and tissue repair.

psychological inventory—Standardized measure of a psychological construct, typically included as part of a questionnaire.

qualitative data—Information gathered in narrative form, typically resulting in rich, detailed descriptors.

qualitative research—A research strategy that focuses on collecting, analyzing, and interpreting nonnumerical data such as language.

quality improvement (QI)—A systematic approach to the analysis of professional practice performance that involves identifying areas that need improvement; implementing policies, procedures, and practices needed to rectify the problem; and reassessing to identify if the problem has been solved.

quantitative data—Measurements characterized as numbers or values.

quantitative research—A research strategy that focuses on collecting and analyzing numerical data to describe characteristics, find correlations, or test hypotheses.

randomized controlled trial (RCT)—One of the most rigorous prospective research designs in which participants are randomly assigned either into experimental treatments or interventions or into control groups.

reaction time—Speed of response to a light or sound, as when pressing a single button after seeing a signal (simple) or choosing one of multiple buttons to press depending on which signal one sees (choice).

receptor—Specialized nerve ending that is found at the end of a sensory neuron and detects changes in the environment.

rehabilitation—Physical treatment, exercise, and educational or counseling sessions that lead a person to regain function and a personally acceptable level of independence.

rehabilitative therapeutic exercise—Processes and treatments designed to restore skills or functions that were previously acquired but have been lost due to disease, injury, or behavioral traits.

relatedness—A sense of connectedness with others in an activity.

relative risk (hazard ratio)—A measure of the association of the exposure–disease relationship typically employed in cohort studies, commonly calculated as a ratio of the probability of a disease occurring in the exposed group (e.g., inactive people) versus the probability of the disease in the nonexposed group (e.g., active people).

research—Systematic inquiry that gathers data with the goal of establishing generalizable inferences and knowledge.

retention—The testing of motor skills following an interval of time after instruction and practice to determine if learning has occurred.

reversibility—A training principle referring to the loss of improvements in fitness once physical activity or exercise levels decline (see *detraining*).

role—Either formal or informal, the required or expected behavior of a member of a team.

sagittal plane—The anatomical plane of motion using both anterior and posterior and superior and inferior directions.

sarcopenia—Loss of skeletal muscle.

self-efficacy—One's belief in their ability to succeed in a situation-specific task.

self-esteem—A global evaluation of one's value as a person.

self-perception—A global term used to describe an evaluation of oneself or one's ability.

septa—Connective tissue between muscles and muscle groups.

serial order—The sequencing and timing of individual movements while performing an action or motor skill.

Six Thinking Hats technique—Provides guidance for examining a question from six different perspectives (i.e., hats), which can help identify and avoid potential confirmation bias.

skill-related fitness—Fitness developed through physical activity experience that is characterized by development of motor skill competence. Common skill-related fitness components are agility, balance, coordination, reaction time, and speed.

specificity—Strength and conditioning principle that training should be as similar as possible to the specific characteristics of the sport or training goal.

sport analytics—Study of sport data using advanced statistics, data visualization, and data management, which sport scientists use to inform coaches and athletes about training and strategy.

sport performance—Multidisciplinary professional services and research using exercise science–related subdisciplines such as strength and conditioning, sport nutrition, sport science, and data analytics.

standard deviation (S)—The average deviation of each score in the data set from the mean of the data set.

standard error of estimate—Standard deviation of prediction errors associated with a given regression equation.

strength (muscular)—Maximal moment of force exerted by a muscle group, typically measured in isokinetic conditions.

stroke volume—Amount of blood pumped out of one side of the heart per beat.

structure—In biomechanics, refers to the physical characteristics (e.g., shape) and composition of a body or one or more of its components.

superior—Anatomical directional positions toward the head.

synergy—The combined, cooperative activation of muscles to effectively complete a motor task.

systematic review—A comprehensive review of scholarly literature that uses repeatable methods to find, select, and synthesize all available evidence for a particular topic or question.

telemedicine—Health care and medical services provided using teleconferencing or telephone calls, thereby negating the necessity for the patient and caregiver to be physically present.

tendons—The connective tissue that connects muscle to bone.

therapeutic exercise—Evidence-based prescription of specialized and individualized movements performed to restore or develop physical capacities that have been lost due to injury, disease, behavioral patterns, or aging.

tidal volume—Amount of air inhaled or exhaled per breath.

torque—Common English term for "moment of force" that is the rotation effect of a force about an axis of rotation.

training exercise—Evidence-based prescription of movements performed for the express purpose of improving athletic, military, work-related, or recreation-related performance.

transfer—The testing of motor skills that is different from the skills practiced or performed in the practice context or situation to determine if learning has occurred.

transverse plane—The anatomical plane of motion using both medial and lateral and anterior and posterior directions.

true experimental—An experiment conducted where the researcher manipulates the independent variable to observe its effect on some behavior or condition.

underwater weighing—Procedure in which a person's body weight is measured while they are completely submerged in water in order to determine body volume for body composition assessment.

variation of training—Usually refers to minor variations planned in exercise programs to improve psychological and physiological responses. Additionally, a genetic element modulates individuals' responses to training.

vector—A complex variable requiring measurement of both size and direction.

ventilation—Process in which gases are exchanged between the atmosphere and the alveoli of the lungs.

ventilatory threshold—Point during a graded exercise test at which ventilation begins increasing at a faster rate than $\dot{V}O_2$ does.

wash-out period—The length of time in a randomized controlled trial when patients are not receiving any treatment before the start of the study.

wearable sensor technology—Small sensors attached to an athlete's body, equipment, or apparel that provide important performance measurements.

Z-score—A score expressed in standard deviation units. Z-scores have a mean of 0 and a standard deviation of 1.

INDEX

Note: The italicized *f* and *t* following page numbers refer to figures and tables, respectively.

ABOUT THE EDITOR

Duane V. Knudson, PhD, FNAK, FISBS, FACSM, RFSA, is a Regents' Professor and University Distinguished Professor in the department of health and human performance at Texas State University, where he teaches biomechanics and research methods. He earned his doctorate at the University of Wisconsin–Madison and has held tenured faculty positions at three universities. His research areas are in the biomechanics of tennis, stretching, qualitative movement diagnosis, the learning of biomechanical concepts, and research quality.

Knudson has authored more than 168 peer-reviewed articles, 26 chapters, and three books: *Fundamentals of Biomechanics*, *Qualitative Diagnosis of Human Movement*, and *Biomechanical Principles of Tennis Technique*. He also coedited the top-selling text *Introduction to Kinesiology*.

He has received numerous state, regional, national, and international awards for his research and leadership, and he has been elected fellow of four scholarly societies, including the prestigious National Academy of Kinesiology. He has served as department chair, associate dean, and president of the American Kinesiology Association and as president of the International Society of Biomechanics in Sports. Photo courtesy of Texas State University.

ABOUT THE CONTRIBUTORS

Scott M. Battley, MS, CSCS, SCCC, CPSS, is a sport scientist within the athletic department at Texas A&M University. As a sport scientist, he helps manage and interpret the data that is obtained during the athlete monitoring process for multiple teams. He is also a doctoral student within the department of kinesiology and sport management. Battley holds the Certified Strength and Conditioning Specialist & Certified Performance and Sports Scientist designation from the National Strength and Conditioning Association and is certified as a strength and conditioning coach by the Collegiate Strength and Conditioning Coaching Association. He received a bachelor of science in applied exercise physiology with a minor in sport management in 2013 and a master of science in sport physiology in 2014 from Texas A&M University. Photo courtesy of Texas A&M University.

Jennifer L. Caputo, PhD, CSCS, is a professor of exercise science at Middle Tennessee State University. She teaches undergraduate and graduate courses in exercise science, along with directing theses and dissertations. She is a founding editor of the peer-reviewed journal *Educational Practices in Kinesiology*. Dr. Caputo holds the Certified Strength and Conditioning Specialist designation from the National Strength and Conditioning Association (NSCA) and is a licensed medical bone densitometer operator. She received her doctoral degree in exercise physiology from the University of North Carolina at Greensboro; she also holds a master's degree in sport and exercise psychology. Photo courtesy of Jennifer Caputo.

Julie Cavallario, PhD, ATC, is the graduate program director of the master of science of athletic training program and an assistant professor in the department of rehabilitation sciences at Old Dominion University in Norfolk, Virginia. Dr. Cavallario has previously worked clinically as an athletic trainer at the collegiate level and as a staff member of the Commission on Accredi-

tation of Athletic Training Education (CAATE). She has authored a multitude of peer-reviewed journal articles in athletic training education and sports medicine. Photo courtesy of Old Dominion University.

Broderick L. Dickerson, MS, CSCS, CISSN, is a doctoral student and graduate teaching assistant at Texas A&M University. He is also a research assistant in the Exercise & Sport Nutrition Laboratory and has assisted with numerous research projects that entail nutritional supplementation, competitive gaming, and exercise in acute and chronic settings. Dickerson holds the Certified Strength and Conditioning Specialist designation from the National Strength and Conditioning Association and the Certified Sports Nutritionist certification from the International Society of Sports Nutrition. He received his bachelor's and master's in kinesiology with an emphasis on human performance from Stephen F. Austin State University. Photo courtesy of Texas A&M University.

James L. Farnsworth II, PhD, LAT, ATC, is an assistant professor at Texas State University. He teaches courses in research methods, statistics, evidence-based practice, and rehabilitation; he also directs theses and culminating projects in athletic training. Dr. Farnsworth is a licensed athletic trainer in Texas and a measurement specialist. He received his doctoral degree in kinesmetrics from Middle Tennessee State University; he also has a master's degree in postprofessional athletic training from Ohio University and a bachelor's degree in athletic training from the University of North Carolina at Wilmington. Photo courtesy of James Farnsworth.

Drew E. Gonzalez, MS, CISSN, SCCC, CSCS,*D, TSAC,*D, EP-C, is a doctoral student and graduate teaching assistant within the department of kinesiology and sport management at Texas A&M University. Gonzalez is also a research assistant in the Exercise & Sport Nutrition Laboratory and has assisted with and led several research projects with respect to sport nutrition, pharmacokinetics, exercise training, and tactical athlete nutrition and health. Gonzalez holds the Certified Strength and Conditioning Specialist and Tactical Strength and Conditioning Facilitator certifications, both with distinction, by way of the National Strength and Conditioning Association (NSCA). In addition, he is certified as a sports nutritionist through the International Society of Sports Nutrition, is certified as a strength and conditioning coach by the Collegiate Strength and Conditioning Coaching Association, and is certified by the Society for NeuroSports. Gonzalez completed his bachelor's and master's degrees at Texas State University in exercise science. Photo courtesy of Texas A&M University.

Xiangli Gu, PhD, is an assistant professor of kinesiology at the University of Texas at Arlington, where she is the director of the Movement and Physical Activity Epidemiology Laboratory. She has strong interests in promoting physical activity and motor performance across the human life span. Her research has taken a programmatic approach to understand health disparities through the behavioral and neuropsychological levels of assessment, as well as interventions to increase physical activity and motor performance. Her research has received over $1 million in funding from agencies and organizations such as the United States Department of Health and Human Services, Department of Labor, private foundations, and SHAPE America. She has published more than 70 peer-reviewed journal articles and five book chapters, and she has delivered over 120 research presentations at international and national conferences. She has served as section editor for the journal *Measurement in Physical Education and Exercise Science.* As one of the Texas Association for Health, Physical Education, Recreation & Dance (TAHPERD) Scholars of the Year, she has also served as the chief editor for *TAHPERD Journal.* She was inducted as research fellow of the Society of Health and Physical Educators (SHAPE America) in 2017 and received the SHAPE America Southern District Scholar Award in 2020 and Joy of Effort Award in 2022. She has served on the advisory boards for the CDC (Division of Population Health) and SHAPE America to develop the National School Recess Guide and Strategies. She is a huge advocate of physical activity as a preventive measure and embraces evidence-based decision-making in prac-

tice, research, and policy by collaborating with local school and community partners. Photo courtesy of The University of Texas at Arlington.

Lindsay E. Kipp, PhD, is an associate professor at Texas State University in the department of health and human performance. She teaches undergraduate and graduate courses related to sport and exercise psychology. She also supervises independent studies and master's theses. Dr. Kipp's research primarily focuses on positive youth development through sport and physical activity. She received her doctoral degree in kinesiology with a specialization in sport and exercise psychology from University of Minnesota. Her master's degree is also in kinesiology with a specialization in sport and exercise psychology from Illinois State University. Photo courtesy of Texas State University.

Richard B. Kreider, PhD, FACSM, FISSN, FACN, FNAK, serves as a professor in the department of kinesiology and sport management and as the director of the Exercise & Sport Nutrition Laboratory at Texas A&M University. Dr. Kreider has conducted numerous studies on nutrition and exercise as well as provided exercise physiology, sport nutrition, and sport science support to several NCAA Division I athletic programs. Dr. Kreider is a fellow of the American College of Sports Medicine; an active member of the National Strength and Conditioning Association (NSCA); a co-founder, board member, and fellow of the International Society of Sports Nutrition; the founding editor-in-chief of the *Journal of the International Society of Sports Nutrition;* a fellow of the American College of Nutrition, and an elected fellow of the National Academy of Kinesiology. Photo courtesy of the School of Education and Human Development, Texas A&M University.

Duck-chul Lee (D.C. Lee), PhD, FACSM, is a professor and the director of the Physical Activity Epidemiology Laboratory at Iowa State University. His research focuses on the health benefits of physical activity, fitness, and exercise training on clinical biomarkers, chronic disease prevention, and longevity using comprehensive epidemiological approaches including large cohort studies and randomized controlled trials of exercise. He has published over 120 peer-reviewed research articles in medical and exercise journals. He serves as a principal investigator on research projects supported by the U.S. National Institutes of Health (NIH) and several research organizations. He is a fellow of the American College of Sports Medicine (ACSM) and serves as a reviewer for NIH grant applications and major medical journals. He received several research awards for his high-impact research publications featured by major media (e.g., NBC and ABC News, the *New York Times*, CNN, and BBC). Photo courtesy of Iowa State University.

Matthew T. Mahar, EdD, FACSM, FNAK, is a professor and the director of the School of Exercise and Nutritional Sciences at San Diego State University. He has taught courses in measurement and evaluation, statistics, research methods, exercise testing and prescription, assessment of physical activity and fitness, exercise physiology, and introduction to exercise science. His research interests include the promotion and measurement of physical activity and fitness, identification of valid and reliable youth fitness testing methods, and analysis of the effects of classroom-based physical activity programs on physical activity and on-task behavior. He is a member of the FitnessGram Advisory Board and developer of the Energizers classroom-based physical activities. He received his doctoral degree in measurement and research in exercise science and his master's degree in exercise science from the University of Houston, and his undergraduate degree in physical education from SUNY Cortland. Photo courtesy of Matthew Mahar.

Natalie L. Myers, PhD, LAT, ATC, is a clinical research scientist with Memorial Hermann's Rockets Sports Medicine Institute. She collaborates with orthopedic surgeons and sports medicine professionals to develop evidence-based research related to shoulder and elbow health conditions. She contributes to the didactic education of the Memorial Hermann's orthopedic and sports physical therapy residences and DI sports physical

therapy fellowship programs. She received her doctoral degree in rehabilitation sciences from the University of Kentucky and her master's and bachelor's degree in athletic training from California University of Pennsylvania and Elon University, respectively. Photo courtesy of Natalie Myers.

Kathy Simpson, PhD, is a professor emerita of the department of kinesiology and the former director of the Biomechanics Laboratory at the University of Georgia. She received a doctorate in biomechanics from the University of Oregon. Her research has focused on determining how people adapt their movements to varying demands (e.g., impact forces on foot prostheses) and how these adaptations affect the functions and structure of the lower extremity and spine. She has applied this research to areas such as improvement of sport performance in athletes with lower-extremity amputations, movement technique adaptations of individuals who have had spinal fusion surgery, and performance of daily activities by individuals with a hip- or knee-joint replacement or a lower-extremity prosthesis. Dr. Simpson has served as the biomechanics section editor for *Research Quarterly for Exercise and Sport*, co-chair of the biomechanics interest group in the American College of Sports Medicine, member of the executive board of the American Society of Biomechanics, and chair of the Biomechanics Academy of the American Alliance for Health, Physical Education, Recreation and Dance. Photo © Morgan Nolan.

Chad Starkey, PhD, AT, FNATA, is a professor and the director of the athletic training division at Ohio University in Athens and the interim director of the School of Applied Health Sciences and Wellness. A graduate of West Virginia University, he received his master's and doctoral degrees from Ohio University. He has served as a commissioner for the Commission on Accreditation of Athletic Training Education, on the board of directors for the Board of Certification, and as chair of the Education Council of the National Athletic Trainers' Association. He has authored several textbooks focused on sports medicine, orthopedic diagnosis, and therapeutic modalities. Photo © Lauren Bowers.

Katherine T. Thomas, PhD, is a retired professor of health and kinesiology. She studied factors, particularly skill, that influence physical activity in children. She has been funded to study health and physical activity in women of color and has received more than $1.4 million in external funding for research, service, and teaching. She has published several books, refereed journal articles, and written chapters in scholarly books. She served as a founding co-chair of one of the state teams for Action for Healthy Kids, co-chaired the committee that wrote the Iowa Association of School Boards model local wellness policy, and served as co-chair of the task force that successfully proposed Iowa legislation to examine physical activity and nutrition in schools. She has been involved in two USDA demonstration projects as social scientist and project co-director. She remains focused on informing stakeholders and decision makers in order to influence practice in schools and communities and to ensure that those practices are evidence-based. Photo courtesy of Katherine Thomas.